# CLIMATE AND WATER

# ADVANCES IN GLOBAL CHANGE RESEARCH

VOLUME 16

# CLIMATE AND WATER

## Transboundary Challenges in the Americas


Edited by

Henry F. Diaz

*Climate Diagnostics Center,*
*Oceanic and Atmospheric Research,*
*NOAA, Boulder, CO, U.S.A.*

and

Barbara J. Morehouse

*Institute for the Study of Planet Earth,*
*University of Arizona, Tucson, AZ, U.S.A.*



KLUWER ACADEMIC PUBLISHERS
DORDRECHT / BOSTON / LONDON

A C.I.P. Catalogue record for this book is available from the Library of Congress.

ISBN 978-90-481-6386-1 ISBN 978-94-015-1250-3 (eBook)
DOI 10.1007/978-94-015-1250-3

Published by Kluwer Academic Publishers,
P.O. Box 17, 3300 AA Dordrecht, The Netherlands.

Sold and distributed in North, Central and South America
by Kluwer Academic Publishers,
101 Philip Drive, Norwell, MA 02061, U.S.A.

In all other countries, sold and distributed
by Kluwer Academic Publishers,
P.O. Box 322, 3300 AH Dordrecht, The Netherlands.

*Printed on acid-free paper*


Softcover reprint of the hardcover 1st edition 2003

# TABLE OF CONTENTS

*Acknowledgments* ix

*Foreword* xi

*Preface* xiii

*Contributing Authors* xvii

***Section A: Water Policy, Institutions, and Society***

1. Climate and Water in Transboundary Contexts: An Introduction 3
   *Henry F. Diaz and Barbara J. Morehouse*

2. Boundaries in Climate-Water Discourse 25
   *Barbara J. Morehouse*

3. Experience and Role of the Organization of American States in Transboundary River Basin Management in Latin America 41
   *Jorge Rucks*

4. Coping with Climate Variability: Municipal Water Agencies Agencies in Southern California 59
   *Denise Lach, Helen Ingram, and Steve Rayner*

5. Impacts of Climate Fluctuations and Climate Changes on the Sustainable Development of the Arid *Norte Grande* in Chile 83
   *Hugo Romero and Stephanie Kampf*

6. Water Resource Management in Response to El Niño-Southern Oscillation (ENSO) Droughts and Floods: The Case of Ambos Nogales 117
   *Terry Sprouse and Lisa Farrow Vaughan*

7. Transboundary Water Problems as "Jurisdictional Externalities": Increased Economic Efficiency Through Institutional Reforms 145
*Charles W. Howe*

8. Impacts in Mexico of Colorado River Management in the United States: A History of Neglect, A Future of Uncertainty 163
*David H. Getches*

9. Reservoir Management in the Interior West: The Influence of Climate Variability and Functional Linkages of Water 193
*Andrea J. Ray*

***Section B: Climate, Hydrology, and Ecosystem Processes***

10. Biomes, River Basins, and Climate Regions: Rational Tools for Water Resources Management 221
*Henry F. Diaz*

11. The Transboundary Setting of California's Water and Hydropower Systems: Linkages between the Sierra Nevada, Columbia, and Colorado Hydroclimates 237
*Daniel R. Cayan, Michael D. Dettinger, Kelly T. Redmond, Gregory J. McCabe, Noah Knowles, and David H. Peterson*

12. The Role of Transboundary Agreements in the Columbia River Basin: An Integrated Assessment in the Context of Historic Development, Climate, and Evolving Water Policy 263
*Alan Hamlet*

13. Climate Doesn't Stop at the Border: U.S.-Mexico Climatic Regions and Causes of Variability 291
*Andrew C. Comrie*

14. Climate and Climate Variability in the Arenal River Basin of Costa Rica 317
*Jorge A. Amador, Rafael Chacón, and Sadí Laporte*

15. Nonlinear Forecasting of River Flows in Colombia Based Upon ENSO and Its Associated Economic Value for Hydropower Generation 351
*Germán Poveda, Oscar J. Mesa, and Peter R. Waylen*

16. Climate Variability and Climate Change, and Their Impacts on the Freshwater Resources in the Border Region: A Case Study for Sonora, Mexico 373
*Victor O. Magaña and Cecilia Conde*

17. Land Cover Changes and Climate Fluctuations in the Upper San Pedro River Basin in Sonora, Mexico 393
*Héctor M. Arias*

# ACKNOWLEDGMENTS

The editors wish to thank all of the contributing authors for their dedicated effort in helping us assemble this book and for their patience in seeing this project to completion. We are very pleased with the results, and we hope that they are as proud of the outcome as we are. We wish to thank the reviewers for giving generously of their time, for their expert evaluations, and for their constructive suggestions. Jon Eischeid was very helpful in helping to assemble the many figures and tables. Diana Miller lent invaluable assistance throughout the editing process and in helping to develop the camera-ready manuscript. We are grateful to the Office of Global Programs of the National Oceanic and Atmospheric Administration for their financial support in enabling us to organize the meeting in Santa Barbara, California, and to Jeff Dozier and other members of the Donald Bren School of Environmental Science and Management of the University of California at Santa Barbara for hosting the conference that led to the development of this book.

# FOREWORD

*"Climate is what we expect; weather is what we get."*
*Robert Heinlein*

This book addresses one of the most perplexing, but one of the most important, challenges in managing water resources in the twenty-first century: How can our increasingly sophisticated understanding of climate be incorporated into everyday decisions about water management and allocation? The authors focus specifically on how to incorporate climate information into *transboundary* water management, where political, legal, social, and cultural factors can, in themselves, render seemingly straightforward allocation decisions intractable.

From the western United States to Central America to South America, the chapters examine how climatic variability has already played a role in the development of transboundary water management systems. These case studies use many different contexts to examine whether water managers and political leaders will be able to adjust policies and legal frameworks to incorporate improved understanding of climate variability and climate change. As virtually all of the chapters recognize, the ability to translate information about climate into management decisions will become crucial as competition for scarce water resources increases.

The barriers to making better use of climate variability and climate change information are not insignificant. In many arid transboundary river basins, such as the Colorado or the Rio Grande, years of contentious wrangling have polarized decision making. Laws, compacts, and treaties—many of which were developed in the early twentieth century—have become sacrosanct, even as their limitations are recognized to lie at the core of some of the region's difficult transboundary water disputes. Injecting information on climate variability, especially in the form of short- and midterm projections that still have some degree of uncertainty, will, as the authors recognize, often be an uphill battle.

In some of the Central American and South American basins discussed in this book, there may be more room for short-term progress, as water is more abundant—at least relative to demand—and the transboundary legal and institutional framework is less restrictive. The greater challenge in these basins may be establishing and funding the infrastructure and capacity to carry out the required monitoring and analysis of climate and hydrological conditions.

Underlying all these challenges, however, is water users' and water managers' inherent skepticism of climate predictability. Not unlike the way the local forecast shapes the layperson's faith (or lack of it) in weather prediction technology, one or two erroneous predictions of El Niño or La Niña effects could be enough to sour a water management entity on incorporating climate variability into its decision-making process, especially if doing so would in any way upset existing allocation or operational arrangements.

One example may illustrate the difficulties that can be expected. Despite two or more years of climate scientists' predictions that the Upper Rio Grande is entering a severe drought phase, in the spring and summer of 2002, irrigation and most municipal use from the reservoir system in New Mexico was scarcely reduced. The river below Albuquerque dried up, resulting in more loss of the endangered silvery minnow and regenerating contentious federal litigation. As of September 2002, the Elephant Butte Reservoir in southeastern New Mexico was at 15% of capacity. In "normal" years, it averages 58% of capacity in September. If the climate scientists are right, and the coming winter fails to provide a sufficient snowpack in the Colorado headwaters of the Rio Grande to replenish the reservoir, 2003 will bring an almost unprecedented water management (and litigation) crisis in this portion of the basin.

Could this situation have been avoided if water managers and water users had taken more heed of the climatologists' predictions? Why didn't they? Did they believe the predictions, but fail to respond because of insurmountable political, legal, social, or economic barriers? Was the transboundary litigation threat from Texas a more important, or at least more believable, factor than climate predictions?

This situation nevertheless illustrates the potential opportunities for using climate information to foster better water management decisions. Had the predictions been sufficiently heeded, the water management entities might have taken early steps to avoid the crisis—for example, using groundwater banking, increasing use of federal farm bill programs that support agricultural water conservation through improved efficiency or temporary fallowing of land, and preparing the public for aggressive municipal water conservation.

These are the kinds of challenges and opportunities this book so capably examines, in an interesting variety of transboundary situations. The authors convincingly demonstrate that we will need to use good climate science if we are to move toward sustainable management of water in the Americas.

Mary E. Kelly
*Austin, Texas*

# PREFACE

In the summer of 2000, scientists representing six countries in the Americas, several international organizations, including the Organization of American States and the International Boundary and Water Commission of the United States and Mexico, met in Santa Barbara, California, to discuss the impact of climatic variations on water resources and water resources management in the Americas, with a focus on border regions that are shared among the countries in the Western Hemisphere. The meeting was convened in recognition that a large number of institutions dealing with freshwater resources throughout the Americas are affected by growing populations and economic development that have made access to adequate water supplies a key policy goal. Even modest variations in supply or demand can have large social consequences, especially for certain regions of the world where adequate access to potable water represents a fundamental, but largely unmet, human need.

As the management and control of water resources has become a key economic force in most parts of the world, understanding the potential impacts of current and future climatic variations on water supply has become critically important. Global change researchers interested in climate impacts on water resources recognized from the outset that water resource management problems might be most tractable on a regional scale, where governments and institutions might take advantage of the experiences of different communities in helping to plan for present and future water needs.

Geographic areas where political boundaries exist are affected by the differing interests, perspectives, and goals of the people living on either side of those boundaries. Both sides may be influenced by similar weather and climate variability on a broad range of time scales, including subseasonal floods and dry spells, interannual variability connected with the El Niño/Southern Oscillation (ENSO), and decadal and longer-term climate variations such as severe and sustained drought. However, impacts on human and natural systems may be radically different on either side of a delineated boundary, and thus may call for very different policy and management responses.

From a societal perspective, interactions among different sectors of society are becoming increasingly complex and more demanding of multidisciplinary approaches that combine considerations of climate variability, hydrology, ecosystem management, and the management of water resources for multiple human uses, particularly on regional scales and in the context of various kinds of boundaries. At the same time, there is a need to weigh historical patterns of water use against the potential for radical changes in water availability arising from decadal and longer-term climatic variations.

These kinds of complex interactions between human demands on water resources, and the natural variability of climate and hydrologic systems, prompted us to begin working toward development of integrated approaches. The kinds of approaches we envision are ones that bring together the knowledge and expertise of researchers in the relevant physical sciences dealing with the water cycle with that of experts in water resource management issues. We therefore undertook to plan and convene a symposium that would provide an opportunity for interaction across political, institutional, and disciplinary boundaries. The proceedings of this symposium form the basic subject matter of this book..

Three key aspects in water resources management call urgently for comprehensive regional assessments of the interactions between climate variations and water resources. First, in many parts of the world, the degree of exploitation of the water has reached such extremes as to magnify greatly the practical effects of climate variation on water supply. Second, a plethora of often conflicting regulatory and other constraints related to environmental protection, and interregional and international water transfers, mandated through a variety of legislative acts, international treaties, and other obligations, circumscribes the range of management options available to regional water resources managers. Third, in some cases, areas that have been traditionally regarded as regional "entities" based on a set of cultural or historical antecedents have been undergoing socioeconomic (demographic and political) changes that may be modifying the traditional linkages that have defined each as a "geographic region." These trends call for new kinds of analysis and integration, and a more comprehensive analytical framework than has been available in the past.

Major advances achieved over the past decade in understanding climate variation, together with the emergence of new capabilities in climate analyses and predictions, present a unique opportunity for the development of integrated assessments. Such assessments are useful for addressing questions such as how society adjusts to the impacts of climate variability and change, how climate information enters (or fails to enter) into decision-making processes—where the critical needs are for climate information and predictions—and what the potential benefits of improvements in, or improved uses of, climate information and forecasts might be.

From a management perspective, uncertainty in water supply has in the past been characterized as a random variable best known through the short historical record on surface discharges. Recent advances in the study of climate and climate variability have demonstrated that some general features of climate can be predicted with reasonable confidence well before they can be detected in the water cycle. In addition, studies of long-term climatic records have shown that projection of climatic variability from the recent historical record can be seriously misleading. In many cases water management and water policy systems throughout the world have not yet assimilated these advances in the understanding of climate variability.

The Santa Barbara symposium undertook to explore linkages between climate and water resources management, specifically in the context of sociopolitical boundaries. We were interested in addressing the nexus of water scarcity versus carrying capacity in the context of the manner in which existing water supplies are utilized. For example, land use and land cover are important components of the interrelationships between the physical and human systems' impacts on water use and the water cycle, which in turn affect water quality and the health of terrestrial and aquatic ecosystems. Obviously, we could not address all the important topics associated with the general subject matter of our meeting. The effects of hydrographic modifications on characteristics of the physical habitat of aquatic ecosystems, for example, is a very relevant topic that was not explicitly addressed in the symposium.

Boundaries are among the most fundamental tools used by society to organize knowledge and action. Indeed, in addition to delineating countries, states, provinces, watersheds, basins, climate regions, ecosystems, etc., boundaries are also widely used for purposes such as locating where to introduce perturbations into climatic and hydrologic models and for metaphorically delineating the points of differentiation among academic and research disciplines. At the most basic level, boundaries are functional markers for sets of rules governing how the entities on either side of the line interact with each other. Boundaries may also mark important political, economic, social, and cultural differences, as well as points of crucial flows and interactions.

Although natural processes do not begin and end at boundaries, boundaries often have a significant influence on decisions made and actions taken in anticipation of or in response to natural events, as well as on how these events, decisions, and actions affect the area and its inhabitants. Sensitivity and vulnerability to climatic and hydrologic stresses in transboundary settings may differ considerably based on relative access and ability to use information, levels of technology, institutional framework, access to resources, and other factors. Likewise, integrating climatological and hydrological data into transboundary water resource management may involve problems associated with differences in availability and comparability of data, control over sources of water supply, latitude to manage demand, and authority to make—and act on—decisions.

A pivotal goal of the conference was to gain insights into how these dilemmas were being addressed in different transboundary contexts, both within and between nation-states. Individuals invited to participate in the Santa Barbara symposium were selected on the basis of their research on topics such as analysis of historical records of precipitation and snowpack; analysis (including use of numerical models) of the relationships between different types of climatic variations at the regional scale; the impacts of regional climate variations such as El Niño and La Niña; and paleoclimatic reconstructions that allow extension of a region's natural record of climatic variations into the more distant past. From a social-science perspective, participants' research interests ranged from the effects of climatic conditions and events on water resource management in transboundary settings, and its implications regarding societal vulnerability to particular climatic stresses, to how political boundaries increase or decrease vulnerability and/or resiliency, and how organizational and professional structures may encourage or discourage adoption of new information, such as long-range climate forecasts, into decision-making processes. Consideration of how the specific socioeconomic context and institutional framework on each side of a boundary facilitate or constrain options for averting or mitigating climate impacts on water supply and demand is central to these analyses.

Overarching goals of the conference were, first, to improve utilization of knowledge about the hydrologic cycle in water resources management, and second, to improve integration of long-range climate forecasts, issued at different time scales, into policy, infrastructure, and related decision-making processes.

In summary, we organized the Santa Barbara symposium to explore critical issues related to regional water resources and climatic change as a contribution to national and international efforts aimed at fostering an interdisciplinary framework for analyzing physical aspects of the water cycle, in the context of regional water management systems. We also had as a major goal improving the application of climate information and forecasts, on different time scales, by entities concerned with assuring sufficient supplies of water, at appropriate levels of water quality, for use by human and natural systems. The meeting was designed to highlight the nature and extent of conflicting interests involving water, and to help set the stage for application of scientific tools to addressing problems of equity and efficiency in water use. We hope our aims have been realized.

*H. Diaz*
*B. Morehouse*

# CONTRIBUTING AUTHORS

Jorge Amador Astua
Centro de Investigaciones Geofísicas y Escuela Física
Universidad de Costa Rica, P.O. Box 2060
San Jose, COSTA RICA

Héctor M. Arias
Gabinete de Estudios Ambientales, A.C.
Blvd. Navarrete #125 Local 21, Colonia Valle Verde
Hermosillo, Sonora, 83240, MEXICO

Daniel R. Cayan
Climate Research Division, Scripps Institution of Oceanography
9500 Gilman Dr. Dept. 0224
La Jolla, CA 92093-0224, USA

Rafael E. Chacón
Centro de Servicio de Estudios Básicos de Ingeniería (Area de Hidrología)
Instituto Costarricense de Electricidad (ICE)
San Jose, COSTA RICA

Andrew Comrie
Department of Geography, Harvill Bldg., Box 2
The University of Arizona
Tucson, AZ 85721-0076, USA

Cecilia Conde
Centro de Ciencias de la Atmósfera, UNAM
Mexico City, D.F. 04510, MEXICO

Michael Dettinger
U.S. Geological Survey and Climate Research Division
Scripps Institution of Oceanography, 9500 Gilman Dr. Dept. 0224
La Jolla, CA 92093-0224, USA

Henry F. Diaz
NOAA/OAR/CDC, 325 Broadway, Boulder, CO 80305, USA

Lisa Farrow Vaughan
NOAA Office of Global Programs
1100 Wayne Avenue, Suite 1225
Silver Spring, MD 20910, USA

David H. Getches
University of Colorado School of Law, Campus Box 401,
Boulder, CO 80309-0401, USA

Alan Hamlet
Department of Civil and Environmental Engineering
170 Wilcox Hall, University of Washington
Seattle, WA 8195-2180, USA

Charles W. (Chuck) Howe
Department of Economics and Environment and Behavior Program
Institute of Behavioral Science, Campus Box 468
University of Colorado at Boulder
Boulder, CO 80309-0468, USA

Helen Ingram
Warmington Chair, School of Social Ecology
202 Social Ecology 1, University of California at Irvine
Irvine, CA 92697, USA

Stephanie Kampf
Department of Civil and Environmental Engineering
University of Washington
Box 352700
Seattle, WA 98195, USA

Noah Knowles
Climate Research Division, Scripps Institution of Oceanography
9500 Gilman Dr. Dept. 0224
La Jolla, CA 92093-0224, USA

Denise Lach
Center for Water and Environmental Sustainability
Oregon State University
Corvallis, OR 97331, USA

Sadí Laporte
Centro de Servicio de Estudios Básicos de Ingeniería (Area de Hidrología)
Instituto Costarricense de Electricidad (ICE)
San Jose, COSTA RICA

Gregory J. McCabe
US Geological Survey, Denver Federal Center
Denver, CO 80225, USA

Victor Magaña Rueda
Centro de Ciencias de la Atmósfera, UNAM
Mexico City, D.F. 04510, MEXICO

Oscar J. Mesa
Programa de Posgrado en Approvechamiento de Recursos Hidráulicos
Universidad Nacional de Colombia, Recursos Hidráulicos
AA 1027 Medellín, COLOMBIA

Barbara Morehouse
Institute for the Study of Planet Earth
The University of Arizona
Tucson, AZ 85721, USA

David H. Peterson
US Geological Survey, 345 Middlefield Road, MS 870
Menlo Park, CA 94025, USA

Germán Poveda
Programa de Posgrado en Approvechamiento de Recursos Hidráulicos
Universidad Nacional de Colombia, Recursos Hidráulicos
AA 1027, Medellín, COLOMBIA

Andrea J. Ray
NOAA/OAR/CDC
325 Broadway, Boulder, CO 80305, USA

Steve Rayner
School of International and Public Affairs, Columbia University
Palisades, NY 10027, USA

Kelly T. Redmond
Western Regional Climate Center, Desert Research Institute
2215 Raggio Parkway, Reno, Nevada 89512, USA

Hugo I. Romero
Departamento de Geografía, Universidad de Chile
Marcoleta 250, Santiago, CHILE

Jorge Rucks
Unit of Sustainable Development and Environment
Organization of American States
1889F Street NW, Washington, DC 20006, USA

Terry W. Sprouse
Water Resources Research Center, University of Arizona
350 N. Campbell Avenue, Tucson, AZ 85721, USA

Peter R. Waylen
Department of Geography, University of Florida
Gainesville, FL 32611-7315, USA

SECTION A

# Water Policy, Institutions, and Society

1

# CLIMATE AND WATER IN TRANSBOUNDARY CONTEXTS

*An Introduction*

Henry F. Diaz* and Barbara J. Morehouse**

**Climate Diagnostics Center, NOAA/OAR, 325 Broadway, Boulder, Colorado 80305*

***Institute for the Study of Planet Earth, The University of Arizona, Tucson, Arizona 85721*

In many parts of the world, population growth, urbanization trends, and economic development and restructuring have elevated water resource issues to unprecedented importance. Even modest variations in supply or demand can generate large social consequences. As management and control of water resources has grown as a political and economic force in most parts of the world, understanding the potential impacts of current and future climate conditions on hydrologic processes and water supplies has become ever more critical.

The often-stated observation that global processes happen in local places is especially salient with regard to the climate-hydrology-water resources nexus. Thus, the imperative grows to connect global and synoptic-scale climate with regional-scale hydrologic processes and regional to local-scale water resource management practices. At the regional scale, for example, different communities may hold very different expectations with regard to environmental issues such as water use. These expectations, in turn, may be related to different strategic approaches, such as full present-day exploitation versus conservation or preservation for the longer term. Contradictions between different strategies may be most tractable at regional scales. Recognition of the importance of regional-scale efforts to address such issues is becoming increasingly evident (see, e.g., Intergovernmental Panel on Climate Change [IPCC] 1998, 2001; Messerli and Ives 1997).

Geographical areas where political boundaries exist are affected by the differing interests, perspectives, and goals of the people living and operating on either side of those boundaries. Individuals on both sides are likely to be influenced by similar weather and climate patterns, on a broad range of time scales ranging from subseasonal (e.g., flood events) to seasonal (e.g., dry spells), annual and interannual (e.g., El Niño/Southern Oscillation



*Henry F. Diaz and Barbara J. Morehouse (eds.), Climate and Water*, 3-24.

[ENSO] events), and decadal or more (e.g., extended dry or wet periods). However, responses to the impacts of these externalities on regional and local resources can be—and often are—radically different. The increasingly complex interactions among different sectors of society regarding the myriad human needs and demands for water, in juxtaposition with the large range of natural climatic variability and its fundamental controls on hydrologic systems, calls for research approaches that emphasize integration across disciplines and interaction with the people the research is intended to benefit (see, e.g., Pulwarty 2001; Miles et al. 1998; Liverman and Merideth 2002).

Resolution of competing demands in contexts of water scarcity likewise demands consideration of historical patterns of water use against the potential for radical changes in water availability arising from decadal and longer-term climatic variations. From this perspective, three areas of water resource management urgently call for comprehensive regional assessment. First, in many parts of the world, extreme exploitation of water magnifies the practical effects of climate variation on water supplies. In many cases, the water being exploited crosses one or more internal or international boundary (Varady and Morehouse, in press). Second, a plethora of often conflicting regulatory and other institutional structures and mandates constrains environmental protection efforts and interregional and international water transfers. These structures and mandates, typically manifested in the form of legislative acts, international treaties and protocols, and administrative rules and procedures, circumscribe the range of management options available to water managers. Third, in some cases, areas that have traditionally been regarded as regional entities, based on various criteria ranging from historical patterns and precedents to expressly articulated economic development strategies, are undergoing dramatic economic, social, and political changes. Such changes may reinforce (for example, socioeconomic ties in the U.S. state of Arizona and the Mexican state of Sonora) or potentially weaken (e.g., the Columbia River Basin) regional ties. In other cases, socially defined "regions" may be emerging where none existed before (e.g., the La Plata River Basin in South America). All such changes have significant implications for management of shared natural resources and call for new kinds of analyses, and a more comprehensive, integrated analytical framework than has been available in the past.

Ideally, this kind of integrated framework includes not only individuals specializing in climatology, hydrology and hydroclimatology, ecology, engineering, and related fields, but also social scientists and economists, water managers, policy makers, and representatives of key user communities. In reality, differences in governance structures and social relations more generally may pose significant barriers to full integration across all of these interests. These barriers are nowhere more apparent than in research addressing transboundary problems. From a management perspective, uncertainty in water supply typically has been characterized as a random variable best known through short historical records of surface discharges. However, recent advances in the study of climate and climate variability demonstrate that some general features of climate can be predicted with reasonable confidence well before they can be detected in the water cycle. Studies of long-term climate records show that projection of climate variability using recent historical records—for example, in constructing analogues to refine predictions—can be seriously misleading. Water management and water policy systems around the world have not yet, in many cases, assimilated these advances into their understanding of climate variability and its applications in decision making.

Today, more than ever, comprehensive treatment of water and climate issues requires consideration not only of multiple temporal and spatial scales but also of the influences of boundaries, border regions, ecotones, interstices, and other "gray areas" that may separate physical and human systems. Clearly, natural processes and impacts neither begin nor end at boundaries. Yet boundaries often have a significant influence on decisions made and actions taken (see Morehouse 1995), whether in anticipation of or response to natural or human events. Boundaries may also, of course, have a significant influence on how these events, decisions, and actions affect the area and its inhabitants. Sensitivity and vulnerability to climatic and hydrologic stresses in transboundary settings may, however, differ considerably from one side of the boundary to the other, based on relative access and ability to use information, levels of technology, institutional structure, access to resources, and other factors. Likewise, considerable challenges often exist with regard to integrating climatological and hydrological information into water resources management. For example, differences in availability and comparability of data, data collection and archiving techniques, interpretation, and communication add further complications to management issues ranging from control over sources and production of water, latitude to manage water demand, and authority to make—and act on—decisions.

Climate, hydrology, and water resources research activities have traditionally been carried out independently of each other. Studies of droughts and floods have provided the primary bridge among these disciplines. With the emergence in the 1980s of global change as a major focus of scientific and public discourse, however, interest has escalated in integrating research across disciplinary boundaries. At the same time, interest has grown in developing a better understanding of how boundaries may facilitate or impede not only the production of scientific knowledge, but also rational decision making and action in border areas.

As is indicated in the chapter by Morehouse, our definition of transboundary regions, for purposes of this volume, is a flexible one. It includes consideration not only of political and administrative boundaries, but also climatic and hydrologic boundaries, as well as social boundaries (e.g., science versus public opinion and policy), and boundaries used in models. The chapters by Rucks, Sprouse and Farrow, Getches, Diaz, and Hamlet expressly focus on international boundaries, while the chapters by Howe, Ray, Cayan et al., Amador et al., Poveda et al., Magaña and Condé, and Arias focus on issues dealing largely with within-country jurisdictional and hydrologic boundaries. The chapter by Lach et al. reveals influences of professional boundary factors in decision making, while Romero and Kampf show, in a case study for Chile, how societal and international boundaries—such as resource allocation, global economic pressures, and history—are interrelated in terms of the allocation of scarce water supplies. Comrie combines discussion of scientific boundaries with examination of climate research issues on the U.S.-Mexico border.

Several chapters focus on the intermountain and southwestern United States, and California. Water issues are paramount in this region, which is characterized by predominantly semiarid to arid conditions. Here, unprecedented transformation of river systems (Worster 1985; Stegner 1987; El-Ashry and Gibbons 1988; Wilkinson 1992; Reisner 1993; Collier et al. 1996) provides a unique context for examining the effects of climate on water supply and of institutional capacity to cope with climate impacts on water supply and demand (U.S. National Research Council [NRC] 1991; National Assessment Synthesis Team [NAST] 2000). Recognition of the critical problems looming in the U.S. West has heightened in recent years (Nichols et al. 2001; Western Water Policy Review Advisory Commission [WWPRAC] 1998), but so far this recognition has not resulted in needed institutional changes. Indeed, prevailing rhetoric merely reprises a long history of similar efforts to come to terms with the finite nature of water resources in the U.S. West (U.S. Bureau of Reclamation [USBR] 1975; U.S. Office of Technology Assessment [OTA] 1983). The Colorado River Compact of 1922 is but one of the long string of institutional efforts aimed at solving the basic dilemma of insufficient water to meet all demands under all climatic conditions in the region, which also includes the Colorado River Delta area of Mexico. Thus, despite decades of studies and recommendations by many people, the U.S. West and adjacent areas of Mexico are as vulnerable today to water scarcity as they were in the past. The absence of any severe decade-long or multi-decadal, watershed-wide drought since the population boom years began in the 1920s has thus far reinforced perceptions of invulnerability to water scarcity. Studies, such as the Severe Sustained Drought Initiative (Powell Consortium 1995), suggest this perception may be ill-founded, though impacts will likely be unevenly distributed both geographically and societally.

Conducting transboundary research in climatology, hydrology, and climate impacts is often challenging, for a number of reasons. One of the most problematical barriers is sheer lack of data, particularly longer time series that allow scientific identification and assessment of trends. Lack of consistency in data collection over time and space, and in archiving techniques, are also problems in many areas. Political barriers may be as difficult to surmount, for many countries consider basic scientific data to be closely linked to questions of national sovereignty and thus subject to rigorous control in terms of both collection and dissemination. Even when data may be accessible, obtaining the data can be prohibitively expensive. Furthermore, sufficient funding to carry out complex border-area research may be difficult to obtain, especially for the multiyear time periods that may be required to produce meaningful results.

In recognition of the multiple layers of complexity associated with social processes, it is becoming increasingly common for environmental research and assessment activities to include social science perspectives that go beyond straightforward economic analysis. It is becoming more common for such research to include methods and theories drawn from political science, anthropology, sociology, and human geography (see, e.g., *Climate Research*, Vol. 21, No. 3, 2002, for a collection of papers on research carried out under the NOAA-funded Climate Assessment for the Southwest—CLIMAS—project). The goal of such assessments is to develop a deeper and richer understanding of the myriad interactions between environment and society so as to identify areas of risk and vulnerability, and to enhance resilience and adaptability among those affected. It is often the case that the nature and intensity of specific societal risks and vulnerabilities are influenced by multiple stressors; hence, it is essential not only to understand what these stressors are, but also to assess the relative contribution of each factor to overall vulnerability and risk. This kind of knowledge provides the foundation for evaluating how well various coping mechanisms might perform with regard to building adaptive capacity within the sector or population of concern (see, e.g., Pulwarty and Redmond 1997).

In response to the lack of a substantial body of scientific research that examines the implications of boundaries and border areas for integrated climate-hydrology-human factors research, we organized a special symposium on the topic. This symposium, entitled Climate, Water, and Transboundary Challenges in the Americas, was held July 16–20, 2000. The meeting brought together climatologists, hydrologists, and social scientists representing North, Central, and South America to discuss research activities, research needs, and challenges associated with trying to build a better understanding of the interactions among climate, hydrology, water resources, and society. Individual sessions examined transboundary equity, sustainability issues, transboundary indicators, interactions between climate variability and transboundary water management, case studies, the roles of organizations and of integrated assessments in water resource planning and management, and using climate information for water resource management in border areas. The chapters included in this volume reflect the variety of approaches and perspectives represented at the symposium and provide insights into the opportunities and challenges posed by boundary contexts.

Section 1, Water Policy, Institutions, and Society, provides insights into the issues faced by researchers, decision makers, and inhabitants of regions where boundaries pose additional challenges to water management institutions, policies, and practices. The authors examine boundary issues from an array of disciplinary and geographical perspectives, and their narratives provide a rich body of insight into how to think about—and meet—the challenges and opportunities posed by boundary contexts.

Morehouse establishes an intellectual framework for this book by articulating why a focus on boundaries and boundary-related issues is important for water management in the context of climate variability and change. She emphasizes that understanding boundary-drawing processes, and boundary functions and rules, is essential for efforts to improve institutions and decision-making processes. She notes that, while boundaries summarize complex information about the rules and structures that we use to understand and manage our world, they may also pose serious impediments to important transboundary interactions and may constrain capacity to conceptualize complex processes operating at different scales. Taking a boundary perspective, she suggests, requires close attention to issues of both temporal and spatial scale, jurisdiction and sovereignty, rationality and efficiency, equity, and transaction costs. Morehouse recommends initiation of climate-water resource research that explicitly includes consideration of the roles played by boundaries of all kinds, as well as of the opportunities and constraints posed by the existence of boundaries.

Rucks discusses the role of the Organization of American States (OAS) in facilitating advancement in scientific understanding of physical processes within key international river systems in Central and South America and in facilitating use of such knowledge to improve water management, decrease vulnerability, and contribute to sustainable development. He notes that Central and South American countries lack deep understanding of transboundary water issues, and that political obstacles impede extensive cooperation in developing integrated transboundary water management policies. The author highlights two case studies in which interactions between water resource issues, societal factors, and climate come together to produce significant challenges, and where OAS has been actively involved.

The first case study involves a five-country initiative on the Río de la Plata system, involving Argentina, Brazil, Paraguay, Uruguay, and Bolivia. The basin is home to 50% of the combined populations of these five countries, and thus is of major importance to all parties. Here, upstream activities ranging from agricultural expansion to hydropower development and rapid urbanization are raising concerns with regard to both downstream flood regimes and related sedimentation problems, as well as preservation of an internationally known wetland area, the Pantanal. Initiation of the Hydrologic Warning Operations Center for the Plata Basin, located in Argentina, constitutes an important step in establishing infrastructure and organizational capacity essential to monitoring hydrologic trends and producing warnings and forecasts. The center has proved its worth in coping with both flooding and low-water conditions, and serves as a focal point for development and communication of transboundary, regionally tailored information.

The second case study Rucks describes is a project underway in the San Juan River Basin, which encompasses portions of Costa Rica and Nicaragua. In this basin, the main issues revolve around water quality, habitat preservation, control of exotic species, and the need for water resource management improvements. Broader goals articulated under the initiative include promoting economic development in the border areas, conservation of natural resources and protection of biodiversity, ensuring the sustainable use of water resources, and protecting drainage basin integrity. Rucks notes that significant challenges exist in this region, including insufficient financial and human resources, inadequate natural resource management capacity, high rates of poverty and population growth, and limited stakeholder participation. He identifies the top-priority need in this region to be a fully functional information system that covers the entire basin. Development of the system requires, among other things, understanding regional climate processes and impacts, and comprehensive, basinwide information about precipitation and runoff, water quality, erosion, and sedimentation processes. Getting the information needed requires capacity building and strengthening of existing institutions (e.g., policies and procedures for sharing information), improving infrastructure, reconciling competing water uses, and reducing poverty levels among residents of the basin.

Rucks concludes by noting ongoing efforts to address and resolve the many issues involving water resources in Central and South America. International meetings are providing opportunities for countries to establish necessary foundations for international cooperation aimed at improving transboundary water management.

The following chapter, by Lach, Ingram, and Rayner, takes us to Southern California, long a hotbed for interbasin transfers to meet the region's voracious water appetite. The authors examine the institutional characteristics of municipal water agencies (MWAs) and assess the capacity of these agencies to adopt the kinds of innovations required to meet challenges posed by continued population growth and climate impacts on water resources. Lach et al. note at the outset that MWAs are "archetypes of conservative organizations" that emphasize reliability far above quality and cost and that they go to great lengths to avoid customer complaints and political visibility. The authors observe that these organizations have long operated in a context of inadequate supplies to meet expanding demands, but that, in the past, interbasin transfers provided a means of resolving such dilemmas. However, with resistance increasing among residents of the water exporting basins, such opportunities are diminishing.

The risk-averse philosophy of the MWAs, according to the authors, leads to a preference for incremental change over substantive innovation. Where innovation occurs, it tends to be when it is perceived as remaining within the boundaries of accepted practice, or when a parallel or small-scale innovative practice is transferred to operational status within the larger organizational context. This process reveals yet another boundary context, that which differentiates internal from external factors driving innovation. Outside organizational boundaries, drivers of innovation include environmental and health regulations, court suits, public mobilization to influence agency decisions or practices, pressures from other agencies, and influences from the academic, public, and private research sectors.

The authors conclude by noting that, with less ability to continue relying on interbasin transfers to meet growing water demand, and with the prospects of ferocious competition over those opportunities that remain, MWAs in Southern California will need to introduce new innovations. Yet adoption is likely to be slow, with only marginal impacts at the beginning on decision making and practice. Such innovations are most likely to arise from external research activities, or from experimentation by internal technicians. The agencies that will fare best, according to Lach et al., are the ones that can take advantage of incremental innovations with regard to interbasin transfers and at the same time develop new management techniques. They will also be the ones that are able to capitalize on new sources of water made available through trade arrangements such as the North American Free Trade Agreement (NAFTA), or to finance expensive desalinization plants to exploit the waters of the Pacific Ocean. The winners are likely to be the more well-endowed organizations, raising the specter that inequities between agencies and their customers will increase with regard to availability, quality, and cost of water.

With Romero and Kampf, we return to South America, this time to the Norte Grande of Chile, which includes the Atacama Desert, one of the world's driest, to examine the interrelationships among copper mining activities, water resources, and climate, and how the interactions of these factors are generating social impacts. In this extremely arid area, most of the water used was deposited millennia ago and much of the area receives virtually no modern-day recharge. At the same time, the area experiences dramatic spatial and temporal climate variability and is strongly affected by ENSO conditions.

This area of northern Chile, the authors note, exhibits many of the same characteristics found in other Latin American countries, where exploitative activities have devastated landscapes and resources, mired communities in poverty, and produced social segregation. Contemporary globalization processes demand that strategies be developed to address social, economic, and environmental needs. Climate variability and change must be taken into account in these strategies. Given these conditions, the authors emphasize, the need for an effective water management strategy is urgent.

The primary economic activity in this region, copper mining, demands a great deal of water, both for direct processing and for indirect uses such as urban services. With demand approaching available water supply capacity, new water management strategies must be devised. In some cases, water can be redistributed within a sector (e.g., from an abandoned to a new mine) or between sectors (e.g., from agriculture to urban use). Romero and Kampf observe that water markets have been introduced to facilitate such transfers. Subsidization of transboundary water imports from Bolivia, where the waters of the Silala River were recently privatized, has also been explored.

Adding to the general dilemma of the imbalance between supply and demand, short-term solutions have revolved around prioritizing water allocations so that mining and urban needs are met first. This policy is threatening agriculture, as well as environmental services and human uses that have been in existence for thousands of years. Protected areas have only loosely defined boundaries, and are pressured by indigenous communities claiming portions of the protected lands. Given that the claimants cannot legally request water rights in these areas, these areas and resources, and the ecological services they provide, face even greater threats.

Transboundary rivers shared by Chile and Bolivia constitute important components in the search for water in northern Chile. Given the rise in demand for scarce water, the authors caution, new conflicts are sure to emerge unless better strategies are devised. Romero and Kampf find hope in that border integration between Chile and Bolivia is becoming increasingly strong, and that both countries have set aside conservation areas in the highlands. They call for a transition from a dividing boundary to an integrated area with corridors, for creation of common resource areas, and for water exchange treaties and agreements to conserve common natural resources. The authors conclude by stressing the importance of shifting from valuations that emphasize economic growth at all costs to ones that stress sustainability, including social equity and environmental protection.

Sprouse and Farrow provide a tangible example of the challenges posed by boundaries in managing shared water resources. Specifically, they examine links between climate variability and water resource issues in the twin cities of Nogales, Sonora, and Nogales, Arizona (commonly known as Ambos Nogales). The two cities share the watershed of the Santa Cruz River, and they rely to a significant extent on surface flows and groundwater recharged by the river and its tributaries to support agricultural, municipal, and industrial uses. Droughts and floods are serious concerns in both communities, and, as the authors emphasize, both communities would benefit from transboundary institutions that allow for coordinated planning and response to climatic stresses. ENSO conditions exert considerable influence on the area, and recent advances in ENSO forecasting hold promise for initiating planning efforts to avert or mitigate detrimental impacts to human and natural systems. Sprouse and Farrow's examination of the forecast for a strong 1997–98 El Niño event and how the event actually transpired provides insights into how transboundary institutional arrangements might unfold. Binational planning efforts were initiated through the Border Liaison Mechanism, which is an arm of the International Boundary and Water Commission (IBWC), to prepare for potential flood events. While the forecast and anticipated impacts proved to be exaggerated (some flash flooding did occur in the area), the initiative demonstrated the viability of binational cooperation to enhance preparedness.

Looking to the future, the authors recommend building on this experience to improve adaptation capacity in the two cities. One avenue would be to establish a binational water planning district. Another would involve institutionalizing systematic planning for droughts and floods and construction of appropriate infrastructure to store water as a buffer against dry times and as a way to reduce flood damages. They also recommend creation of a binational ENSO task force, perhaps through the Border Liaison Mechanism, to establish a network of entities involved in transboundary resource management, early warning, and emergency management. The task force could also promote education and awareness about ENSO in the region, and serve as a clearinghouse for regionally relevant ENSO information. Activation of a border task force to develop transboundary climate diagnostics and forecasting capacities, the authors observe, would go a long way toward addressing constraints posed by current practices that limit such activities to areas within national boundaries.

In the next chapter, Howe examines boundary issues from the point of view of the effects of jurisdictional externalities. He observes that such externalities are likely to arise when the physical boundaries of water systems fail to conform to institutional boundaries, and that resolution of issues arising from such externalities can be especially challenging due to the unidirectional course of water flows. He reviews examples illustrating the characteristics of jurisdictional externalities and several cases where institutional mechanisms have successfully addressed problems associated with these externalities. He notes that current institutions may be legal but inequitable, inefficient, and economically costly. His solution to the dilemmas posed by jurisdictional externalities is threefold. First, reinstitutionalize river basin commissions (a program designed to do just this was eliminated in the United States early in the Reagan administration). Second, expand the role of water markets in connection with, and within the areas covered by, these commissions. Third, substitute flexible allocation rules for the fixed rules that are currently embedded in most river basin institutional arrangements. From the perspective of climate impacts, the third recommendation is particularly interesting for its suggestion to incorporate continuous information about the current state of the water system as well as climate forecasts. Howe suggests that a substantial increase in net payoffs would arise if decision makers were to switch from relying on current inflexible rules for water allocation to more flexible rules based on more complete use of information about climatic and hydrologic variability in the watershed.

The impacts of U.S. management of the Colorado River form the basis for an analysis by Getches of what the future holds for the biologically rich and environmentally threatened Colorado River Delta. The delta was historically one of the great desert estuaries of the world. Increased diversion and use of the river over the years, however, diminished and at times completely eliminated freshwater flows into the area. The environmental impacts were huge. Periods of high precipitation and runoff between 1983 and 2000, however, stimulated re-creation of wetlands and revitalization of riparian habitat. Getches warns that a return to previous (lower) levels of flow will again threaten the area. Much of the threat arises from legal arrangements that allocate water utilization rights allowing full consumption of all water available in "normal" years without regard to environmental impacts. The author observes that the river has long been a source of contention between the United States and Mexico, as well as among U.S. riparian states and with Indian tribes. With ten dams, the river is one of the most managed and controlled in the world. It sustains a U.S. population of more than 20 million people and irrigation of some 1.7 million acres of agricultural land.

Getches reviews the history of the Law of the River, the collection of treaties and other legal instruments that govern management of the river and use of its waters. Being on the upstream end of the river, the United States has historically been at a considerable advantage over Mexico in determining how the river and its waters would be managed. However, whether the United States will continue to be able to prevail under conditions of climatic variability and change is questionable.

Getches sees endangerment of the delta as the next crisis in U.S.-Mexico relations in the Colorado Basin. Yet, he emphasizes, only relatively small amounts of water would be required to ameliorate most environmental impacts to the delta. Thus neither country would need to make large sacrifices. Even under these relatively advantageous conditions, however, it is unlikely that agreement would be reached quickly or easily. Overallocation of Colorado River water in the United States, current water appropriation practices, and institutions that emphasize water consumption over environmental protection are among the primary barriers to reaching an accord. The author reviews recent efforts to establish a binational framework for protecting the delta region, in the face of these entrenched institutional structures and practices. The efforts have not yet been successful.

Noting that climate variability constitutes "the elephant in the closet," Getches warns that several manifestations of variability could exacerbate problems in the delta. One is persistence of the historical pattern of large annual variation in flows. A second is the possibility of severe sustained drought, and a third is long-term climate change affecting timing and volumes of flow in the river. Finding water to provide assured flows to the delta is the crucial issue, and one that needs to be addressed before climate-driven crisis, which could harden positions, occurs. He particularly warns that intransigence among the U.S. riparian states, if allowed to drive U.S. policy, could derail resolution of the problem. The author identifies a number of potential sources of water, ranging from purchase and transfer of some existing rights to improving water conservation practices and replenishing flows by using sources such as effluent and groundwater. With sufficient capital, Getches observes, water deliveries could begin. The author also cites institutional and legal changes that could facilitate protection efforts. He considers the IBWC to be the appropriate organizational mechanism for addressing the problems, particularly under the umbrella of Minute 306, which recognizes the two countries' shared interest in preserving the ecology of the delta.

A tributary to the Colorado River, the Gunnison River, provides the focal point for Ray's examination of spatial linkages between the Gunnison

Basin and areas outside its hydrologic boundaries, how climate variability could affect reservoir management plans, and how climate information could improve water managers' ability to achieve their goals. Ray catalogs water problems in the Gunnison Basin as ranging from water quality (especially salinity), environmental protection, and interstate obligations, including those associated with the Colorado River Compact.

ENSO conditions exert a strong effect on snowpack in the watershed, influencing variability in runoff to the river and thus streamflow. Boundary issues, she observes, pose additional challenges to management of the river, for while the basin is well defined hydrologically, its hydrologic boundaries do not correspond well with patterns of water allocation and use. Indeed, as Ray notes, the basin is "linked with places and problems far outside its hydrologic boundaries." Thus, as water uses increase overall, relationships among such uses in the Gunnison Basin must be taken into account, as must linkages with water and other resource uses elsewhere. The author emphasizes that, through influencing water supply and demand, climate constitutes a physical link connecting the basin with these other places.

In this context, cross-scale issues, both temporal and spatial, emerge as important variables in water management. Among the cross-scale issues Ray identifies are salinity control, ecosystem sustainability, Colorado's legal water allocations under the Colorado River Compact, transbasin diversions to population centers east of the Rocky Mountains, hydropower generation, and recreation. U.S. National Park Service ownership of federal reserved water rights in the Black Canyon of the Gunnison River, Ray notes, sets important preconditions for broader management of the river's waters (see NRC 1996).

Climate and hydrologic information, the author emphasizes, is crucial to planning and decision making within this complex institutional, socioeconomic, and physical arena. Ray reviews current uses of forecasts, and details other forecast products that hold potential for improving water management. One potentially useful product is use of seasonal forecasts to develop improved snowpack (and hence runoff) forecasts. Another is use of seasonal forecasts, done in March or April, for summer flow conditions that could affect demand as well as supply. Such forecasts could be used, for example, to assure instream flows through the Black Canyon stretch of the Gunnison. Subseasonal forecasts could also prove useful, she advises. For example, such forecasts could be used for predicting the timing of peak flow periods, which in turn would be useful for making decisions about whether or not augmentation of the natural peak might be needed. Ray further suggests that anticipating and coping with cross-scale effects of climate variability could also be aided by climate forecasts.

Section 2, Climate, Hydrology, and Ecosystem Processes, develops the scientific foundations underpinning the societal issues addressed in Section 1. The authors in this section bring a wealth of information about the state of climate and hydrologic science and its application to real-world water management problems.

In examining water resource management from the broad geographical perspectives of biomes, river basins and climate regions, Diaz establishes the foundation for the chapters that follow. He argues that geographical areas such as biomes or major river basins, which tend to correlate with large-scale climate features such as ENSO, are the most logical units of analysis and management. Focusing on the U.S.-Mexico border, he notes that this area exhibits a high degree of year-to-year climate variability. Overlying this is a pronounced warming trend over the past 20 years of 2–3°F (~1.5°C) as well as a trend of above-average precipitation. Over longer time frames, he notes, significant changes in vegetation structure have been found, related in part to human activities but also to climatic changes. In this context of high temporal and spatial variability, Diaz observes, human decisions seldom reflect—but should involve—careful consideration of the potential impacts of interactions between climatic shocks and human actions, and of the interconnectedness of natural systems.

In this context, reconciling contradictions between natural and political boundaries is essential. Thus, according to Diaz, rational water resource management requires holistic consideration of the entire drainage of a river system, from headwaters to delta and including all tributaries. Using the Colorado River as an example of this holistic approach, Diaz suggests that combining improved climate prediction capabilities with forecast-sensitive institutional mechanisms for assuring continued water supply to the delta offers a way to ensure the survival and viability of that biologically rich area.

Cayan, Dettinger, Redmond, McCabe, Knowles, and Peterson, examine hydroclimatic linkages between California water and hydropower in the Sierra Nevada with the Columbia River and Colorado River watersheds. The authors observe that California's water supplies and electrical power are closely linked to the hydroclimates of all three watersheds, due to the prevalence of transboundary water supply arrangements and hydroelectricity trading. California engages in large intrastate and interstate water transfers to move water toward demand. At the same time, the state imports about 20% of its total electrical consumption; these imports, which are crucial to meeting its needs, are obtained about equally from sources in the Northwest and Southwest United States. It is important to note that California's ability to balance supply and demand is highly influenced by the hydroclimates of the Columbia and Colorado River Basins. Thus, dry conditions in one sending basin may pose challenges to the state; shortages in two or more of the watersheds, the authors observe, are "particularly threatening."

Using natural discharge estimates for the study areas, the authors identify crucial factors such as the length of time it takes for the yearly water supply to accumulate in each watershed, and patterns of clustering in annual low- and high-discharge series relative to climatic conditions during those times. Their findings indicate that, though the climates differ among the three watersheds, correlations exist with regard to water-year hydrologic variations. Indeed, Cayan et al. find that the three watersheds "share hydrologic extremes much more than would be expected by chance." In examining climate patterns, the authors observe that, while ENSO and the Pacific Decadal Oscillation (PDO) have a strong influence on wet/dry years in the Columbia River system, this is not reliably the case for the Sierra or Colorado watersheds. On the Colorado, they emphasize, summertime precipitation is highly influential. Given that summer precipitation remains difficult to forecast with any confidence, it is not surprising that, for the three watersheds, the authors consider Colorado River water managers to be at the greatest information disadvantage. They conclude by emphasizing that differences in the seasonality of greatest contribution to wet or dry years in each of the three watersheds influence how well—and when—managers can foresee eventual water-year discharge.

Hamlet discusses the historical development of the Columbia River Basin, and uses this background as providing a context for water resources management today. This in turn provides a backdrop against which the impacts of climate variability and potential future changes in climate might affect water resource availability and use. Hamlet shows how the evolving management goals in the U.S. portion of the basin, and current scenarios for the impact that future climate change may have on the hydrology of the basin, may affect international agreements with Canada in the management of the Columbia River flows.

Comrie discusses a project to delineate climatic regions in the U.S.-Mexico border region as a means of overcoming the constraints to transboundary climate research posed by political boundaries. He then explores monsoon variability in these regions. Common patterns of seasonality and variability, rather than simple precipitation totals, are used to determine cohesive areas, and two techniques are employed to determine the proper locations of the boundaries of each region.

Comrie identifies nine coherent regions, each of which has its own precipitation seasonality characteristics. For example, Region 1 encompasses the monsoon-dominated area along the Sierra Madre Occidental in northwestern Mexico through eastern Arizona and most of New Mexico. Here, there is a strong mid- to late-summer precipitation maximum embedded in a pattern of considerably less precipitation over the rest of the year. Region 2, by contrast, is dry all year but has marginally greater precipitation in the winter. This region lies to the west of Region 1 and encompasses the Mojave Desert and lower Colorado River Valley.

The monsoon-dominated region, Region 1, then forms the basis for Comrie's analysis, based on four subregions within Region 1, of spatial and temporal variability of monsoonal precipitation. Using calculations of climate anomalies by subregion, he finds significantly different 500-millibar (mb) circulation patterns under wet and dry summer monsoon conditions. It appears, he notes, that shifts in circulation at this level in the atmosphere (~5,900 m, or ~19,500 ft, in summer), relative to the geographic locations of the different subregions, either directly or indirectly influence seasonal precipitation.

The author concludes by observing that improving the spatial coverage of meteorological stations might improve definition of regional boundaries and thus analyses of climate processes and impacts. Also needed, he observes, is further research into the role of the Hadley circulation and the Intertropical Convergence Zone (ITCZ) in influencing the shape and strength of subtropical circulation over the region.

With Amador, Laporte, and Chacon, the focus shifts to the Arenal River Basin in Costa Rica. The Arenal River system drains parts of north-central Costa Rica. The Arenal Dam and Reservoir, located at the southern end of the Guanacaste Mountains, produces nearly a quarter of the electricity used in the country. Water from the system is also used for agricultural irrigation in Guanacaste Province, which with an average rainfall of just over 1,400 mm each year, is one of the driest regions of the country.

The Arenal Basin is characterized by complex topography and high climatic variability, with ENSO predominating in the northwestern area and changes in the Caribbean low-level jet, some of which are influenced by ENSO, dominating in the southeast. Average rainfall over the basin varies from 5,000 mm in the southeast to 2,000 mm in the northwest. Variability is clearly evident over distances of only 30 to 40 km, between the northwestern and southeastern portions of the basin. Temporal variability on time scales ranging from intraseasonal to interannual is also evident. Using pentad precipitation data, the authors have developed a better understanding of seasonal distribution of precipitation over the basin, as well as areas of strong spatial and temporal contrast in precipitation patterns. The authors find that the midsummer dry period, locally known as the "veranillo," has a weak influence on the western part of the reservoir during July and August. By contrast, intense trade winds associated with summer development of a low-level jet over the Caribbean produce, in interactions with local topography, increased precipitation in the eastern area of the reservoir. El Niño conditions, together with other factors, are associated with decreases in reservoir levels. By contrast, La Niña conditions tend to produce increased rain in the basin, especially in the western and northern parts of the reservoir. The authors stress the importance of considering ENSO influences in the context of other climatic factors, such as Caribbean low-level winds and the frequency of cyclone formation in the Caribbean region. The authors note that climate studies in Central America are challenging due to the region's location amidst warm ocean pools where intense convective activity occurs and due to interactions between trade winds and topography that produce significant local climate variability.

Due to rapid growth in electricity demand in Costa Rica, developing tools to manage hydropower resources is becoming ever more essential. Developing and disseminating useful climate information are increasingly recognized as essential to this effort. Yet, the authors caution, forecasters should not place inordinate reliance on ENSO signals as primary inputs, nor should analogs or multiple regression models depend only on tropical Pacific sea surface temperatures (SSTs). Rather, they recommend, forecasts should be based on tools that are more solidly based on physical data, such as those provided by numerical models producing multi-parameter outputs.

Poveda, Mesa, and Waylen also examine links between climate variability and hydroelectric power generation. Focusing on the El Peñol project on the Naré River in Colombia, the authors report on the development of the Multiple Adaptive Regression Splines (MARS) model, which incorporates an array of climate and forecast variables to produce streamflow predictions. Given that hydroelectricity constitutes 75% of total energy production in Colombia, droughts can have serious economic impacts on operation of power generating facilities reliant on streamflows. The situation is further exacerbated by a lack of intraregional energy trading, meaning that any shortfall in hydroelectric power supplies must be met by using more expensive alternative sources.

Peaks and troughs in precipitation and streamflow over the basin correspond to movement of the Intertropical Convergence Zone. ENSO influences are also strong in the study region: El Niño–related droughts negatively affect power production, and cause fluctuations in other economic sectors such as agriculture and health services.

Reservoir managers need forecasts 6 to 12 months in advance in order to maximize the economic efficiency of their power generation activities. The MARS model, which incorporates climatic variables, was developed to improve reservoir management. The model's study site, the El Peñol Reservoir, is the largest in the country and supplies about 14% of the country's hydropower supply.

The authors note that initial results from MARS, a nonlinear regression model, show that, in tests, the model performed best for low-flow seasons and least well (though not intolerably so) for high-flow seasons. They observe that the low-flow season is dominated by human decisions regarding drawdowns, while the high-flow season is when climatic influence is greatest. Poveda et al. also observe that the model seems to be better attuned to high streamflows associated with cold ENSO events, than to the drought conditions generated by the ENSO warm phase.

In testing the model's ability to make real-time forecasts based on historical and current macroclimatic variables, Poveda et al. observed that forecasting skill became lower as the forecast period expanded into the 6- to 12-month horizon. At the same time the greatest value of the forecast to reservoir managers lies in precisely this time frame. Nevertheless, the authors found that, running the model by using data for the years 1977–92, operating cost savings over the long term were generated, and appeared to be in the neighborhood of 35 to 40%. More generally, the authors found, the model offers an opportunity to mitigate economic losses and social unrest associated with the impacts of climate-influenced hydroelectric power shortages.

Magaña and Condé examine climate impacts on water supplies in Sonora, Mexico. The authors note that, in Mexico, economic development is constrained by inequitable distribution of water, a situation that climatic variability and change may exacerbate. Regional plans and programs generally view water supply as being a constant or unlimited resource to be used to satisfy increasing human demand. Hydrologic data, however, contradict this mindset: According to data for 96 aquifers collected by the Mexican National Water Commission, annual demand is about one and one-half times the amount of annual recharge. Some of the important water sources in Sonora are transboundary, most notably the San Pedro, Santa Cruz, and Colorado Rivers.

Population in the border states continues to expand rapidly, with more than 10.5 million people living within the 100 km U.S.-Mexico border zone and 4.3 million of these individuals living in Mexico. The population is 90% urban. Continued growth, the authors caution, is likely to lead to intensified competition for available water supplies, increased demand for better-quality water, and increases in the cost of water. A recent initiative to gradually remove water subsidies, especially for agriculture and the mining industry—both of which are important in Sonora—pose additional need for adjustments.

Magaña and Condé note that Sonora has two quite different hydrologic areas. Inland, the mountains provide most of the water for the state, while the coastal valley constitutes the main source in the southern region of the state. Two rivers account for 87% of drainage in the state. The runoff is stored in 14 dams, of which 5 account for 98% of total storage capacity. Almost all of these dams are near the end of their useful life. Storage facilities can provide only 12% of the state's total water demand, 94% of which is generated by the agricultural sector. Although some dams also generate electricity, the predominant form of power generation is from thermal sources.

Sonora, with an average annual precipitation of 428 mm, receives higher precipitation in the areas along the coast than it does inland. Along the Arizona-Sonora border, precipitation averages less than 100 mm a year. The area has a bimodal precipitation regime, characterized by winter frontal storms arriving from the north and monsoon precipitation in the summer. Summer precipitation is highest in July and August; reservoir levels increase most notably in early spring and early fall, lagging slightly behind the precipitation peaks. Yet averages do not tell the whole story, for the region is influenced by interannual and interdecadal patterns of climate variability, with ENSO being a particularly strong influence. The authors identify correlations between climatic variability and changes in water supplies in selected Sonoran reservoirs. More frequent and stronger El Niño events occurring in the past two decades, they note, have produced increases in reservoir water levels in Mexico; however, in the northern border states, including Sonora, water levels in recent years have dropped substantially and are linked with climatic variability. The results of their research indicate that information about interannual climate variability must be integrated into water planning and management.

Magaña and Condé also consider the potential impacts of climate change on water resources. Results of their analyses suggest that Mexico may see either no change or a decrease in summer precipitation, but an increase in winter precipitation of perhaps 10 to 20%. This, they observe, may enhance water availability in the wintertime, in the northern border areas. However, expectations are that snow-dominated watersheds will experience earlier snowmelt and thus earlier peak flows, as well as reduced summer flows. The authors suggest that increasing winter storage for use during the summer season may be a feasible adaptation strategy. They conclude by recommending production of improved forecasts and integration of climate information and forecasts into water management, together with construction of new infrastructure, maintenance and upgrading of existing facilities, and promotion of efficient technologies to increase reuse of water. Together, these measures could go some way toward addressing the current and anticipated imbalances between water supply and demand in Mexico's northern border area.

The chapter by Arias also focuses on the northern border area of Sonora, but with a sharp focus on interactions between climate variability and land cover changes in the Upper San Pedro River Basin. The San Pedro River originates near the copper mining community of Cananea, Sonora, and flows northward through agricultural and ranching lands, crossing the international border at Naco, just south of Bisbee, Arizona. Analysis of remotely sensed imagery indicates that natural vegetation in the upper reaches of the river underwent significant change during the time period 1973 to 1986, primarily due to conversion of grasslands into mesquite groves and of riparian gallery forests into agricultural lands. This transformation occurred during a period, from the late 1940s to the end of the 1970s, of below-average precipitation in the area, as is indicated by data from six rain gages having at least 35 years of recorded data. The driest years during this time period occurred between 1953 and 1964, and between 1973 and 1983. Arias postulates that the observed changes in land cover are at least partly caused by a prolonged and severe dry period that beset the area during this time period.

Specifically, Arias finds that, in 1973, two-thirds of the basin was covered by grassland, and almost 20% of the land cover was in oak woodland. Less than 1% of the area showed human impacts from urbanization, mining, and other uses. During the succeeding 25 years, grasslands were reduced to just over half, while coverage by mesquite groves increased from about 2% to 14%. Over the same time period, human uses rose to almost 4% of land cover. The rate of change slowed after 1986, but his analysis indicates that a new category of land cover had emerged: abandoned cropland, which he attributes to the virtual elimination of agricultural subsidies. Over the 25-year period, 1973 to 1997, Arias found that the greatest changes involved reduction of grasslands, desert scrub, and riparian forest, and increases in mesquite and cropland (including abandoned cropland). He concludes that replacement of grassland by mesquite may be accelerated in another dry period, and that measures to control mesquite establishment and growth are required to conserve the grasslands. Such conservation has been identified as a major objective in the basin. His findings provide insights important to understanding the combined impacts of land use/land cover change and climate variability/change on the hydrology of the Upper San Pedro River Basin.

## REFERENCES

Bureau of Reclamation, 1975: Executive Summary of *Critical Water Problems Facing the Eleven Western States.* Washington, D.C.: U.S. Dept. of the Interior, 85 pp. (Available through the U.S. Government Printing Office, Washington, D.C.)

Collier, M., R.H. Webb, and J.C. Schmidt. 1996. Dams and Rivers: A Primer on the Downstream Effects of Dams. USGS Circular 1126, Tucson, Arizona, 94 pp.

El-Ashry, M.T. and D.C. Gibbons. 1988. *Water and Arid Lands of the Western United States.* Cambridge: Cambridge University Press, 415 pp.

Intergovernmental Panel on Climate Change (IPCC). 1998. *The Regional Impacts of Climate Change: An Assessment of Vulnerability.* Edited by R.T. Watson et al., London: Cambridge University Press, 512 pp.

Intergovernmental Panel on Climate Change (IPCC). 2001. *Climate Change 2001: Impacts, Adaptation, and Vulnerability. Contribution of Working Group II to the Third Assessment Report of the Intergovernmental Panel on Climate Change.* Edited by J.J. McCarthy, O.F. Canziani, N.A. Leary, D.J. Dokken, and K.S. White. London: Cambridge University Press, 1032 pp.

Liverman, D.L. and R. Merideth. 2002. Climate and society in the U.S. Southwest: The context for a regional assessment. *Climate Research* 21(3): 199–218.

Messerli, B., and J.D. Ives. 1997. *Mountains of the World: A Global Priority.* London: The Parthenon Publishing Group, 495 pp.

Miles, E., N. Mantua, A. Hamlet, K. Gray, R. Francis, and P. Mote. 1998. Preliminary results of an integrated assessment of ENSO impacts on the Pacific Northwest. Paper presented to the 25th Anniversary Symposium of the School of Marine Affairs, University of Washington, Seattle, Washington, May 8, 1998.

Morehouse, B.J. 1995. A functional approach to boundaries in the context of environmental issues. *Journal of Borderlands Studies* 10(2): 53–73.

National Assessment Synthesis Team (NAST). 2000. *Climate Change Impacts on the United States. Overview Report.* The U.S. Global Climate Change Research Program (USGCRP). Cambridge, U.K.: Cambridge University Press, 154 pp.

National Research Council (NRC). 1991. *Managing Water Resources in the West Under Conditions of Climate Uncertainty.* Washington, D.C.: National Academy Press, 344 pp.

National Research Council (NRC). 1996. *River Resource Management in the Grand Canyon.* Washington, D.C.: National Academy Press,

*Natural Resources Journal.* 2000: Water Issues in the U.S.–Mexico Borderlands. Vol. 40(4), Albuquerque, New Mexico: The University of New Mexico School of Law.

Nichols, P.D., M.K. Murphy, and D.S. Kenney. 2001. *Water and Growth in Colorado: A Review of Legal and Policy Issues.* Boulder, Colorado: Natural Resources Law Center, University of Colorado School of Law, 191 pp.

Office of Technology Assessment (OTA), U.S. Congress. 1983. *Water-Related Technologies for Sustainable Agriculture in U.S. Arid/Semiarid Lands.* Washington, D.C.: U.S. Government Printing Office, 412 pp.

Powell Consortium. 1995. *Severe Sustained Drought: Managing the Colorado River System in Times of Water Shortage.* Papers reprinted from the *Water Resources Research Bulletin* 31(5), October 1995. Tucson: Arizona Water Resources Research Center, University of Arizona.,

Pulwarty, R.S. 2001. Integrated assessments. Paper presented at Fire and Climate 2001 Workshop, Tucson, Arizona, February 14–16, 2001.

Pulwarty, R.S., and K.T. Redmond. 1997. Climate and salmon restoration in the Columbia River Basin: The role and usability of seasonal forecasts. *Bulletin of the American Meteorological Society* 73: 381–397.

Reisner, M. 1993. *Cadillac Desert: The American West and its Disappearing Water.* New York: Penguin Books.

Stegner, W. 1987. *The American West as Living Space.* Ann Arbor: The University of Michigan Press, 89 pp.

Varady, R.G., and B.J. Morehouse. In press. Moving borders from the periphery to the center: River basins, political boundaries and water management policy. In, Fort, D., and R. Lawford (eds.). *Science and Water Resource Issues: Challenges and Opportunities.* American Geophysical Union Water Resources Monograph Series..

Western Water Policy Review Advisory Commission (WWPRAC). 1998. *Water in the West: Challenge for the Next Century.* Available from the *National Technical Information Service*, Springfield, Virginia.

Wilkinson, C.F. 1992. *Crossing the Next Meridian: Land, Water, and the Future of the West.* Washington, D.C.: Island Press, 376 pp.

Worster, D. 1985. *Rivers of Empire: Water, Aridity and the Growth of the American West.* New York: Pantheon Books.

2

# BOUNDARIES IN CLIMATE-WATER DISCOURSE

Barbara J. Morehouse
*Institute for the Study of Planet Earth, The University of Arizona, Tucson, Arizona*

**Abstract** "Boundary" is a deceptively simple term for a very complex ensemble of ideas, expectations, and theories. From a social science perspective, boundaries may (among other things) spatially define the limits of authority, jurisdiction, and sovereignty; map the edges of identity; define market and service areas; filter flows; or legitimize surveillance. By contrast, in scientific disciplines such as climatology and hydrology, boundary metaphors may facilitate description (e.g., atmospheric or oceanic structure), locate the point at which specified conditions (or changes in those conditions) begin and end, or define the shape and size of the spaces into which data are aggregated. Unlike politically defined boundaries, the definitions and functions of these boundaries are derived through scientific inquiry. Water resource management is one of the areas in which the two discourses of water and boundaries meet and sometimes collide. Climatological, hydrological, and social science research would benefit from a more sharply focused examination of how different uses of "boundary" as concept and as material reality might influence the nature and range of water resource management options deemed feasible to implement under various types of climatic conditions.

## 1. WHY SHOULD WE THINK ABOUT BOUNDARIES?

We tend to take boundaries for granted, and why not? Boundaries not only are ubiquitous on the landscape, but, as metaphors, are pervasive in social discourse around the world. Yet people have not always viewed their worlds in terms of mapped boundaries. Indeed, some languages do not even contain a word for boundaries (Morehouse 1996). Boundaries as we know them today are an outgrowth of the desire of European monarchies, begin-



*Henry F. Diaz and Barbara J. Morehouse (eds.), Climate and Water*, 25-40.

ning in the seventeenth century, to consolidate power over lands and subjects, and of the efforts of Enlightenment scientists to categorize and map the world and its contents (see, e.g., Jones 1959; Kristof 1959; Prescott 1978). The spaces encompassed by boundaries constitute individual geographies, each with its own location, shape, size, contents, and meaning (see, e.g., Paasi 1996).

We continue to rely on these concepts today as a means of establishing order out of chaos, whether we are engaged in microbiology or astronomy, microscale groundwater recharge or global change modeling. We rely on them for many reasons, not the least of which is because they symbolize, in very simple form, complex information about the structures and rules that govern our world. Yet boundaries drawn for one purpose, such as establishing the jurisdictional space—and limits—of a government agency or the dividing line between two sovereign states, may be irrational within other contexts such as watershed management (where watersheds are bisected by boundaries) or research into larger-scale climate or hydrologic processes.

## 2. BOUNDARIES, SCIENCE, AND SOCIAL PROCESS

Much of the literature about boundaries focuses on various aspects of international boundaries and border areas (for a review of the literature, see Newman and Paasi 1998), or internal political boundaries (see, e.g., Morehouse 1996). There is, however, a body of work that considers boundaries from an array of perspectives ranging from the very personal (the body) to the very abstract (e.g., psychology). Further, island biogeography grapples with boundary and edge issues, as does climatology (e.g., the atmospheric boundary layer), and conservation biology (in terms of ecological edges and transition areas). Use of Geographic Information System (GIS) and remotely sensed technologies poses boundary challenges as well, for error is greatest at the edges where two or more maps or images must be juxtaposed.

Regardless of whether boundaries are viewed from social, natural, or physical science perspectives, all boundaries to one degree or another represent an attempt to assert power (or in the case of science, authority) over specific subject matter. In some cases, particularly those based on scientific methods, this assertion of authority is associated with efforts to bring order out of chaos. In others, particularly in the social realm, boundaries represent explicit or implicit power relations between entities wishing to articulate and enforce differentiation in material or metaphorical space. Over time, most boundaries and the institutions associated with them come to be embedded in society and to be seen as permanent fixtures on the landscape. The boundaries and their embedded institutions, in turn, influence the structure of internal and transboundary relations, patterns of flows and interactions, collection and analysis of data, and ideas about what kinds of cross-border interactions may or may not be possible or desirable. These conditions hold true, at least to some extent, at all scales from international boundaries to micro-local jurisdictions and individual property.

The various rules governing border decisions, interactions, and behaviors arise in response to articulated needs, and persist because they continue to address one or more concerns of those whose responsibility it is to maintain the integrity of the area (and its contents) that is encompassed within the boundaries. At the same time, boundary rules may seriously impede important transboundary interactions. Efforts to manage natural resources in transboundary settings, to conduct research on natural processes occurring at supra-boundary scales, and to efficiently share information across boundaries are among the most compelling examples of situations in which boundary rules can impede progress.

## 3. FUNCTIONS SERVED BY BOUNDARIES

Broadly speaking, boundaries may serve in the following types of functions: identity, inclusion, exclusion, control of flows, jurisdictional authority, surveillance, and territoriality (including assertion of market and service areas) (see, e.g., Paasi 1996; Sibley 1995; Morehouse 1995; Sack 1986). In science, boundaries facilitate description, locate the beginning and end points of designated conditions, and define the shape and size of research domains for purposes such as data collection, aggregation, and analysis.

### 3.1. Identity

Identity at all levels from the individual to the supranational increasingly has come to be articulated in terms of material or metaphorical boundaries. Merely glancing at conflicts in today's world, not to mention the strong drive toward affirmation and protection of cultural identity, reaffirms the importance accorded by society to clearly delineated territory. Having one's own identity geographically defined and depicted on a map, be it a map of the world or of one's own neighborhood, in turn, affords at least the promise of internal cohesion and capacity to actively participate in external affairs. Today, of course, the extensive activity surrounding the drawing and maintenance of boundaries coexists with processes of globalization, which challenge the utility of boundaries at all scales. The tension between the drawing/enforcement of boundaries and the elimination of territorial distinctions will continue to influence the political and physical landscape for many years to come.

Boundaries delineating geographies of social and cultural identity can provide a valuable structure for asserting control over internal affairs and warding off unwanted outside influences. Native Americans, for example, went from resisting the establishment of reservation homelands to vociferous defense, based on common tribal identity, of the lands they now occupy. Fomenting nationalistic fervor based on an idea of common identity in a shared homeland has been successfully employed in many countries, including Iran, Afghanistan, and Mexico, to rouse citizen support for governmental policies and actions or to rally them in times of disaster or strife.

It is not a far stretch to transpose geopolitics of identity to issues of transboundary cooperation in managing common-pool resources such as water and coping with climate impacts (see, e.g., Blatter and Ingram 2001). Substantial reluctance may exist with regard to cooperating across boundaries on such issues, particularly in cases where substantial power imbalances and economic disparities pose real or perceived threats to sustaining a separate national or group identity. For the U.S.-Mexico border region, the International Boundary and Water Commission (IBWC) serves as a mediating institution that facilitates interactions between the two countries on boundary and water issues. The IBWC, while operating binationally, nevertheless reflects the value attached to the international boundary, for it is composed of distinct U.S. and Mexican sections, each of which reports to its own federal government bureau: the U.S. section reports to the State Department, and the Mexican section is part of the Mexican Foreign Ministry.

Even in cases where the parties are relatively equal, as in the case of the United States and Canada, achieving satisfactory levels of cooperation may require substantial negotiation, particularly in cases where Canada believes its identity may be co-opted. The International Joint Commission (IJC), like the IBWC, its counterpart on the U.S.-Mexico border, provides an institutional mechanism for overcoming these barriers, within the carefully restricted parameters of boundary maintenance and resolution of transboundary water issues. The IJC, however, like the IBWC, maintains separate sections reporting to their own federal governments.

Related to the notion of identity, human territoriality has been defined as a "powerful geographic strategy to control people and things by controlling area" (Sack 1986, p. 5). Far from being biologically motivated, however, human territoriality is "socially and geographically rooted" (Sack

1986, p. 2). Controlling an area and its contents frequently involves controlling and policing the boundaries that contain those contents. Boundaries not only define material or imagined space but also give form to territorial identity and provide a context for defining membership within the bounded area. The territories thus defined and controlled, like the defining boundaries, are not simply created as one-time structures. Rather, as emphasized by LeFebvre (1991) and Soja (1989), these structures continue to exist by virtue of, but only so long as they are maintained by, social process.

Territoriality functions to legitimate valuation of human and natural resources and provides a foundation for engaging in social relations with entities outside the territorial unit. Territoriality also comes into play when companies define market areas, and when private- and public-sector entities delineate service areas.

Viewed from the perspective of climate and natural resource management, territoriality provides an essential framework for understanding how and why natural processes are important to inhabitants and for justifying investments in infrastructure, institutions, and research needed to maximize opportunities and resilience while minimizing vulnerability.

## 3.2. Sovereignty/Jurisdictional Authority

Sovereignty and jurisdictional authority are closely related to identity and territoriality. Sovereignty refers to the capacity of the governing entity (or entities) to shape and preserve the identity of the society contained within the territorial boundaries, and to control what occurs within the territory and at the territorial borders, free of undue influence from external sources. Sovereignty is typically associated with nation-states but is also a crucial factor in relations between subnational entities and the state. For example, native peoples in the United States and Canada consider themselves to be sovereign entities having a special relationship with the governing nation-state. U.S. states and Canadian provinces, likewise, have a degree of sovereign status granted them through their federal constitutions.

Jurisdictional authority refers to the geographical area controlled by entities such as cities; water management areas; agencies at the federal, state, or local level; and transboundary bureaus such as the IBWC and the IJC, which have been authorized to govern or administer specifically delineated territories as representatives of a higher-ranking entity. For example, in Arizona, the Department of Water Resources (ADWR) is charged with managing the state's water supply, while the Arizona Department of Environmental Quality (ADEQ) has primary responsibility for water quality regulation. The jurisdiction of each, broadly defined, encompasses the entire state. However, much of the power afforded by law to ADWR relates to its authority to strictly manage groundwater in legally designated portions of the state called Active Management Areas (AMAs). These are areas identified as far back as the late 1940s as facing critical problems due to excessive overpumping of groundwater.

Today, there are five AMAs, all located in the most heavily populated north-south central corridor of the state. Within the boundaries of the AMAs, ADWR has considerable authority to set and enforce rules aimed at balancing demand with renewable supply. Outside the AMA boundaries, ADWR has no such authority. Likewise, the agency's regulatory power over surface water supplies is relatively minimal (Norton 1999), and legal authority to engage in conjunctive management of surface and groundwater supplies remains fraught with controversy (Glennon and Maddock 1994). This geographical pattern of jurisdictional authority has significant implications for how growth and development unfold throughout the state.

## 3.3. Inclusion and Exclusion

Boundaries serve an important function in determining what is supposed to be "inside" and what is supposed to be "outside." In the context of social process, boundaries have become an almost indispensable tool in bids to control what is to be allowed to enter and exist within the delineated territory, and to define that which is "not us," and therefore must be relegated to the other side of the line. The combination of rules of inclusion and exclusion constitutes a powerful means of maintaining identity and territorial control, and for sustaining sovereignty. It also contributes substantially to the enforcing body's capacity to engage in surveillance within, at the boundaries, and outside the bounded area. One need only look at the surveillance technology arrayed on the U.S.-Mexico border to grasp how important this function has become.

In natural resource management, rules of inclusion and exclusion provide, for example, a means for protecting and maintaining control over resources (e.g., water supplies) within the territorial unit while externalizing undesirable elements (e.g., exotic species, hazardous waste, etc.). In science, research often entails building a "fence" around what will be included in a study, with all else relegated to the categories of externalities or irrelevancies. This is a necessary process, and one that allows for rational collection and analysis of data, ideally leading to significant advances in knowledge within defined parameters. In some cases, such as climate modeling, "forcing" conditions at the boundary of the model allows for controlled perturbation of the system's parameters, and thus experimentation on how the perturbation affects the factors contained within the boundaries. In others, creating a boundary between two factors, such as surface water and groundwater, may facilitate structured hydrologic research and experimentation.

Institutional separation of groundwater and surface water, however, can lead to significant problems in water resource management, as has been the case in Arizona. Here, surface water and groundwater are considered to be separate under the law (see Glennon and Maddock 1994). Efforts to modify the legal framework in order to recognize hydrologic interaction between surface and groundwater has thus far foundered on the particulars of case law precedent and judicial decisions. The need to conjunctively manage surface and groundwater is especially critical in areas where the water table is highly responsive to climatic variability. Here, boundaries may make no sense from either a hydrologic or a management point of view, but institutional entropy impedes making the necessary changes.

An interesting inclusion/exclusion dilemma is posed, for example, by a stretch of jurisdictional boundary of the Tucson AMA running along the northeast edge of the AMA near the town of Oracle. The water system that serves Oracle is within the AMA boundaries and is thus subject to ADWR requirements, but the service area is 10 miles away, outside the AMA. Although the water provider can meet current water demand, concern has been expressed recently regarding the possibility for future water shortages if anticipated massive residential development occurs. Inside the AMA boundaries, new developments are required to demonstrate availability of enough good-quality potable water to serve estimated demand for 100 years; outside the AMA boundaries, no such requirement exists.

One potential, but politically unlikely, solution is to incorporate the lands where proposed development is anticipated within the AMA. The developers would probably fight any such proposal. Another solution would be to extend the assured water supply requirement to the entire state, thus eliminating the inside/outside problem. In a recent audit of ADWR, the Auditor General of Arizona (Norton 1999) recommended that this be done. Again, however, it would be very difficult to enact legislation on the issue, because rural interests could be expected to strongly oppose any such initiative. Indeed, it may take a severe extended drought, plus extensive urban development, even to gain recognition of the need for institutional change within the official discourse of state water policy.

## 3.4. Filtering Flows

The function of boundaries in filtering flows differs from the function of inclusion/exclusion in that, in this case, rules of selectivity (rather than blanket allowances or prohibitions) are in operation. The primary question asked with regard to the filtration of flows is, "who or what can cross and under what circumstances?" It is an important function, for although banning all movement across geographical boundaries is very difficult to enforce and not particularly desirable, controlling individual flows may be essential for any number of reasons. For example, Arizona water law generally prohibits the transfer of water out of a non-AMA area into an AMA. This law was enacted to address the concerns of rural counties where such transfers posed a serious threat to the continued viability of local economies. By contrast, under conditions of drought, water may be transferred between non-AMA areas—but only so long as detailed rules are followed. Among the boundary filtering rules is one that states that the water must be moved by train or motor vehicle; another rule stipulates that the water must be required for stock watering or city water service. Invoking a boundary rule of exclusion, the law expressly prohibits water from being used in an AMA (reported in University of Arizona Water Resources Research Center 2000).

Natural processes do not respect political boundaries; they might not even respect natural boundaries identified and delineated through scientific research. This is certainly the case with climate, and in many cases, with hydrologic flows as well. Nowhere are these realities brought into sharper relief than in the case of communities juxtaposed at political boundaries. For example, heavy rains may occur in both southern Arizona and northern Sonora. If these rains lead to flooding in, for example, Nogales, Sonora, the effects are likely to be experienced equally or even more in Nogales, Arizona. This is due to regional topographic characteristics that produce northward flows of water in streambeds and arroyos (Ingram et al. 1995). Furthermore, droughts as well as floods have significant cross-border impacts (Morehouse et al. 2000). Along the U.S.-Mexico border, border communities must work together, with the assistance of the IBWC/CILA (Comisión Internacional de Límites y Agua) and other organizations, to address these kinds of problems. Recognition that this kind of boundary rule is being invoked may be crucial in efforts to resolve problems involving transboundary flows, whether of contaminants or of other phenomena.

## 4. ADDRESSING BOUNDARY ISSUES THROUGH INTEGRATED ASSESSMENT

In recent years, integrated assessment (IA) and IA modeling (Easterling 1997; Rotmans and Dowlatabadi 1998; Meyer et al. 1998; Parson 1995; see also "Thematic Guide to Integrated Assessment Modeling," (http://sedac.ciesin.org/mva/iamcc.tg/), have emerged as a promising framework for transcending geographical, disciplinary, and science-society boundaries through a greater emphasis on collaborative research. Interest in integrated assessments may be traced to recognition of the need for new tools to address large issues that cannot be effectively addressed through narrowly focused science inquiries. Global change research and modeling efforts, and the results of such research reported by the International Panel on Climate Change (IPCC) and others (see, for example, IPCC 2001; USGCRP 2000) have underlain much, though by no means all, of the progress made in integrated assessment methodology and empirical research.

The Climate Assessment for the Southwest (CLIMAS) Project, funded by the U.S. National Oceanic and Atmospheric Administration's Office of Global Programs (NOAA-OGP), is an example of an ongoing integrated assessment, in this case focused on natural climate variability and its impacts on the region's human and natural systems (see http://www.ispe.arizona.edu/climas/). Social science research and stakeholder outreach adhere closely to political boundaries in terms of what is included and excluded from the project and consider climate impacts at more local scales. The natural science research component, by contrast, includes larger geographical areas, but tends to exclude the factors specifically associated with social process.

The project recognizes the need for boundaries that identify its geographical focus, but these boundaries are multiple and overlapping. From a social science perspective, the boundaries coincide with the political boundaries of Arizona and New Mexico, but also stretch over the boundary into the northern Mexico border region. On the natural science side, the region takes in an area stretching from the Gulf of California on the south to the Colorado River Basin on the north, and includes parts of eastern California as well as the southern tip of Nevada. This larger area allows for analysis of important physical processes that affect human and natural systems within the socially defined region. One of the significant challenges for researchers involved in the project is to find ways to resolve problems generated through use of different scales of inquiry and due to boundary-related disjunctures in availability, quality, and attributes of social and physical/natural data.

# 5. THE CHALLENGES OF INTEGRATING CLIMATE AND WATER RESOURCE MANAGEMENT ACROSS BOUNDARIES

Integrating climate and water resource management across boundaries is possible, but not easy to accomplish. Issues of temporal and spatial scale, jurisdiction and sovereignty, scientific rationality, technological efficiency, transaction costs, and social equity are likely to arise, as are political and economic considerations. All of these issues cannot be addressed simultaneously, but how they come to be prioritized and acted upon may spell the difference between success and failure. At the outset, scientists need to have a clear understanding of how boundaries, with all of their political, economic, social, cultural, and scientific baggage, influence relations and dynamics at the boundary interface. This information provides essential background information for working closely with decision makers, stakeholders, and community members to set research and implementation agendas for assessment of climatic variability and its impacts.

## 5.1. Temporal and Spatial Scale

Identifying the appropriate temporal and spatial scales for research and for dissemination of information is always challenging. In boundary areas, the problem can be exacerbated by the existence of the boundary itself, especially when the territorial unit delineated by the boundary (country, state, reservation, jurisdictional unit) serves as the limit for collection of data, mapping of key variables, or other essential practices. Selection of appropriate spatial scale is essential, as is harmonizing disparities in data arising from data collection practices as well as from the influence of the boundary itself on data collection practices.

Consideration of temporal scale is equally important, for shorter-term trends in human and physical systems may exclude important longer-term patterns. For example, recent research on climate based on analysis of the Pacific Decadal Oscillation (PDO) (see, e.g., Mantua et al. 1997) indicates a correlation between the positive phase of the PDO and multidecadal-scale patterns of wetter than average climate conditions. Such conditions have implications for water management as illustrated by the fact that an extended wet period was in effect in the early 1920s when the crucial calculations were done in dividing the waters of the Colorado River among the basin states and, later, with Mexico. The negative phase of the PDO has been correlated with the most severe drought experienced in the southwestern United States and in northern Mexico within the historical record of more than 100 years, that of the 1950s.

Similarly, long-term trends may mask important short-term variations that may have profound impacts on human or natural systems. The recent wildfire in the vicinity of Los Alamos, New Mexico, may be understood in terms of climate patterns over just the past 3 years, combined with poor judgment and longer-term fire suppression practices that caused a large fuel-load buildup. In this case, the policies in force within the jurisdictional boundaries of the U.S. Forest Service and National Park Service combined with shorter-term but larger-scale climate processes to produce a fire that threatened not only the local forest, but also the town of Los Alamos and the Los Alamos National Laboratory. Subsequent concerns about mudslides, flooding, and massive erosion during the summer monsoon season, have only added to the actual and potential damages associated with the event.

## 5.2. Jurisdiction and Sovereignty

Concern about maintaining control not only over data, but over how those data are analyzed and used, may make it difficult to work at optimal temporal and spatial scales. Such situations may arise out of issues associated with the desire to maintain jurisdictional control or political sovereignty. Working with native peoples in the U.S. Southwest has brought home to CLIMAS researchers the need to carefully negotiate agreements with specific tribes regarding what information will be collected, by whom, relating to what area or resources, and for what purpose (Austin et al. 2000). Likewise, methods for reporting results and archiving data must be agreed upon before work can proceed.

The collection of data by satellites and other remote sensing devices may be useful in overcoming some of these problems, although ownership and use of such data have been contested among groups, such as some Native American tribes, that are concerned about preservation of culture and sovereignty. Ground-based measurements continue to be essential for data validation as well as for long-term comparative studies. Such measurements are crucial to assessment of how climate and hydrologic processes play out locally and how they generate particular kinds and intensities of social and ecological impacts. Gaining access to data collection sites may present challenges to researchers who must cross protected boundaries of all kinds. Further, dissemination of information may generate border issues. NOAA's Climate Prediction Center (CPC), for example, issues official climate forecasts for the United States. However, these forecasts stop at the international boundaries for political reasons. Many people, particularly those living near the international boundaries, have a strong interest in forecasts and other climate information that spans the boundary. Currently, such information can be obtained only from sources, such as the International Research Institute (IRI), whose products do not stop at the U.S.-Mexico border.

## 5.3. Scientific Rationality and Technological Efficiency

Integrating climate and resource management across boundaries, or even "erasing" the boundaries, is essential to transcending partial and imperfect knowledge about the physical characteristics and dynamics of climate and hydrologic processes and about the impacts of those processes on human and natural systems. Addressing issues of geographical or temporal scale is a valuable means for assuring scientific rationality in the study of complex interactions among the physical processes, and between physical and social processes, associated with climate and water resource management. Expansion of inquiry across the boundaries of study areas, including the boundaries embedded in climate and hydrologic models, and across disciplinary boundaries, may also facilitate technological efficiency with regard to formulating and disseminating climate information and forecasts as well as using such information and knowledge to improve water resource management. The past two La Niña episodes, for example, resulted in serious drought conditions along portions of the U.S.-Mexico border. The impacts of the drought were very significant for binational water resources, particularly on the Rio Grande/Río Bravo. The capacity to meet treaty requirements for delivery of water became a point of significant concern for both Mexican and U.S. water managers.

A newly implemented U.S. National Drought Mitigation Center Monitor (http://enso.unl.edu/monitor/) offers a means for anticipating and planning for such events in the future. However, scientific rationality about drought impacts on water resources and technological efficiency in water management require that comparable data be available for the entire watershed, on both sides of the international boundary. Yet such data are not available at either the temporal or spatial scales required. Only now are these issues beginning to be addressed

## 5.4. Politics, Economics, and Social Equity

The capacity of an entity to effectively incorporate climate considerations into water resources management may hinge very directly on the availability of sufficient financial, scientific, educational, and human resources. Capacity may also be facilitated or constrained by the degree of flexibility, resilience, and adaptability inherent in internal political structures, as well as by the nature and amount of external political pressure being exerted on internal regimes. In cases where coping capacity, resilience, and adaptability are low, issues of social equity are likely to be particularly problematical. There may be a strong impetus to allocate scarce resources to those entities who have the most power and influence, rather than to those who stand to be most adversely affected by such scarcity.

## 5.5. Transaction Costs

Transaction costs may be defined to refer to the actual or potential "price" of following a particular course of action. In cases such as that of integrating climate information into water resource management practices, willingness to change ways of doing business may hinge very directly on the decision maker's assessment of the relative risks and benefits of doing so. It is very important to ascertain how individuals define risks and benefits, prioritize them, and assign relative values to them. In transboundary settings, the risks and benefits—and the values assigned to them—may be quite different. Reconciling these differences may be an essential first step toward changing institutional rules and practices. In turn, institutional changes may need to be achieved before decision makers and managers will assume the risk of changing their practices.

In the realm of climate and water resource management, this latter step may be perceived to be fraught with peril, in no small part because climate forecasts in particular are issued in probabilistic terms and by their very nature are uncertain. Understanding the nature of these types of transaction costs must underlie recommendations for institutional change. In addition, transaction costs associated with educating managers about how to use climate and hydrologic information and forecasts, developing infrastructure to accommodate new ways of doing business, undertaking research to fill knowledge gaps, and acquiring data must also be taken into account. Again, in transboundary settings, each of these factors may be very costly, and may pose considerable barriers to achieving the desired level of change.

## 6. TARGETING RESEARCH AND OUTREACH EFFORTS IN BOUNDARY CONTEXTS

Knowing one's audience is essential to success in disseminating new knowledge, technologies, and ways of managing water resources based on enhanced use of climate information. Different entities have differing needs. A pilot survey of stakeholders in the U.S. Southwest, for example, revealed that some individuals—notably those involved in daily operations—cited a desire for mainly local-scale, short-term information. Those with broader missions (such as budget and infrastructure planning and general administration) were interested in longer-term, larger-scale information. Typically, stakeholders wanted the information presented in less technical terms than scientists customarily use in communicating information, and they wanted the information to be set in a broader context (see Benequista and James 1999).

Among those entities to whom information about climate and water resource management can be usefully communicated are politicians and government bureaucrats at all levels from the national to the local, communities located on or near boundary lines of all sorts from international to local jurisdictional, and nongovernmental organizations. Important targets also include private-sector entities whose interests are transnational and whose operations are sensitive to climate variability, such as power companies, water companies, and high-volume water users (e.g., electronics manufacturers and agribusiness operations). In each case, preliminary research establishing the nature and extent of sensitivity to climate, actual or potential vulnerabilities to adverse climate conditions or events, and institutional factors that influence the degree of resilience to climate impacts, is essential. Such research provides a foundation for determining what sorts of climate information are needed, the appropriate temporal and spatial scales of the information, and how that information is formatted and transmitted.

Integrated assessments that emphasize collaboration across disciplinary boundaries and full incorporation of key stakeholders into the research and outreach process provide a useful framework for carrying out this kind of research. If properly designed, integrated assessments facilitate investigation of research questions from multiple perspectives and allow for dynamical interaction and feedback among the various segments of a project, thus enhancing the potential for maximizing the utility of the resulting knowledge and products to society. Furthermore, integrated assessments, by explicitly valuing multiple perspectives, provide a solid philosophical basis for grappling with issues that span—and transcend—boundaries of all kinds. Climate assessments being carried out in the Americas through organizations such as NOAA and the InterAmerican Institute (IAI) are excellent examples of ways in which this kind of work can be approached productively.

## 7. LOOKING TO THE FUTURE

Incorporating climate considerations into water resource management requires a thorough understanding of how the drawing of boundaries, definition of boundary functions, and establishment/enforcement of boundary rules favor certain ways of not only managing water resources, but also of understanding the underlying physical and social dynamics involved. This understanding applies not only to the political boundaries that separate countries, states, or local jurisdictions, but also to those boundaries that define markets areas, service areas, private property limits, natural areas, computer model parameters, atmospheric and hydrologic characteristics, and divisions between disciplines.

Advances in assessment of complex issues, such as climate impacts on water resources, could be enhanced by the emergence of a subsector of the integrated assessment community devoted to addressing these and other boundary questions in a focused but multifaceted way. Particularly needed are studies explicitly designed to compare climate and water resource issues within different boundary contexts. Such studies are necessary for testing different theoretical and methodological approaches, and for distinguishing broader trends and patterns from the highly idiosyncratic characteristics of individual locations. Comparative studies also have the potential to stimulate innovation in the production and communication of climate and hydrologic information tailored to specific target audiences.

## 8. REFERENCES

Austin, D., S. Gerlak, and C. Smith. 2000. *Building Partnerships with Native Americans in Climate-Related Research and Outreach*. CLIMAS Report Series CL2-00, Institute for the Study of Planet Earth, The University of Arizona, Tucson, Arizona.

Benequista, N., and J.S. James. 1999. *Pilot Stakeholder Assessment Report*, February, 1999. http://www.ispe.arizona.edu/climas/archive.html.

Blatter, J., and H. Ingram. 2001. *Reflections on Water: New Approaches to Transboundary Conflicts and Cooperation*. Cambridge, Massachusetts: Massachusetts Institute of Technology Press.

Easterling, W.E. 1997. Why regional studies are needed in the development of full-scale integrated assessment modeling of global change processes. *Global Environmental Change* 7(4): 337–356.

Glennon, R.J., and T. Maddock, III. 1994. In search of subflow: Arizona's futile effort to separate groundwater from surface water. *Arizona Law Review* 36(3): 567–610.

Ingram, H., N.K. Laney, and D.M. Gillilan. 1995. *Divided Waters: Bridging the U.S.-Mexico Border*. Tucson: University of Arizona Press.

IPCC. 2001. Intergovernmental Panel on Climate Change, United Nations Environment Program. http://www.usgcrp.gov/ipcc/WG3SPM.PDF.

Jones, S.B. 1959. Boundary concepts in the setting of place and time. *Annals of the Association of American Geographers* 49(3): 241–255.

Kristof, L.K. 1959. The nature of frontiers and boundaries. *Annals of the Association of American Geographers* 49(3): 269–282.

LeFebvre, H. 1991. *The Production of Space*. Translated by Donald Nicholson-Smith. Oxford: Basil Blackwell.

Mantua, N.J., S.R. Hare, Y. Zhang, J.M. Wallace, and R.C. Francis. 1997. A Pacific interdecadal climate oscillation with impacts on salmon production. *Bulletin of the American Meteorological Society* 78: 1069–1079.

Meyer, W.B., K.W. Butzer, T.E. Downing, B.L. Turner, II, G.W. Wenzel, and J.L. Wescoat. 1998. Reasoning by analogy. In, S. Rayner and E.L. Malone (eds.), *Human Choice and Climate Change, Vol. 3, Tools for Policy Analysis*. Columbus, Ohio: Battelle Press, pp. 217–289.

Morehouse, B.J. 1995. A functional approach to boundaries in the context of environmental issues. *Journal of Borderlands Studies* 10(2): 53–73.

Morehouse, B.J. 1996. Conflict, space and natural resource management at Grand Canyon. *The Professional Geographer* 48(1): 46–57.

Morehouse, B.J., R.H. Carter, and T.W. Sprouse. 2000. The implications of sustained drought for transboundary water management in Nogales, Arizona and Nogales, Sonora. *Natural Resources Journal* 40: 783–817.

Newman, D., and A. Paasi. 1998. Fences and neighbours in the postmodern world: Boundary narratives in political geography. *Progress in Human Geography* 22(2): 186–207.

Norton, D.R. 1999. Performance Audit, Arizona Department of Water Resources. Report to the Arizona Legislature by Douglas R. Norton, Auditor General, April 1999, Report No. 99-8, 32 pp.

Paasi, A. 1996. Inclusion, exclusion and territorial identities: The meanings of boundaries in the globalizing geopolitical landscape. *Nordisk Samhällsgeografisk Tidskrift* 23: 3–17.

Parson, E.A. 1995. Integrated assessment and environmental policy making. *Energy Policy* 23: 463–475.

Prescott, J.V.R. 1978. *Boundaries and Frontiers*. London: Croom Helm.

Rotmans, J., and H. Dowlatabadi. 1998. Integrated assessment modeling. In, S. Rayner and E.L. Malone (eds.), *Human Choice and Climate Change, Vol. 3, Tools for Policy Analysis*. Columbus, Ohio: Battelle Press, pp. 291–377.

Sack, R.D. 1986. *Human Territoriality: Its Theory and History*. Cambridge: Cambridge University Press.

Sibley, D. 1995. *Geographies of Exclusion: Society and Difference in the West*. London: Routledge.

Soja, E.W. 1989. *Postmodern Geographies: The Reassertion of Space in Critical Social Theory*. London: Verso.

University of Arizona Water Resources Research Center. 2000. 2000 Legislative session enacts varied water-related legislation. *Arizona Water Resource* 8(6): 7.

USGCRP. 2000. *Climate Change Impacts on the United States: The Potential Consequences of Climate Variability and Change*, Overview. U.S. Global Change Research Program (USGCRP), National Assessment Synthesis Team, Cambridge, U.K. and New York: Cambridge University Press.

3

# EXPERIENCE AND ROLE OF THE ORGANIZATION OF AMERICAN STATES IN TRANSBOUNDARY RIVER BASIN MANAGEMENT IN LATIN AMERICA

Jorge Rucks
*Unit of Sustainable Development and Environment, Organization of American States, 1889 F Street, N.W., Washington, D.C. 20006*

**Abstract** The governments of the American states recognized the importance of the integrated management of water resources when they approved the Action Plan for Sustainable Development of the Americas at the Summit of the Americas for Sustainable Development (Santa Cruz de la Sierra, Bolivia, December 2001), thus giving a new official thrust to this issue. The heads of state and presidents of the American countries delegated to the Organization of American States (OAS) General Secretariat the responsibility for supervising this action plan. In order to do so, the General Secretariat has made use of its comparative advantage in dealing with groups of countries in the Americas by working on the management of transboundary river basins. The Unit for Sustainable Development and Environment (USDE)[1] of the OAS supports specific agreements that bring together different countries with similar objectives to work together in the management of the key transboundary basins, sharing experiences and supporting the development of national and regional policies. In this chapter, I describe some of the activities of the OAS in support of transboundary water resources management in the Americas.

[1] Also known by its Spanish name, Unidad de Desarrollo Sostenible y Medio Ambiente (UDSMA).



*Henry F. Diaz and Barbara J. Morehouse (eds.), Climate and Water*, 41-57.

# 1. INTRODUCTION

The quantity and quality of water resources associated with transboundary water systems and basins in the Americas normally rely on the will and capabilities of two or more countries. About 44 transboundary river basins in Latin America contain border areas, and therefore have political relevance. For example, the Amazon and the La Plata River Basins include part of all the South American countries with the exception of Chile. One of the biggest aquifers in the world, the Guarani Aquifer, with an area of more than $1.2 \times 10^6$ km$^2$, is a transboundary water system shared by four countries: Argentina, Brazil, Paraguay, and Uruguay.

The La Plata River Basin, with an area of about $3.2 \times 10^6$ km$^2$, encompasses the most important industrialized areas of Latin America, with high rates of population growth. Droughts and floods with significant negative social and economic consequences periodically impact the La Plata River Basin. To the north, the Amazon Basin, covering an area of over $5 \times 10^6$ km$^2$, is shared by nine countries and is home to many critical natural ecosystems; such diversity in natural and sociopolitical systems presents a major challenge to efforts to harmonize development interests and biodiversity protection. In addition, in the Caribbean region, socioeconomic levels are partly determined by the manner in which the countries—each with highly variable degrees of capabilities—are able to manage their water resources and basin landscapes. In Figure 1, the major water resources projects with OAS involvement are summarized.

# 2. TRANSBOUNDARY WATER MANAGEMENT

To date, the countries of the Americas have not developed general conventions or agreements to provide an integrated or transboundary approach to water resources management. Lack of a deep understanding of the variety, complexity, and importance of the issues related to transboundary water management have allowed for only partial successes in limited international agreements. This lack of general principles and global legal instruments is one of the causes that significantly restrict the sustainable management of transboundary waters. This is of course a major issue, but it also provides a clear indication that the complexity of the problems and interests involved call for shared objectives and common actions to build specific agreements for system-wide management. Thus, the difficulty of managing transboundary water systems and basins can itself become a catalyst for countries to work together.

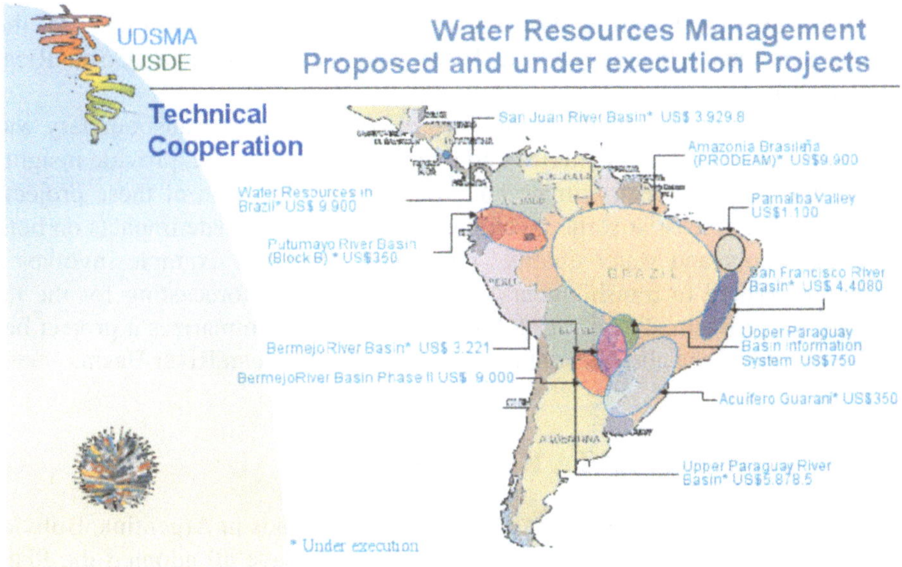


*Figure 1*. Overview of OAS water resources management projects in Central and South America. Dollar amounts given in thousands.

Although much progress has been made at the technical level, obstacles at the political level still form a barrier to extensive cooperation on transboundary integrated water resources management. For example, the existing treaties among Latin American countries do not provide for the creation of river basin committees. Nevertheless, South American countries have been cooperating to improve management of their shared water resources. Two major multilateral agreements have been signed: The Treaty for Amazon Cooperation among Bolivia, Brazil, Colombia, Ecuador, Guyana, Peru, Suriname, and Venezuela; and the Treaty for the La Plata River Basin among Argentina, Bolivia, Brazil, Paraguay, and Uruguay. Several countries have signed bilateral treaties.

The Central American countries impacted by Hurricane Mitch in 1998 have an increased awareness of the need for a more integrated approach to water resources management in their river basins. As a consequence, under the coordination of the Central American Commission for Environmental Development (CCAD), and the Central American Regional Committee of Water Resources (CRRN), several countries of the region are undertaking bilateral or trilateral projects in transboundary basins. Some examples worthy of mention are the Rio Paz (Guatemala-El Salvador), the Rio Lempa (Guatemala-Honduras-El Salvador), the Rio Coco (Honduras-

Nicaragua), the Golfo de Fonseca (Nicaragua-El Salvador-Honduras), the Rio San Juan (Nicaragua-Costa Rica), and the Rio Sixaola (Costa Rica-Panama).

Two examples of international efforts to manage transboundary watersheds, one in South America and one in Central America, provide insights into the links between climate and water resources. Each of these projects recognizes, at least at some level, the importance of climate impacts on both water supply and water quality. The South American example involves a center devoted to transboundary data collection and forecasting for the La Plata River Basin. The Central American example summarizes a project between Costa Rica and Nicaragua to manage the San Juan River Basin.

## 3. THE LA PLATA RIVER BASIN

The Rio de la Plata Basin (Fig. 2) drains lands in Argentina, Bolivia, Brazil, Paraguay, and Uruguay. These countries have all adopted the Plata Basin Treaty, which established a coordinating committee and framework for cooperating in integrated, transboundary development of the basin. This treaty, structured in a classic top-down diplomatic approach, has been credited with helping the signatory countries avoid open conflict and for facilitating construction of hydroelectric dams on the international river network (Varady et al. 1999).

The Plata Basin, fifth largest in the world, encompasses $3.1 \times 10^6$ $km^2$, including large parts of Brazil, Uruguay, Argentina, and Bolivia, as well as all of Paraguay. The basin includes the Paraná River, one of the longest rivers in the world and the one with the largest flow in the world. This river alone drains almost half of the entire basin; 59% of the river's watershed lies in Brazil. The Paraguay River Basin is the next largest sub-basin; roughly a third of the area drained by this river is in Paraguay and another third is in Brazil; the remainder of the watershed is distributed between Argentina (15%) and Bolivia (19%). The Uruguay River, with a much smaller watershed area, constitutes about 12% of the total drainage in the Plata Basin; 43% of this river's watershed lies in Brazil, 41% in Uruguay, and 16% in Argentina (Tucci and Clarke 1999). These rivers contribute to the La Plata River's mean annual flow rate of 24,000 $m^3/s$.

*Figure 2*. Map illustrating the approximate locations of the Amazon, Paraná, and La Plata River Basins.

Physical conditions in the Plata Basin vary substantially from north to south and from west to east. Elevations range from 4,000 m in the western Andean Cordillera to 200 m in the south. Likewise, mean annual rainfall varies across the basin from 200 mm along the western boundary of the basin to 1,800 mm in the maritime uplands along the coast of Brazil. In general, rainfall is greatest in the upper Paraguay and Paraná River Basins. In terms of river flows related to the high rainfall in the upper stretch of the

Paraguay River, the shallow channel slope of much of the river has led to formation of the Pantanal wetland, one of the world's largest. The Pantanal, which encompasses an area of approximately 140,000 km$^2$, produces a difference of 4 months in flood peaks north and south of the wetland. It also influences annual peak water levels through reducing variance in annual peaks that would otherwise occur due to variation in annual precipitation amounts. On the Upper Paraná River, changing land use from natural forest to field cropping of soybeans has affected flow regimes in the basin. Downstream from the Paraguay and Paraná Rivers, navigation and flood control are the predominant concerns, as past floods have caused significant loss of life and damage to property, notably since 1970 (Tucci and Clarke 1999).

In part, increases in flood impacts since 1970 may be traced to changes in land use. As Tucci and Clarke (1999) noted, annual floods were relatively low during the period 1950 to 1973, giving rise to a belief that it was safe to build in areas subsequently shown to be at severe risk of flooding. That flooding is a paramount concern may be seen from the damages of the largest flood of the twentieth century. This flood occurred on the La Plata and Paraná Rivers in 1983 when, for a year and a half, the flood level of the Paraná River was above street level in Santa Fe, Argentina (the town is protected by dikes). The cost of the flood in the city of União da Vitória, located on the Iguazú River, ran to US$78 million. The state of Santa Catarina sustained damages equal to 8% of its gross state product for that year (Tucci and Clarke 1999). More generally, uneven temporal and spatial distribution of flow, and degradation of water quality (especially in and near São Paulo and Curitiba) are major concerns.

Historically and currently, water use in the Plata Basin has been driven by national goals, largely focused on generating economic benefits. Efforts to establish an international framework for managing the river began in 1967 and 1968, culminating in the signing of an agreement designed to promote rational development of the basin by the five countries. Due to many diplomatic conflicts in the years 1969 to 1979, little was accomplished. In the early 1980s, efforts were renewed, with Paraguay, Brazil, and Argentina signing an agreement covering the operation of the hydropower plant at Itaipú. Other binational agreements concerning hydropower operations followed. The growth of Mercosur, an alliance of South American countries that began as a commercial trade agreement, has led, in recent years, to initiation of a number of water basin projects in the region.

Sustaining international cooperation in management of the Plata River Basin is of foremost importance, given that the basin encompasses heavily populated urban centers and is home to 100 million inhabitants (50% of the combined population of the five basin countries); indeed, the region has the highest level of development in all of South America (Tucci and Clarke 1999). The lands of the basin also support highly productive agricultural and ranching activity, and include forested areas in the river valleys. Furthermore, the basin contributes 70% of the gross product of the five basin countries. There are more than 40 hydropower dams in the upper basin in Brazil alone. River navigation and river port projects are under way as well (Goniazki 1999).

Among the development activities that have generated environmental impacts on the basin are: construction of hydropower operations on the Upper Paraná River in Brazil over the decades 1960–90; deforestation of the Paraná, Uruguay, and Paraguay Basins over the same time period; the introduction, since 1970, of intensive agriculture (primarily in Brazil), and navigation activities; and related conservation efforts on the Upper Paraguay River. Of importance for management of the Paraná River is an increase in flow that has been observed since 1970. It has been speculated that these increases may have been caused by changes in vegetation cover, or by climatic variation (Tucci and Clarke 1999). Hydrological records for the basin reveal increases in rainfall and runoff since 1970. These increases have had important impacts in the basin, including the Pantanal wetlands on the Paraguay River and the Paraná River Basin. Proposed navigation improvements on the Paraguay River are predicted to add to the impacts on the Pantanal wetlands (Tucci and Clarke 1999).

As was noted by Tucci and Clarke (1999), flows in the river system have increased by 19% to 46% since the 1970s. These increased flows have resulted in soil erosion and deposition of sediment in the river channels, increases in river level and frequency of floods, changes in the riverbed and in the riparian environment, decreases in reservoir storage capacity due to sedimentation, increased energy production at the hydropower plants, and changes in water quality due to resuspension of riverbed materials during flood events. Unusually heavy precipitation in the 1990s, as well as in the early 1900s, has been cited as the most important reason for flooding during those time periods. However, there is no consistent evidence that changes in rainfall and runoff dynamics associated with land use changes played an important role in recent flooding. On the other hand, some evidence suggests that, even at times when precipitation was not extremely high, streamflows have been higher than expected (Anderson et al. 1993, cited in Tucci and Clarke 1999). In general, Tucci and Clarke (1999) conclude that streamflow has increased in the Upper Paraguay, Paraná, and Uruguay River Basins and that both precipitation and land use changes have contributed to this increase, though the relative contribution of each component is not yet known. The most significant question for managing the Plata Basin in the future is how permanent this increase will be. While the increase enhances energy production, reservoir life may be reduced due to higher than anticipated rates of sedimentation. Increased water depth improves navigation both in terms of distance upstream and in terms of length of navigation season; however, these same changes may increase flood hazards. This latter concern comes into sharp focus when the services provided by the Pantanal wetland are considered. Plans to increase the length of the navigable waters of the Paraguay River through dredging would result in a reduction in the volume of water retained in the wetland, possibly altering the ecosystem to that of a savannah, with related loss in the area's biodiversity.

One organizational development that is contributing to transboundary management of the Plata River Basin is the Hydrologic Warning Operations Center for the Plata Basin. Located in Argentina, the center was established in the wake of the devastating floods that occurred in 1982–83. The center operates the Hydrologic Warning and Information System for the entire basin.[2] Management of this system requires permanent hydrologic monitoring of basin conditions and forecasting of both potential flood events and low-water conditions. The data are furnished by the riparian countries, as well as by the operators of the binational hydroelectric facilities, and by provincial agencies. The center digitizes the data and archives them in databases, runs models for making hydrologic forecasts, and works with data processing and Geographic Information System (GIS) technologies. Processing of remotely sensed data is also carried out. These activities allow the center to maintain a permanent forecasting and hydrologic warning service, including dissemination of the forecasts to decision makers in the basin. Among the products distributed by the center are daily reports on hydrometeorological conditions, and early warnings about risk conditions. The forecasts and data collection work carried out by the center, as well as the practice of responding to stakeholder needs, proved invaluable in confronting flood conditions that occurred in 1987, 1992, 1995, 1997, and 1998, as well as when marked drops in water levels were experienced in 1985 and 1988. Lead times for these events have ranged from 7 to 40 days. Centers such as this one provide important focal points for advanced development of regionally tailored climate information and dissemination of such information in transboundary contexts.

---

[2] Organization of American States: Implementation of Integrated Watershed Management Practices for the Pantanal and Upper Paraguay River System: Project Description. http://www.oas.org/usde/ALTOPARA/ Accessed September 20, 2001

## 4. THE SAN JUAN RIVER BASIN AND COASTAL ZONE

The impetus for initiation of the San Juan River Basin Project (location illustrated in Fig. 3) emerged from the Central American Action Plan for the Development of Border Zones, issued in conjunction with the XIII Summit of Central American Presidents held in Panama in December 1992. Identification of the San Juan River Basin as a priority area in this plan prompted a request to the United Nations Environment Progamme (UNEP) and the OAS, by the governments of Nicaragua and Costa Rica, to carry out a diagnostic study of the state of the basin. The study was published in 1997 under the title, *Diagnostic Study of the San Juan River Basin and Guidelines for an Action Plan.* Among the goals identified were economic development of the border areas, conservation of natural resources, protection of biodiversity, strengthening the Mesoamerica biological corridor, sustainable use of water resources, and protection of the integrity of the drainage basins. Specifically with regard to transboundary water resource issues in the San Juan River Basin, four major areas of concern were identified in the diagnostic study: degradation of water quality; degradation of habitat in coastal and nonmarine areas, including lakes and streams; introduction of detrimental exotic species; and excessive and/or inappropriate exploitation of resources due to inadequate management and control measures. Climatic conditions can affect all of these areas of concern to a greater or lesser degree.

Among the fundamental causes identified for these problems were inadequate planning and management, weak institutions (particularly insufficient financial and human resources), insufficient natural resource management capacity, limited stakeholder participation in decision making, and the extreme poverty and high population growth rate of those living in the basin.

The ultimate objective of the project is to "ensure the availability of the goods and services provided by water resources for conserving natural ecosystems and social and economic development in order to satisfy present and future demands as agreed by all parties involved."[3] The plan includes components aimed at strengthening a basinwide information system to facilitate availability of information needed for decision making; creation of a well-coordinated, bilateral planning process for the basin; implementation of a gender-oriented public participation process; strengthening of public insti-

---

[3] Organization of American States: Formulation of Strategic Action Program for the Integrated Management of Water Resources and the Sustainable Development of the San Juan River Basin and its Coastal Zone. http://www.oas.org/usde/SanJuan/ Accessed September 20, 2001.

tutions and private organizations; and formulating/implementing environmental education activities.

*Figure 3.* Map illustrating the approximate location of the San Juan River Basin in Central America.

Particularly important to these efforts was the occurrence of Hurricane Mitch in 1998. The devastation caused by this storm elevated the issue of development of integrated basin management as a priority in regional and national political circles (Rucks 1999). In 1998, a program called "Block B" was approved in the amount of US$283,000. These funds were earmarked for integrated management of water resources and sustainable development in the San Juan River Basin and its coastal zone. The objective of the project was to prepare a project plan that included tasks needed to formulate a strategic action plan for managing the basin and coastal area. Among the specific objectives were identifying regional solutions aimed at assuring water quality, resolving erosion and sedimentation problems, and prioritizing actions that could be undertaken in a consensus framework between the countries of Nicaragua and Costa Rica. The objectives also included facilitating public involvement in the management process, and assisting the governments in the region to encourage sustainable development. Scientific research activities, such as defining and evaluating geomorphological and biochemical processes, were also cited. Furthermore, the goals included identifying pilot projects that could be used to evaluate methodologies and techniques as well as resolve sustainable development problems, and rehabilitating degraded areas. Prioritizing activities and analyzing potential mechanisms for recuperating costs, and executing public participation programs was an important task area identified by Rucks (1999). The following summary of the basin's characteristics places the plan in context, and suggests areas where climate impacts are important, and where climate information could contribute to sustainable management of the basin and its water resources.

The San Juan River Basin project (see Fig. 3) encompasses approximately 38,500 km$^2$ in Nicaragua and Costa Rica. The basin comprises four subsystems: Lake Nicaragua (source of the San Juan River), the San Juan River itself, the Reserva Biológica de Indio Maíz (Indio-Maíz River Biological Reserve, and the location of one of Nicaragua's most biologically diverse areas), and the Tortuguero Plains Conservation Area (in Costa Rica, this area is notable for reflecting the greatest biodiversity and ecological fragility in the southern section of the basin) (see footnote 1; Rucks 1999). The river basin has been identified as critical to meeting future development needs on the semiarid Pacific slope of these two Central American states. The river basin is unevenly divided between the two countries, with 64% of the land area lying in southern Nicaragua and 36% in northern Costa Rica. The basin includes important zones of biological diversity—51 protected areas exist within the basin's boundaries, 33 in Costa Rica and 18 in Nicaragua. Almost the entire costal zone is within protected areas. Nicaragua features strong expansion of the agricultural frontier, which is exerting pressure on some of the protected areas, particularly the Reserva Biológica de Indio Maíz.

Total population in the basin is approximately 1.1 million, with 73% of the people living in Nicaragua and 27% in Costa Rica. Population density in the Nicaraguan portion of the basin is about 46 inhabitants per square kilometer, with 55% of this population being rural. In Costa Rica, 68.3% of the population is rural, and average population density is only 22.3 persons per square kilometer. Nevertheless, the population in Costa Rica is more homogeneously distributed than that in Nicaragua. In both countries, poverty rates and quality of life indices for inhabitants of the basin exceed the respective national average; however, quality of life and availability of infrastructure and services are better, on average, in Costa Rica. This difference is associated with strong Nicaragua-to-Costa Rica migration pressure (Rucks 1999). In both countries, the economy of the basin is based on primary-sector activities. Deforestation is an important issue, particularly in Nicaragua.

Elevations in the San Juan Basin range from 500 to 3,000 m above sea level in the Cordillera Volcánica Central, in Costa Rica. Precipitation in the basin varies between 1,500 mm in the west to 6,000 mm in the coastal zone. The flow of the San Juan ranges from 475 $m^3/s$ at the mouth of Lake Nicaragua to 1,308 $m^3/s$ at the river's outlet (Rucks 1999). In addition to river flows, the basin also possesses abundant supplies of good-quality groundwater.

The location of the San Juan River Basin makes it a corridor for tropical weather systems moving from the Atlantic Ocean to the Pacific. Hurricane impacts, combined with threats posed by volcanic activity and seismic events, increase the already high vulnerability of water supplies in the basin. For example, heavy precipitation during Hurricane Mitch in 1998 resulted in flooding of Lake Managua (not located within the San Juan River Basin), and a flow of floodwaters into Lake Nicaragua. Because Lake Managua is heavily polluted, such flows threaten water quality in Lake Nicaragua. The San Juan River Basin project is designed to address this concern through close cooperation with a separate project, operated under the Regional Seas Program for the Caribbean, to prevent contamination arising from flooding of Lake Managua.

Lake Nicaragua is one of the largest stores of freshwater in the Americas and is known for its relatively high water quality. The lake serves multiple purposes, ranging from navigation to energy production, irrigation, a source of potable water, and a destination for tourism and recreation. Possible conflicts over water may occur in the context of activities such as further hydroelectric power development or the long-proposed construction of a canal linking the Atlantic and Pacific Oceans, because 300 to 400 $m^3/s$ of the flow of the San Juan River are already devoted to meeting demand for potable water and for irrigation water for the Pacific region of Nicaragua.

A fully functional, basinwide information system is a top priority in achieving sustainability in the San Juan River Basin. Thus, development of comprehensive knowledge about rainfall and runoff, water quality, erosion, and sedimentation across the entire basin is a key focus of the project. Also targeted are improved information on the physical and biological characteristics of Lake Nicaragua and how these elements will respond to increasing human impact. Although not explicitly mentioned, developing a better understanding of how climatic variability and change affect both the physical and human environment over time and space can also be essential in formulating management strategies and responding to climatic impacts in the management of the basin's resources.

Effective use of such information, however, requires considerable effort with regard to capacity building and strengthening of institutions. The San Juan River Basin Plan reflects the desire of both Nicaragua and Costa Rica to foster formation of councils in critical subbasins that interact with local border governmental entities. The plan also targets building sustainable development planning and management capabilities, improving existing infrastructure, improving equipment levels, and fostering mechanisms that increase basic incomes among local and regional organizations. These innovations, and others, such as development of policies and procedures for sharing information among institutional entities within each country and across the international border, are required to assure integrated water management. A solid understanding of regional climate and its impacts is among the basic needs in this as in other transboundary water management institutions. But utilization of climate information and forecasts requires effective institutional and organizational structures, as well as appropriate and sufficient infrastructure and education. These elements must be in place to assure unimpeded collection, analysis, and flow of climate and climate impacts data. At the national level, it is generally accepted that integrated water resources management is a problem of conflict resolution among users. At the regional level, the issue of who has ultimate control over the water resource adds a new obstacle to the management of the resource.

There is a need to develop enforcement mechanisms that work in both a bottom-up and a top-down manner. In transboundary water management, other actors must be able to participate, particularly the foreign affairs ministries, as well as organizations that are involved in addressing basin-wide problems as a whole, including local, regional, and international organizations, among others. Mechanisms to incorporate the stakeholders for the whole basin have to be established to make decisions that permit resolution of conflicts generated by the different uses of water with the participation of the society in general. Involving the public in such efforts is essential to ensuring social sustainability of the efforts of the integrated water management teams.

# 5. TRANSBOUNDARY WATER MANAGEMENT: THE BROADER CONTEXT

The requirements and principles for water resources management have been widely recognized at least since the United Nations Conference for Environment and Development in Rio de Janeiro in 1992, and have been the basis to start the Inter-American Dialogues in Water Management in 1993 (Miami), 1996 (Buenos Aires), and 1999 (Panama) (see Fig. 4). Academic institutions, governmental and nongovernmental organizations, and the private sector have all promoted these Inter-American Dialogues. In the First Dialogue, aimed at invigorating information exchange and technology transfer, the Inter-American Water Resource Network (IWRN) was created as a network of networks (Fig. 5). The Sustainable Development and Environment Unit of the OAS was selected as the technical secretariat of the IWRN. The Second Dialogue of Buenos Aires established the basic agreements between the different actors related to water resources that were the technical inputs for the Santa Cruz Summit in the area of water resources management (Fig. 5). The Third Dialogue of Panama analyzed the progress achieved in the management of water resources in the Americas, taking into account the results of three regional meetings that were held in South America, Central America, and the Caribbean, and decided to incorporate the Americas in the World Water Vision process. In all these meetings the issue of transboundary water resources management was considered as a priority area for action.

The fourth and most recent Inter-American Dialogue took place in Foz do Iguaçu, Paraná, Brazil in September 2001. The general theme of this event was to search for solutions to the problems of management and water resource supply for various uses, with special focus on arid and semiarid regions, metropolitan regions, transboundary river basins, and regions that are vulnerable to climate changes. The results of this meeting were used as input for the Inter-American Ministerial Meeting for Sustainable Development, called "Santa Cruz + 5," which took place in Santa Cruz, Bolivia, in December 2001.

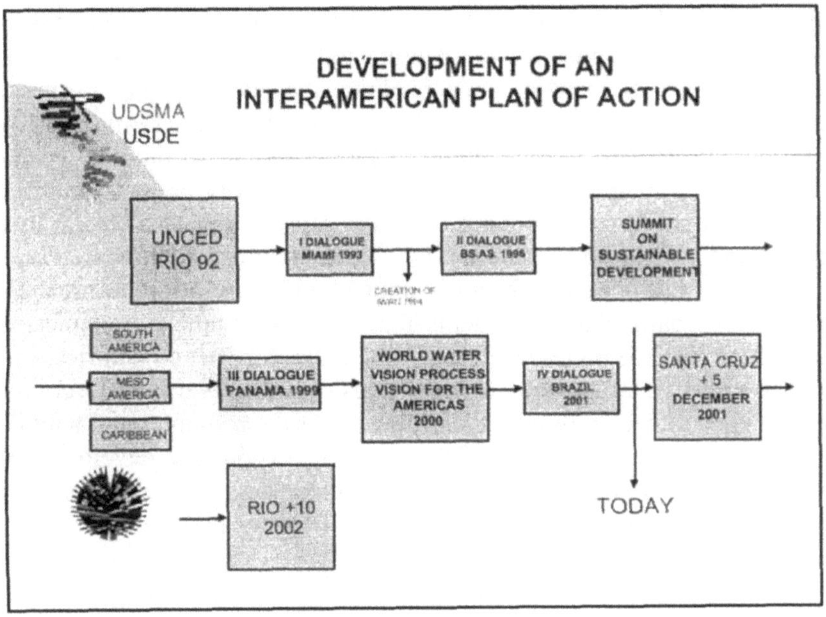


*Figure 4*. Schematic for the development of an Inter-American Plan of Action for the integrated management of transboundary water resources.

## 6. CONCLUSIONS

Proper management of water resources and rational administration of the goods and services that these resources provide to society, including its economic development, are key factors that will determine the sustainability of those development efforts throughout the American continent. Improving knowledge about climate and its impacts, particularly at regional and local scales, is essential to this effort. Because of their large sizes, the numbers of people living in transboundary river basins, and the ecosystems and freshwater resources they contain, special attention must be given to these critical areas, such as the Amazon, the Orinoco, and the La Plata in South America and the San Juan in Central America. Nevertheless, other smaller basins are also of critical importance and must not be left unattended. Agreements among the American states are crucial to the achievement of these objectives.

To facilitate the integrated management of water resources, the countries in the Americas have mandated the OAS's General Secretariat to act in three complementary stages. First, the General Secretariat must support the definition of policies by developing forums and in particular by sponsoring the Inter-American Dialogues for the Administration of Water Resources. Second, the General Secretariat must facilitate the exchange of information, experience, and capabilities, which it does through the Inter-American Water Resources Network. Third, the OAS must continue to provide technical assistance to its member states by carrying out priority projects, especially in transboundary basins. In doing so, it has already gained a wealth of experience and has placed it at the service of the American countries.

Each transboundary basin is unique, and therefore requires the development of particular instruments that rely on common principles. Because of the relative underdevelopment of Latin American countries, in order to be effective, all agreements or treaties regarding water resources management have to deal both with the protection of the resources and with their sustainable use for the benefit of the social and economic development of the communities involved. This will ensure the general protection of the environment and the active participation of all interested parties.

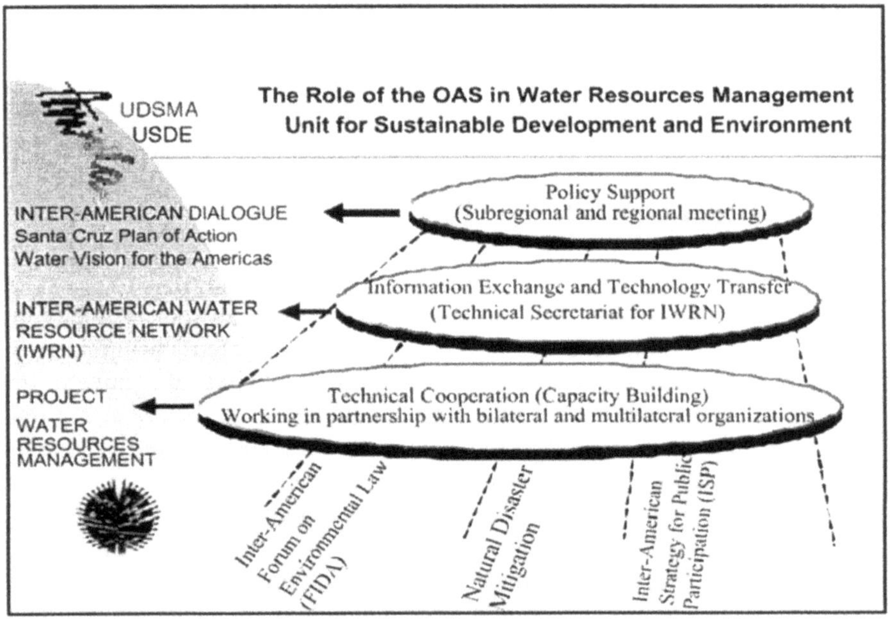


*Figure 5.* A schematic diagram depicting the OAS role in water resources management in the Americas.

In summary, the Americas have an important regional organization (the OAS) that acts simultaneously in political and technical areas for the benefit of the member countries. The Sustainable Development and Environment Unit, the technical part of the OAS, deals with the mandates and plans related to sustainable development and regional water resources management issues. Climate factors must be among the variables considered in formulating and implementing development plans and in addressing water resource issues.

## 7. REFERENCES


Anderson, R.J., Jr., N. da Franca Ribeiro dos Santos, and H.F. Diaz. 1993. An analysis of flooding in the Paraná/Paraguay River Basin. LATEN Dissemination Note #5, The World Bank.

Goniazki, D. 1999. Hydrologic Warning and Information System for the Plata Basin. *Proceedings of the Third Inter-American Dialogue on Water Management*, Panama City, Panama, March 15–21, 1999. http://www.iwrn.net/D3_Proceedings.pdf. Accessed September 20, 2001.

Rucks, J. 1999. Manejo integrado de recursos hidricos en cuencas transfronterizas: el caso de la cuenca del Río San Juan Costa Rica-Nicaragua. *Proceedings of the Third Inter-American Dialogue on Water Management*, Panama City, Panama, March 15–21, 1999. http://www.iwrn.net/D3_Proceedings.pdf. Accessed September 20, 2001.

Tucci, C.E.M., and R.T. Clarke. 1999. Environmental issues in the Plata Basin. *Proceedings of the Third Inter-American Dialogue on Water Management*, Panama City, Panama, March 15–21, 1999. http://www.iwrn.net/D3_Proceedings.pdf. Accessed September 20, 2001.

Varady, R.G., L. Milich, and R.E. Ingall. 1999. Post-NAFTA environmental management in the U.S.-Mexico border region: Openness, sustainability, and public participation. *Proceedings of the Third Inter-American Dialogue on Water Management*, Panama City, Panama, March 15–21, 1999. http://www.iwrn.net/D3_Proceedings.pdf. Accessed September 20, 2001.

# 4

# COPING WITH CLIMATE VARIABILITY

*Municipal Water Agencies in Southern California*

Denise Lach,* Helen Ingram,** and Steve Rayner***

**Center for Water and Environmental Sustainability, Oregon State University, Corvallis, Oregon 97331*

***Warmington Endowed Chair of Social Ecology, University of California at Irvine, Irvine, California 92697*

****School of International and Public Affairs, Columbia University, Palisades, New York 10027*

**Abstract** This chapter reviews the current water situation in Southern California, and explains the predilection, capacity, and strategies of municipal water agencies to innovate. Drawing on data collected in Southern California, as well as in the Pacific Northwest and the Potomac Basin, the values held by municipal water agencies are discussed, as well as the incentives, sources, and processes for innovation. Finally, the chapter relates these findings to the future of equitable distribution of water resources in the far west border region of California, Nevada, Arizona, and northern Mexico.

## 1. INTRODUCTION

On May 31, 2000, a superior court judge in the state of California called a temporary halt to development of the largest housing project in Los Angeles County history. The judge ruled that the developers of the 22,000-home Newhall Ranch had not proved that they could supply enough water, especially during periods of drought (Kelly 2000). This court ruling is emblematic of growing concern for the ability of municipal water agencies (MWAs) to provide adequate supplies to growing populations, taking into account climate variability. Limiting growth based on the availability of water has been complete anathema in Southern California, where more and



*Henry F. Diaz and Barbara J. Morehouse (eds.), Climate and Water*, 59-81.

bigger is a way of life. And, since Southern California already relies heavily on water from outside the region—including Northern California, other western states, and Mexico—there exists high potential for increased international conflict growing out of competition for larger shares of transboundary surface and groundwater supplies.[1]

The purpose of this chapter is to examine how Southern California MWAs are dealing with water shortages exacerbated by population increases and climate uncertainty. To avoid heightened international tensions over shared water resources, MWAs could adopt various techno-scientific innovations to better manage available supplies and to encourage conservation. How well such organizations are able to meet these challenges is of obvious importance to controlling levels of domestic and international conflict over water. Although the need for innovation may be rising, MWAs are archetypes of conservative organizations. They are highly averse to risk, and value invisibility and the absence of controversy. On the whole, water utility managers ascribe to what might be termed the "dentistry theory of administration." That is, they aspire to painless service. Yet, current and predicted conditions of growing demands in the face of limited supplies suggest the need for significant, and perhaps painful, change within the water sector.

## 2. STUDY RESULTS

To ascertain the conditions under which water agencies adopt techno-scientific advances, an interdisciplinary team conducted approximately 100 semistructured ethnographic interviews, about 40 of which took place in Southern California and Sacramento. Interviews were also conducted in the Columbia River Basin of the Pacific Northwest and the Potomac River/Chesapeake Bay area near Washington, D.C. The research team included an anthropologist, sociologist, political scientist, and water resources engineer. Key questions included agency mission and how officials knew they were or were not doing a good job, what innovations had taken place within the organization, and the source of the impetus and ideas for change. The project was funded by the National Oceanographic and Atmospheric Administration (NOAA), and was directed particularly to the use of

[1]An example of the increased competition between the United States and Mexico is provided by the recent example of the Lining of the All American Canal and the sale of water between the Imperial Irrigation District and San Diego County. While portrayed as a win/win strategy for both areas of origin and receipt, there are considerable negative side effects to urban and rural water users in Mexico (Cortez-Lara and Garcia-Acevedo 2000).

probabilistic forecast information about seasonal and inter-annual climate variability in water agency planning and operations. At least in concept, being able to predict wet and dry seasons should be highly useful to conjunctive management of surface water and groundwater demand management. Most of our attention in the interview protocol and in this chapter is directed to innovations in general rather than NOAA products in particular.[2]

## 3. THE PHYSICAL AND ORGANIZATIONAL CONTEXT FOR INNOVATION

Municipal water agencies in California have long been accustomed to addressing inadequate water supplies to meet growing demand. The enormous growth of Los Angeles County, and in more recent decades, San Diego and Orange Counties, has occurred in an area of very low and highly variable rainfall. Similar growth has occurred in Mexico in Baja California and the Valley of Mexico. Due to the semiarid climate of Southern California, only water transfers from locations both north and south of Los Angeles County make such growth possible. For example, transbasin diversions along the U.S.-Mexico border total between 6 and 8 million acre-feet each year (Michel 2000). Such interbasin transfers are the primary cause of many water- and land-based problems, are enormously expensive, and raise serious equity issues. Moreover, they may involve increasing competition between rival localities over limited supplies.

The population grew by 2.8 million people within the Metropolitan Water District (MWD) of Southern California service area between 1987 and 1997 without increasing imported water supplies. Yet, population growth continues unabated in the region and resistance to water transfers from other states in the Colorado River Basin and from Northern California has escalated. Present water delivery strategies, moreover, are unlikely to be sufficient to meet the water supply challenges of the future. And, climate change may increase the frequency and severity of extreme events such as drought and flooding.

Alternatives to interbasin transfers include a variety of management strategies, such as more efficient use of existing surface and groundwater supplies, recycling and reuse, and demand management. Such alternatives

[2] Interviews were conducted with the promise of anonymity. While a list of those water officials in California interviewed is appended to this chapter, the authors have not, and the reader should not associate any comments to any particular interviewees. Conclusions are based upon careful, independent review of interview notes among the investigators.

are already being employed to a meaningful extent in Southern California today. For example, local water districts like San Juan Capistrano have implemented water-pricing strategies that place high premiums on water consumption that exceeds conservation targets.

The major water conveyance facilities in the region are portrayed in Figure 1. About 55% of the water supply in the MWD service area is imported from either the Colorado River Aqueduct (CWA) or the State Water Project (SWP).[3] Currently MWD receives about 1.2 million acre-feet/year, considerably more than the 550,000 acre-feet designated for urban use under the Law of the River managing the Colorado River.[4] The SWP is designed to transport 4.2 million acre-feet/year of water from Northern to Southern California, but currently delivers only about half of the anticipated yield, in large part because environmentalists have blocked construction of planned surface water storage facilities. Moreover, supplies depend upon snowpack in the high Sierra, which is highly variable and sensitive to El Niño/La Niña events (Reynolds et al. 2001). Water deliveries can also be interrupted when pumping affects the environmental health of the bay/delta region.

It is generally acknowledged that there is not enough water for all uses and that Southern California will be forced to cut back its use of Colorado River water. Moreover, even the legally allocated quantity of water is not entirely secure. The Colorado River is drought prone and moderately sensitive to El Niño/La Niña events (Grimm et al. 1997). In response to concerns throughout the region about water allocation, former Secretary of the Interior Bruce Babbitt established interim guidelines for determining when surplus Colorado River water would be available for California, Nevada, and Arizona. These criteria will be in effect until 2015, giving California a greater certainty of supply and a transition period in which to further develop water conservation, recycling, storage, and transfer programs that will provide for separation from an overreliance on the Colorado River.

---

[3] Water is also provided from localized surface and groundwater sources.

[4] The Colorado River is governed by an international treaty with Mexico and several minutes of the International Boundary and Water Commission (IBWC), two major interstate compacts, a decree of the U.S. Supreme Court, various statutes, and contracts between the United States and water and power customers. Collectively these are known as the "Law of the River."

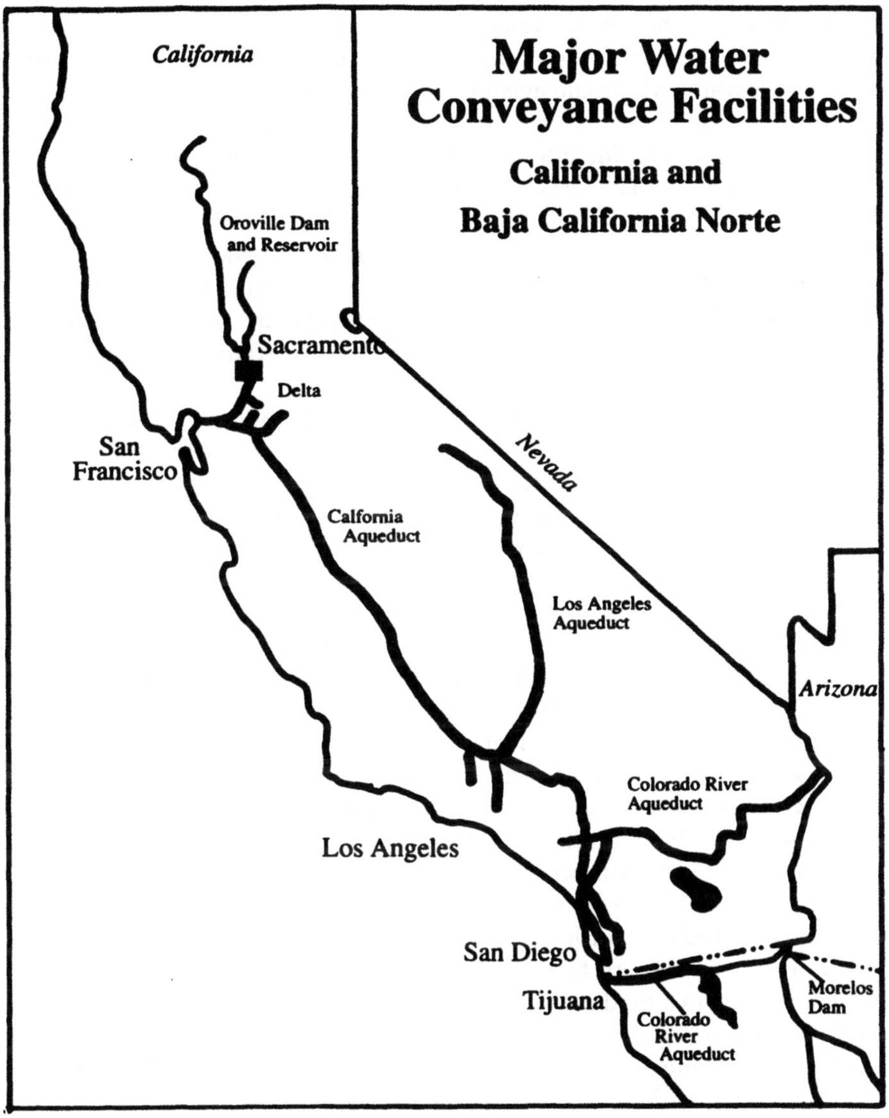


*Figure 1.* Major water conveyances: California and Northern Mexico.

Figure 2 shows an example of how water flows from the two major sources through the MWD to one local water utility in Orange County, the Irvine Ranch Water District (IRWD). The figure traces the flow of water among agencies, and, thus, the degree of dependence of one agency upon another. The first link in the organizational chain is the MWD, established in 1928 as a water wholesale agency consisting of 13 member agencies. Today, the MWD has evolved into a regional water management agency composed of 27 members, including the Metropolitan Water District of Orange County (MWDOC). The MWDOC became a member of the Metropolitan Water District (MWD) in 1951, and buys all of its water from MWD. MWDOC sells the water to its 28 member agencies, including the Orange County Water District (OCWD) and ultimately the Irvine Ranch Water District (IRWD). At the end of the water delivery pipeline, local water districts like the IRWD are highly vulnerable to any shifts in water quantity or distribution strategies.

Seeking more control over the quantity, quality, and sources of water, local agencies are seeking alternative sources such as reclaimed water, groundwater, captured rainwater, and strict conservation among users. The search for locally controlled sources of water has led to competition among the member utilities within the OCWD. The Irvine Ranch Water District, in particular, has been aggressively pursuing options that include reclaimed water technology and rainwater capture to serve the rapidly expanding population in South County. Other member agencies of the OCWD, such as Anaheim, have resisted expansion of the service areas of other member agencies within the OCWD for fear of increases in water rates or declines in water quality for their own residents.

Localized sources of water supply are identified for each agency in Figure 2. They include surface flows, natural and artificial recharge, and reclaimed water. Variations in the reliability, quality, and cost of different sources of water, and competition among agencies, present significant water management challenges to utilities. For example, capturing all the surface water flows, which are of high quality and relatively low cost, is an important priority even though these flows are highly variable. This means regulating surface flows by dams or berms, and storage in reservoirs or recharge basins. Reclaimed water, which is suitable for outdoor watering and toilets, is the most reliable water supply in Southern California. Reclaimed water must be blended with other water sources in order to be potable.

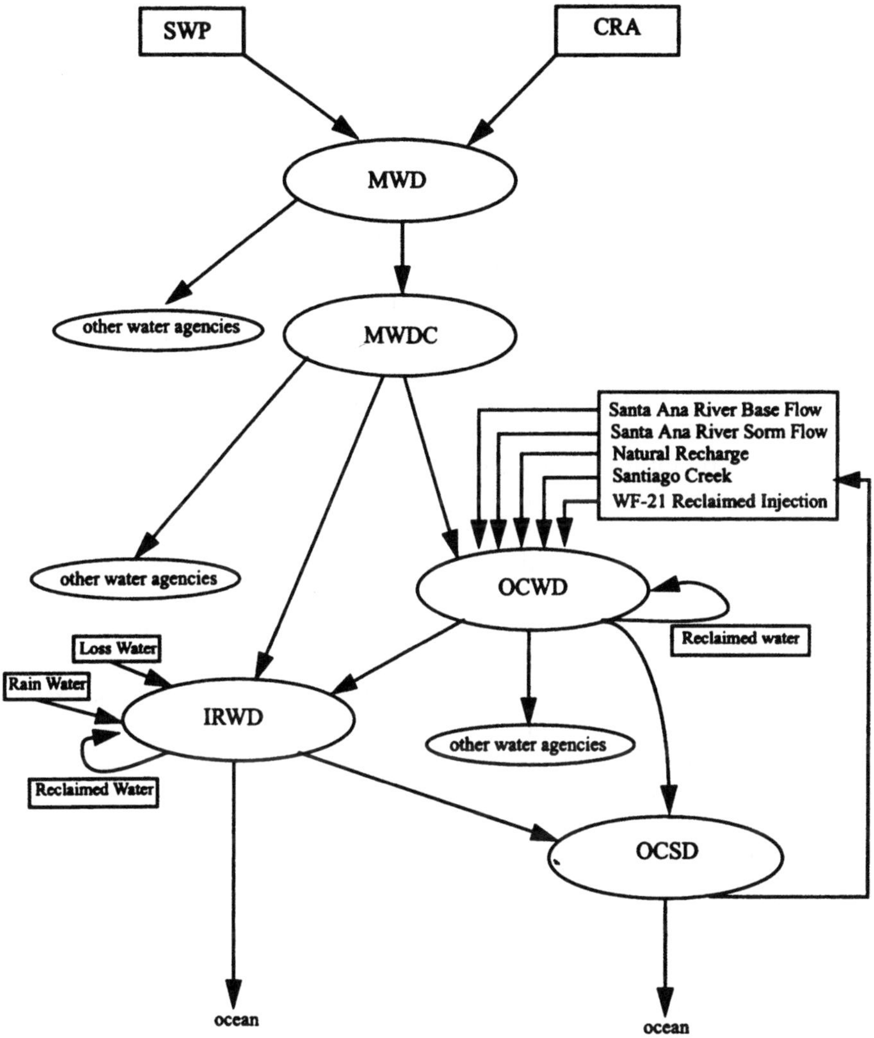


*Figure 2.* Transect of water flows in Southern Calfornia.

Another source of water not identified in Figure 2 is water conservation. Water not used by existing customers of municipal water utilities becomes available to new residents. The Orange County Water District has tried to deal with increased competition among its member agencies through long-term planning and greater independence from imported water (Kennedy 1999). To accomplish this objective, many Orange County MWAs will need to implement highly innovative water management strategies.

## 3.1. Hierarchy of Values in Water Agencies

Water agency officials we talked with clearly express a hierarchy of values that includes reliability, quality, and cost. Reliability ranks far above other agency values according to virtually every water official we interviewed, each articulating some version of the statement made by a municipal water utility director: "We cannot optimize. We must be conservative in our decisions." Most managers told us that because of the great variability of water quality and supply, it is impossible to manage for average conditions, as no such conditions exist. Instead, ensuring water system reliability depends on incorporating the extremes into standard operating procedures or, in other words, "routinizing" the irregular. To do this, water agencies build systems full of redundancies that ensure water is always available when it is needed. In Southern California, this includes finding multiple sources of water.

The emphasis placed on reliability is derived in part from the importance of water as a human resource. Water agencies are the human custodians of this fundamental necessity and treat water supply as a bedrock public service along with fire and police protection. Therefore, to the water officials we spoke with, water is too important ever to become a contentious public or political issue. If water supply or quality ends up on the public agenda, water agencies have failed to perform their function. Any water supply or quality problem that drew attention to the water agency was perceived by the officials as politically costly because they knew that their competence and decisions would be questioned. They preferred to keep a low profile. When we asked water agency officials how they knew they were doing a good job, they consistently told us that success was measured by an absence of complaints. One official we spoke to in the Los Angeles Department of Water and Power said that his agency was doing well when people were not "storming the building," which, interestingly, is surrounded by a moat.

This innate conservatism of water agencies—as manifest in both the redundancy of systems and preference for low visibility—suggests that they are much more likely to embrace incremental rather than innovative change. Incremental change involves an agency action very much like what has occurred previously, except to a greater or lesser degree. Such change does not involve any new conceptual or organizational alterations that may draw unwanted attention to the agency but also does not typically provide large gains in capacity. However, changing physical and organizational contexts may require institutional, organizational, and individual innovation. Innovation involves agency actions that change the way they do business and often even the way they think about what their business is. Growing competition for the benefits associated with water will make water agencies search for ways to improve management of existing sources, including incremental change and innovation. Even if no improvements can substantially relieve the pressures under which water agencies operate, water resource management must appear to have been open to all reasonable alternatives.

## 3.2. Conditions and Sources of Innovation

The hierarchy of values expressed by water agency officials fosters an aversion to any risks that may lead to unreliable water delivery, making these agencies reluctant to innovate; they instead prefer to remain satisfied with incremental change. When we asked for criteria against which adoption of significant change might be measured, we were consistently told that innovations had to deliver results that were no worse than current conditions.

A further requisite for any innovation is that it be compatible with the current craft skills of the organization. Water management is as much craft as science, according to the agency staff we talked with, and we were told frequently that it takes 5 years or so to get used to any water system. The place-specific nature of watersheds, combined with the longevity of water delivery infrastructure, means that operators must be knowledgeable about and able to adjust to idiosyncrasies in the natural and built systems. We found that while many ideas and technologies are copied from one utility to another, and there is an industry standard of good practice, most new techno-scientific changes are vetted by "old timers" at each organization before they are put into practice. If they do not "feel" right, if they are not compatible with the "way things are done here," innovations often remain on the shelf rather than being put into service.

Conservative organizations like these water resource agencies are not likely to innovate without a well-informed understanding of the benefits and details of any change. The existence of well-tested technologies or sectorwide best practices is thus a precondition for innovation. Incremental change may originate from inside the organization (i.e., current employees), and as long as the innovation stays within accepted boundaries of practices, a rich information base of experience is available in efforts to implement change. Innovations that are more challenging to accepted practices and arrangements may receive impetus for action from outside (i.e., best practices of industry), although the actual change will be homegrown in its design and implementation in conservative organizations.

### 3.2.1. Internal Drivers of Innovation

While water management agencies admit to being conservative when it comes to taking risks concerning core goals of reliability, quality, and cost, they are determined not to be thought of as backward or not up to date technologically. Agency identity and professionalism are bound up with keeping current on industry practices and ideas. Moreover, should problems arise, it would be enormously damaging for utilities to be vulnerable to charges of having allowed their technology or organizational procedures to become antiquated. There is a "keeping up with the Joneses" pressure upon water management agencies. If one water utility in an area begins to use Geographic Information Systems (GIS) to map the locations of its service lines, for example, then other utilities in the general area are likely to feel they need GIS capability as well, even though the technology requires expenditures on software, data entry, upkeep, and operator training. One water utility official spoke to us about wanting "bragging rights" to claim that his was the most advanced municipal water utility in the region and perhaps the nation. Individual agency officials we talked with were anxious to avoid being caught in error, and consequently were eager to make improvements when possible.

Despite competition among agencies over reputation for excellence, however, water engineers feel as if they all belong to the same fraternity and treat each other like colleagues. One agency planner told us he was forced to admit that he mistook the possible effects of a new flood control dam upstream upon a downstream structure for which he was designing an enlargement. He treated challenges to his work as professional peer review, and while he found flaws in the criticism to which he was subjected, he willingly altered his analysis.

Water management agency professionals are graduates of universities to which they often maintain ties. They turn to former professors when awarding grants and consulting contracts, and in seeking advice. They attend professional meetings, such as the American Waterworks Association, where they listen to research papers and look at displays of the latest advances in water technology. They read professional journals and newsletters. When technological or organizational innovations are introduced with the sanction of a respected source of professional expertise, they are especially likely to get attention and serious consideration. A senior specialist in planning in the Metropolitan Water District explained how he and his colleagues introduced a substantial innovation into their modeling of water supplies and demand by putting into practice ideas they had heard about at a professional meeting. Their effort was facilitated by a contract to a hydrology professor at a prestigious state university. They cited their training in economics as explanation for their willingness to consider significant modifications of what had been longstanding ways of doing things, which made them more sensitive to the importance of demand forecasting. These resource specialists were also new to the organization. Recruitment of new professionals brings the latest training and knowledge about technology to the agency and was identified as a primary source of internal innovation in a number of organizations we studied.

### 3.2.2. External Drivers for Innovation

Officials also identified several external drivers that they perceive as pushing innovation in water resource agencies. Environmental and health regulations were frequently mentioned as agents for change in Southern California water agencies. We were told that the increasing number of water quality and wildlife protection regulations by both state and federal agencies requires that standard operating procedures of water supply and waste treatment agencies be continually reexamined and altered. Several managers, for example, referred to how water agencies have been forced to make accommodations to Endangered Species Act (ESA)–listed threatened and endangered species such as salmon, bald eagles, and other plants and animals. Water flow requirements for listed riparian species have placed additional limitations on interbasin transfers from rivers that may already be oversubscribed.

Court suits are another external driver. Most of the river rights in Southern California river basins have been adjudicated and a court-appointed water master monitors allocations. The threat of court battles is widely used as a weapon in negotiations among local water agencies. To be on the losing end of such a suit limits the options of a water agency and often dictates significant change. One California source informant explained that while an upstream utility probably had the right to reclaim discharges from their waste treatment plants rather than allowing wastewater to flow downstream into a neighbor's recharge basin, such action would result in an expensive court suit neither party could afford to lose, so the rights weren't pursued.

Other political drivers also exist, in the sense of mobilizing popular support or opposition to water agency decisions or practices more directly. Pressures for change can come through governing bodies and advisory boards of the water agencies themselves as well as other political bodies. Any increase of rates is likely to raise agency visibility and thus increase the potential for public pressure to change. A plan to raise replenishment fees to member utilities, for example, was one of the drivers behind development of a new Municipal Water Utility of Orange County Comprehensive Plan.

Political pressures emanating from one water management institution may drive innovation in others. Initiatives by the IRWD to supply water to its expanding service area has caused the Municipal Water District of Orange County to begin long-term planning to assess what such annexation might mean for water quality and water rates for other municipal water utilities within the district. And, while water utilities in Southern California would like to see more water directed southward from the State Water Project in order to demonstrate the efficiency of current water management strategies, both environmentalists and agencies serving the Central Valley in California are resisting, saying the water is needed in the north. This pressure has increased efforts to enhance demand management and conservation, which might not have otherwise occurred. One source talked about the "sympathetic" droughts that occur in Southern California when real droughts hit the north. Forced to export water even when supplies are scarce, Northern Californians are especially likely to be critical of Southern Californians' imprudent use of water at such times. Consequently, municipal water utilities in the south step up visible conservation activities to forestall campaigns to reduce exports through the State Water Project.

The most innovative techno-scientific changes are likely to enter conservative water agencies from sources outside the organization, such as universities, think tanks, and consultant firms. The missions of most water organizations do not include much emphasis on research, and the kinds of people employed in water agencies are not often research oriented, nor does their job description encourage the development of innovations. Research organizations and consultant firms often hire water resource graduates with advanced degrees. The largest number of water researchers with doctorates outside academic institutions, for example, can be found in environmental and engineering consultant firms, including large worldwide organizations such as Dames and Moore and CH2M Hill, and thousands of smaller firms. These businesses sell not only their services but also often their products. As a consequence, there is a good deal of outside pressure from consultant firms, think tanks, and universities upon water agencies to adopt new technologies. We found that innovations, therefore, tend to be driven by increases in techno-scientific capability in external organizations rather than needs of either the water agency managers or political decision makers. The push comes from the research community rather than any pulls coming from the conservative water industry.

### 3.2.3. The Interaction of External and Internal Drivers

The adoption or use of a techno-scientific innovation is often the result of interactions between external and internal drivers. Pressures from outside for a particular change are most likely to be effective when internal support exists, although on occasion, interactions between outside and inside become so complex that it is difficult to distinguish the sources of the drivers. Figure 3 portrays the interaction of external and internal drivers. One example of how innovation reinforcement occurs when external and internal drivers are intertwined is the development of innovative demand management alternatives during the extended drought in the Southwest between 1987 and 1992.

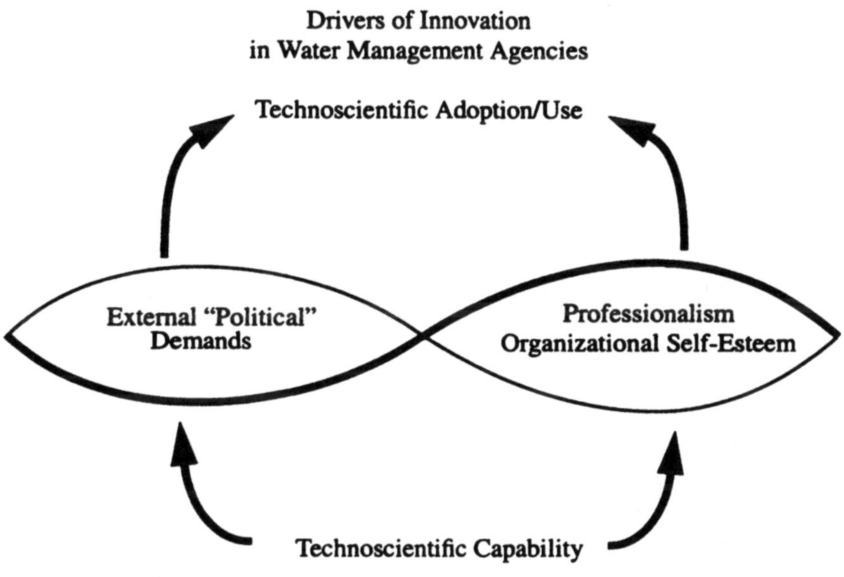


*Figure 3.* Drivers of innovation in water management agencies

The 1987–92 drought was notable for its length and the statewide nature of its impacts. Because of California's size, droughts often do not occur simultaneously throughout the entire state. But during this drought, the Sacramento River system (Northern California) experienced two dry years and four critically dry years; the San Joaquin River system (Southern California) experienced six critically dry years. Most water users served by large suppliers such as the SWP did not begin to experience shortages until about 1990 or 1991, because reservoir water storage provided a buffer during the initial drought years. The SWP, for example, was able to meet all delivery requests during the first 4 years of the drought, but was then forced to substantially cut back deliveries. In 1991, the SWP terminated deliveries to agricultural contractors and provided only 30% of requested urban deliveries. By the third year of the drought, overall statewide reservoir storage was about 40% of average. Statewide reservoir storage did not return to average conditions until 1994, and that was thanks to an unusually wet 1993.

Groundwater extraction increased substantially during the drought, with most new wells being drilled for individual domestic purposes. According to the Governor's Advisory Drought Planning Panel (2000),

> . . . water levels and the amounts of groundwater in storage declined substantially in some areas. Groundwater extractions were estimated to exceed groundwater recharge by 11 million acre-feet in the San Joaquin Valley during the first five years of the drought. Precise surveys of the California Aqueduct identified an increase in subsidence along the aqueduct alignment in the San Joaquin Valley, in response to increased groundwater extractions.

Most of the state's urban water retailers and water wholesalers implemented either voluntary or mandatory demand reduction techniques at some point during the drought. Demand reduction programs were typically accomplished through extensive customer education and outreach programs.

Municipal water agencies, which had previously relied on finding new sources of supplies, began to hire staff whose job it was to encourage conservation. While some new staff members were in public or community relations, others were professionals trained in multiple methodologies for reducing water use. Research and experience extending back to the previous serious drought in the 1970s showed that a significant percentage of household water use could be reduced by placing bricks or water-filled plastic bags in toilet tanks. Plumbing manufacturers began to market low-flow toilets based on this principle, and with the support of water management professionals, many cities made such appliances mandatory in new building construction.

In order to make any impact through this new technology, however, it was necessary to convince homeowners to replace existing, functional appliances with new, low-flow toilets. Since the cost of water had traditionally been low—one of the values water agencies pursued as described in the hierarchy above—there had been little financial incentive for homeowners to make changes without encouragement. Outside pressure from environmentalists, often working through city councils, resulted in a number of new municipal ordinances that subsidized or fully funded the costs to homeowners of replacement appliances. These regulations were fully supported and often initiated by water utilities looking to conservation as a new source of water. Rather than the water utilities raising questions about their ability to provide water by asking users to limit their use, incentives were developed and promoted by third parties to encourage the desired conservation. In response to our questions about innovations, several individuals we interviewed mentioned toilet replacement programs as evidence of the progressiveness of their organization. While such programs are expensive, they are easy for water agencies to justify. Water officials need only say that the program has been tried and found to be popular and successful in many other cities.

### 3.2.4. Stealth and Disguise Pathways for Innovation

There is often a considerable disconnect between managers and technicians within water management organizations, opening up what we call the stealth pathway for actualizing agency policy or behavioral change. Agencies regularly depend on a number of routines or models in their operations. The builders and users of these models are trained professionals who make the assumptions and establish the relationships among variables. Relatively major changes can occur in the assumptions, data, and even models themselves without top executives or decision makers necessarily knowing or caring. So long as everything continues to run smoothly and a high-quality water product is delivered inexpensively, there is little reason for top administrators to question detailed operational issues. A number of modelers and resource specialists told us that higher-level administrators depended upon but did not really understand their work. The right side of Figure 4 illustrates this pathway of change.

In 1992, the MWD in Southern California undertook to integrate demand and supply into a single model. While such models were common in the electric power industry, they were not used by water agencies. Previously, historical data were used to estimate demand separately. Between 1992 and 1996, two new integrated resource planning models were developed in quick succession with the help of a consultant from the University of California at Los Angeles. As a consequence of its use of these models, the MWD was able to convert itself from a pass-through agency, with little prior knowledge of what supplies or demands it might face, into an active overseer of the resource. While top management, which was experiencing rapid turnover at the time, may have been cognizant of the innovative model changes, it is unlikely they knew or cared much about the mechanics of the modifications.

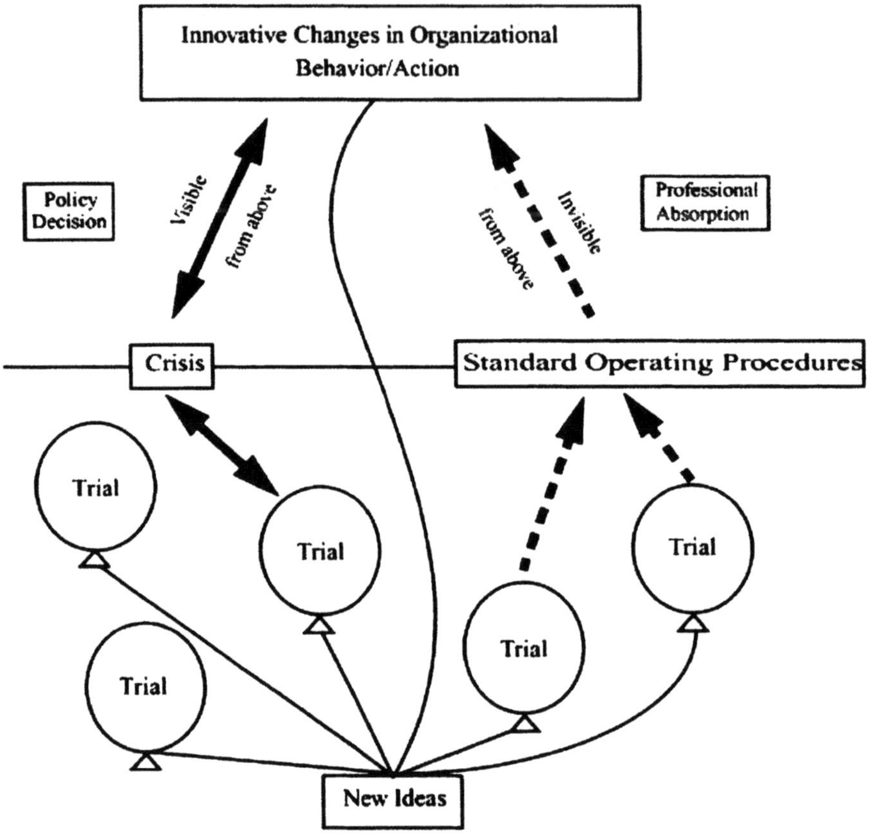


*Figure 4.* Pathways of innovation in water management agencies.

Another route for innovation is the disguise pathway, which is portrayed on the left side of Figure 4. Changes that take this course begin as techno-scientific trials that are adopted and used by a single individual or small group in the organization. Until a crisis occurs, however, no one really depends on this innovation. We found that when water management agencies are under strong internal or external pressure to change behavior or policy directions, they look to these local innovations to find something they can rely upon as a basis and rationalization for changing policy. This also helps them to claim that they had been doing it all along. What was once a kind of useless, rudimentary appendage to the body of water agency activity becomes more robust, capacitated, and meaningful.

The disguise pathway explains why so many of our respondents emphasized how long it took for a water management agency to "bless" the use of a new idea. Change, we were told, comes only if an idea has been routinized into the culture of the organization over time and experience. The gestation time required for a new idea to affect policy was estimated by one respondent to be somewhere between 10 and 20 years. The accepted venue for overt adoption of an innovation is typically the solution to a problem situation that is not adequately dealt with by standard operating procedures. The solutions that water decision makers turn to when what they have at hand will not do the job is likely to be something they picked up, for a variety of reasons, and got familiar with, but never put into full use.

An example of disguise innovation is the conservation rate structure adopted by three small public water utilities during the California drought described above. The Otay Water District, City of San Juan Capistrano Water District, and Irvine Ranch Water District adopted a highly unusual conservation rate structure. Rather than a cost of service structure, which is standard for public utilities, the innovative rate structure funds all current operations by the base rate. The conservation rate structure is therefore revenue neutral. However, each water sector is allocated a conservation use target. For example, residential target use is based on a combination of factors including size of the lot, housing characteristics, and daily calculations of evapotranspiration. Customers using more than the target are assessed sharply escalating penalty fees. When this policy was first implemented as an emergency measure during the drought, some residential customers so far exceeded their target rate that they were billed as much as $300 a month.

This rate structure has continued to be in force, with what sources called a lot of "wiggle room." One utility assesses penalties to only 1% to 2% of its customers. However, work proceeds to define and allocate the conservation water rate to different sectors of water customers, including utility cooling towers, hotels, and hospitals. Because the rate does not have much bite and seems at face value to be fair and science based, water users seldom challenge it. Should serious drought require significant reduction of demand, it is anticipated that the wiggle room in the conservation use targets will be substantially tightened. Rather than across-the-board water rationing, which is traditional in the industry, these utilities will engage in much more advanced demand management methods. The long track record being established in less stressful times, however, will likely make these strategies appear to be business as usual. Even while making a considerable change, the agency appears to be doing no more than conducting business in a professional way as they have always done in the past.

### 3.2.5. Locales of Innovation

Contrary to much of the organizational literature, we found no relationship between the size of an organization and its propensity to innovate. While it would seem logical to expect that only large organizations might have the diversity and the redundancy to innovate, this turned out not to be the case. Instead, we found that the presence of organizational resources was more critical. Rich organizations with wealthy customers, regardless of size, can afford to try out techno-scientific changes while continuing to perform standard operations without much sacrifice. Less well-off agencies were more likely to tell us that they wanted to try something new, but simply lacked the time, people, and money. The age of an organization may well matter. A relatively new municipal water utility, like the Irvine Ranch Water District, that is building its infrastructure from the ground up, can implement strategies such as dual delivery systems, one for potable and the other for irrigation water. People open to new ideas can be recruited to organizations of any size or age, but some moderate rate of turnover would seem to have a positive correlation with trying out new ideas.

## 4. CONCLUSIONS

As the reliability of interbasin transfers is increasingly threatened by population growth and climate shifts, local water agencies that traditionally have been dependent on large water conveyance systems like the MWD are searching for localized sources of water. This is one way to reduce their vulnerability to decreases in the quantity or quality of water for their customers. Continuing growth in the areas from which water has typically been transferred to Southern California is also raising questions about the equity of existing allocation schemes. If Southern California can no longer depend on interbasin transfers for reliable supplies, conservative water agencies are faced with the necessity for developing alternative strategies. This favorable context, however, does not appear to translate directly into the adoption of innovations.

While the challenges facing the water agencies suggests that innovations in water supply and delivery would be welcomed, we found that innovations take a long time to work their way into altered decision-making processes and organizational practice. In the past, conservative organizations relied on redundant systems to manage water supply and quality, accepting incremental changes to the system as necessary. The search for localized water supply sources to decrease dependence on water delivery systems is an example of water agencies creating multiple (redundant) sources to ensure reliability and describes how the agencies are likely to choose incremental change by adopting strategies with which they are familiar. Since the most common strategy in Southern California to increase water availability is through interbasin transfers, there will continue to be fierce competition among MWAs looking to make additional claims upon existing and new sources of supply.

The impact of organizational innovation on actual decision making or practice is likely to be marginal at first. Innovations are tried out alongside existing tools for decision making and practice. It is to be expected that water management personnel will say that they have more confidence in old ways of doing things, and when signals about the efficacy of innovative strategies diverge, they will place their confidence in older technologies. Innovations arising from research in universities, think tanks, and consultant firms are important sources of innovation as they become integrated into professional training and specific applications. Innovations may well come from technical staff within organizations who incorporate specialized changes into standard operating procedures as new models and technology are developed within their field. Top officials and decision makers, however, are unlikely to know whether or how techno-scientific innovations are being used if the changes are included in models and technology they are not familiar with or do not use.

Municipal water agencies that are likely to prosper in the changing context are those that are able to fully exploit incremental adaptations in interbasin transfer strategies as well as develop innovative management techniques. Such agencies must be richly endowed with organizational resources, most likely deriving from a high-income customer base. Organizations like the Irvine Ranch Water District are likely to fare better than utilities representing older urban areas or inner cities, which may lose out in the competition for new supplies and not be sufficiently well off to innovate in the management of present supplies. Current inequities in the availability, quality, and cost of water supplies within Southern California may well increase as all water agencies are driven to compete for scarce supplies. The insatiable thirst for water in Southern California, combined with variable and changing weather patterns that challenge the capacity of existing sources and technologies, however, may create a sophisticated group of water agencies skilled in traditional management strategies and experienced with introducing innovation in conservative organizations. These agencies may be able to reliably provide low-cost and safe water while others are unable to even protect their current sources of water.

The conservative values of all water agencies, combined with localized pockets of wealth, are likely to increase inequity in water allocation in Southern California. For example, global trade treaties like the North American Free Trade Agreement (NAFTA) may provide new sources of water for Southern California agencies and systems that have the capacity to propose and implement large-scale, cross-continent water transfers from domestic sources farther afield than existing sources (e.g., Columbia River or Alaska) or new international sources (e.g., Canada). Or, as one of our respondents told us, as the cost of desalinization technologies declines, wealthier water systems will be able to provide unlimited water to certain communities even as once-reliable sources of ground and surface water are exhausted through existing practices and agreements. When some local agencies are capable of claiming the Pacific Ocean as a water source and others are limited to existing and overallocated sources, innovative adaptations to water shortages and climate variability are likely to become increasingly concentrated in fewer, wealthier delivery systems—as will be the water.

## 5. REFERENCES

Cortez-Lara, A., and M.R. Garcia-Acevedo. 2000. Lining of the All American Canal: Social and economic effects. *Natural Resources Journal* 40(2): 261–280.

Grimm, N.B., A. Chacon, C.N. Dahm, S.W. Hostetler, O.T. Lind, P.L. Starkweather, and W.W. Wurtsbaugh. 1997. Sensitivity of aquatic ecosystems to climatic and anthropogenic changes: The basin and range, American Southwest and Mexico. *Hydrological Processes* 11(8): 1023–1041.

Governor's Advisory Drought Planning Panel. 2000. *Preparing for California's Next Drought*. California Department of Water Resources, Sacramento.

Kelly, D. 2000. Ruling reflects crucial shift in water policy. *Los Angeles Times*, June 14, 2000, A3.

Kennedy, J.C. 1999. *Master Plan Report for the Orange County Water District*, April, Fountain Valley, CA.

Michel, S.M. 2000. Place and water quality politics in the Tijuana–San Diego region. In, Herzog, L.A. (ed.), *Shared Space: The Mexico–United States Environmental Future*. San Diego, California: University of California at San Diego Center for U.S.-Mexican Studies, pp. 233–263.

Reynolds, R., M. Dettinger, D. Cayan, D. Stephens, L. Highland, and R. Wilson. Effects of El Niño on Stream Flow, Lake Level, and Landslide Potential. U.S. Geological Survey Website: http://geochange.er.usgs/gov/sw/changes/natural/elnino. 2001

*Appendix A.* Interview contacts in California.

| | |
|---|---|
| John Anderson<br>Office of Emergency Services | Rafael Fernando, P.E., Sr. Engineer<br>System Operations Branch<br>Metropolitan Water District of Southern California |
| Jonathan Bishop, Chief, Regional Programs<br>Los Angeles Regional Water Quality Control Board<br>California Environmental Protection Agency | Robert Floerke, Assistant Deputy Administrator<br>Office of Spill Prevention & Response<br>Department of Fish & Game |
| Timothy Blair, Sr. Resource Specialist<br>Planning & Resources Division<br>Metropolitan Water District of Southern California | Sonny Fong, Emergency Preparedness Manager<br>Executive Division<br>Department of Water Resources |
| Steve Borroum, Environmental<br>Department of Transportation | Larry Gage, Chief<br>Operations Control Office<br>Division of Operations & Maintenance<br>Department of Water Resources |
| Chester Bowling, Chief<br>Water Operations Division<br>Central Valley Operations Office<br>U.S. Bureau of Reclamation | Brandon Goshi<br>Planning & Resources Division<br>Metropolitan Water District of Southern California |
| Arthur Bruington, Former Manager<br>Irvine Ranch Water District<br>Public Assistance Coordinator, Flood Control | Gregory Heiertz, Director<br>Engineering & Planning<br>Irvine Ranch Water District |
| Kerry Casey, Operations Division<br>Army Corps of Engineers | Gary Hester, Chief Forecaster<br>Hydrology Branch<br>Division of Flood Management<br>Department of Water Resources |
| Francis Chung, Supervising Engineer<br>Division of Planning<br>California Department of Water Resources | Gary Hildebrand,<br>Supervising Civil Engineer III<br>Hydraulic/Water Conservation Division<br>Los Angeles County Department of Public Works |
| Robert Collins, Sr. Hydrologist<br>Water Management Section<br>Sacramento District<br>U.S. Army Corps of Engineers | Chiun-Gwo (Simon) Hsu, Hydrologic Assistant<br>Aqueduct Section<br>Water Supply Division<br>Los Angeles Department of Water & Power |
| Robert Copp, Operations Division<br>Department of Transportation | Bob Joe, Assistant Director for Management<br>Army Corps of Engineers |
| Christopher Crompton, Manager<br>Environmental Resources Division<br>Environmental Management Agency | Katherine Kelly, Chief<br>Office of State Water Project Planning<br>California Department of Water Resources |

| | |
|---|---|
| Gary W. Darling<br>The Resources Agency<br>State of California | Francie Kennedy, Conservation Coordinator<br>City of San Juan Capistrano |
| John Kennedy, Assoc. General Manager<br>Advance Planning<br>Orange County Water District | David Pettijohn<br>Water Supply Division<br>Los Angeles Department of Water & Power |
| Vern Knoop, Chief, Water Supply<br>California Department of Transportation | Terri Porter, Maintenance<br>California Department of Transportation |
| Donald Long, Office Chief<br>State Water Project Analysis Office<br>Division of Operations & Maintenance<br>Department of Water Resources | Paul Pugner, Chief, Water Management and Hydraulic Engineer<br>U.S. Army Corps of Engineers, Sacramento District |
| Robert McVicker, District Engineer<br>Water Replenishment District of Southern California | Gwen Sharp, Hydrogeologist<br>Orange County Water District |
| Craig Miller, Director<br>Recharge & Wetland Operations<br>Field Headquarters<br>Orange County Water District | Joan Sollenberger, Program Manager<br>Transportation Planning<br>Department of Transportation |
| William Mork, State Climatologist<br>Division of Field Management<br>Department of Water Resources | Timothy Sovich, Sr. Engineer<br>Orange County Water District |
| Elizabeth Morse, Meteorologist in Charge<br>National Weather Service<br>National Oceanic & Atmospheric Administration<br>U.S. Department of Commerce | Matthew Stone, Assoc. General Manager<br>Municipal Water District of Orange County |
| Christian Nagler, Chief<br>Watermaster Service Section<br>Division of Planning & Local Assistance<br>Department of Water Resources, Southern District | Karl Seckel, Asst. Manager/District Engineer<br>Municipal Water District of Orange County |
| Don Noble, Maintenance Operations Manager<br>Public Works Department<br>City of Huntington Beach | Daniel Tunnicliff, Chief Engineer<br>Safety & Emergency Response Division<br>Orange County Sanitation District |
| Langdon Owen, Member<br>Board of Directors<br>Metropolitan District of Southern California | James Yannotta, Manager, Operations<br>Los Angeles City Department of Water & Power |

| | |
|---|---|
| Francis Palmer, Chief<br>Ocean Standards Unit<br>Standards Development Section<br>Division of Water Quality<br>California State Water Resources Quality Board | |

*Appendix B.* Interview protocol.

I. Scope and Scale of Organization
   1. What does your organization (or part of the organization) do?
   2. Tell us a little about yourself: your current job/role, experience, etc.
   3. Whom do you interact with to do your job (both inside and outside the organization)?

II. Organizational Norms
   1. How do you know when your organization is doing its job well?
   2. How do other people judge your organization's performance?

III. Sources of Information and Innovation
   1. What do you personally need to know to do your job effectively?
   2. Where do you go to find or generate that information?
   3. How do you select sources of information?
   4. How do you decide what information is reliable?
   5. What do you do when reliable sources disagree?
   1. Can you describe any recent innovations affecting your organization? Probe regarding information used/necessary to determine problem, make innovation, assess impact, etc.

IV. NOAA Forecasts
   1. What would you do differently if you had a reliable prediction about prevailing weather conditions? A 90-day prediction? A year ahead?
   2. Have you ever looked at NOAA's long-term forecasts?
   3. If so, what did you make of them? Did you use them? Why/why not?

5

# IMPACTS OF CLIMATE FLUCTUATIONS AND CLIMATE CHANGES ON THE SUSTAINABLE DEVELOPMENT OF THE ARID NORTE GRANDE IN CHILE

Hugo Romero and Stephanie Kampf*
*Departamento de Geografía de la Universidad de Chile, Santiago, and Centro de Ciencias Ambientales/EULA, Universidad de Concepción, Chile*
**Fulbright Program Fellow 1998–99*

**Abstract** The environmental effects of mining under the conditions of globalization are analyzed at regional and local scales in the context of their impact upon the sustainability of the Chilean "model" for economic development. The extraction of raw materials, such as copper, from mines is still the main basis of the Chilean economy. Mining, developed in the arid regions of the north of Chile in and around the Atacama Desert, is examined here with regard to its effects on the natural and social components of the environment. Major scientific concerns are focused on the availability of water to sustain economic and social needs, given the extreme aridity and the effects of climate change and fluctuations on water sources. Furthermore, most of the water in the area can be considered a nonrenewable resource because it was formed under very different paleoenvironments.

Modern recharge of water is extremely limited there, and modern economic development (mines, tourism, urban growth), as well as the cultural heritage and natural habitats in these regions, depend largely on these fossil water reserves. Accordingly, the water demand crisis urgently requires the development of a strategic policy at a regional scale in order to conserve the remaining resources, to prioritize the uses of water, and to search for alternative sources.



*Henry F. Diaz and Barbara J. Morehouse (eds.), Climate and Water*, 83-115.

# 1. INTRODUCTION

The simplest and best-known definition of sustainable development addresses the need for nations and regions to combine economic growth, social equity, and environmental protection. Sustainable development should also emphasize inter- and intragenerational equity and develop a special concern about transboundary issues.

Ten years after the Rio de Janeiro summit, progress toward environmental conservation and sustainable development is still difficult to recognize in Latin America. Economic growth, one of the main ingredients, has been very slow, and in the worst cases, has been negative for many years for some countries. Globalization has generated large asymmetries between developing countries that export raw materials and natural resources, and industrialized, developed countries that export modern services. International markets and financial flows of capital, which are increasingly needed in developing nations, have been affected by international crises, by policies imposed by international agencies, and by territorial and social fragmentation that have resulted from the geographical selectivity of capital investments.

Numerous analyses show that poverty levels have remained high and even increased in many Latin American countries and that the social polarization between those linked to global business and those dependent on domestic economies has resulted in a pattern of social and spatial segregation at national, regional, local, and urban scales (Romero 2000a; Romero and Rivera 1996).

In environmental terms, economic and social stresses have caused tremendous impacts. Based on the theory of comparative advantages, developing countries have intensively exploited and exported large amounts of their stocks of raw materials. Furthermore, pressed by economic needs and social urgencies, all the countries have changed from inward-oriented towards outward-oriented economic policies; meanwhile, they are seeking inclusion in the world economic system, where most of the available niches for them are reduced to the selling of their natural resources. At the same time, the periodic occurrence of international trade crises has forced the developing countries to increase their exploitation of minerals, forests, lands, fisheries, oil, etc., to compensate for the drop in prices in international markets.

During the recent Asian economic crisis, most of the Latin American countries have ignored the enforcement of environmental legislation and social regulations in favor of economic investments in order to satisfy political exigencies. Despite theoretical assertions, instead of the stated goal of combining economic growth, social equity, and environmental protection, what has been observed is a seemingly permanent conflict among these positions. In terms of transboundary issues, open economies are producing a redefinition of national frontiers, regional borders, and place connections. New and flexible production systems, multilateral trade agreements, and the formation of regional integrated markets like MERCOSUR (Mercado Común del Sur, formed by Argentina, Brazil, Paraguay, and Uruguay, with Chile as an associate member), are leading toward the definition of a new "geography" in which the privatization of the land and its resources generates the most income and political support. Transnational companies also create a new spatial organization, which is increasingly more efficient and independent from the traditional political states.

The development of more advanced transportation and communication systems in developing countries has made remote regions more accessible, led to a reassessment of many natural and cultural resources located at distant places, and shown the weakness of the local sociopolitical institutions. Regional and spatial fragmentation has been the direct result of the development of modern trade and telecommunication networks, which have selectively connected only the areas that are most interesting to international markets.

Globalization requires, however, the increasing integration of places and people, especially because it needs larger markets, access to natural resources, agglomerations of urban populations (to strengthen domestic consumption), vertical and horizontal integration of production chains (including industries and services), and the adaptation of scale economies to increase competitiveness and place utility.

In terms of intergenerational responsibilities, it is clear that in Latin America the current priorities are leaving behind eroded soils, devastated life support systems, deforested slopes, less biodiversity, social segregation, and massive environmental pollution in the form of thousands of tons of air, water, and soil pollutants to future populations. In terms of intragenerational equity, 160 million of the 430 million people in Latin America live below poverty thresholds, and increasing social polarization is predominant everywhere. It has become necessary to improve the share of development benefits among different social groups, to install mechanisms to ensure the participation of the people in the decision-making process, and to effectively democratize the decision-making institutions.

One of the major current challenges is to understand how compatible growing social and economic needs will be in the future with respect to global environmental changes (National Research Council [NRC] 1999). Climate change and fluctuations are among the most important challenges, particularly in arid regions where they primarily affect water availability.

To illustrate the challenges of sustainable development, Chile may be regarded as an excellent example because of its natural conditions and state of economic development. Chile is the longest Andean country (more than 4,500 km in length from north to south, extending latitudinally from 18° to 54°S), and it contains arid, mediterranean, and temperate types of climates. Located in the north of Chile, the Atacama Desert is at the very core of the arid regime: Its mean annual rainfall is less than 3 mm, although amounts increase with altitude and latitude.

Along the whole eastern boundary, which separates Chile from Bolivia and Argentina, the main mountain chain of the Andes reaches over 6,000 m above sea level (asl), with the highest peaks in the northernmost section. From the main Andean chain, many transversal ranges reach the Pacific Ocean, which is only at a mean distance of 150 km westward. Near the ocean, the coastal range forms another north-south mountain chain, which can easily reach above 1,000 m altitude. As a consequence, Chilean landscapes in general are extremely mountainous and include north-south successions of closed basins.

In the arid north of Chile, the main Andes axis extends eastward and contains flat highlands, known as altiplanos, or as the Puna de Atacama. These altiplanos, in turn, can be sources of lakes, lagoons, salars, and transboundary rivers and streams. The Lauca River, for example, is born in swampy areas in the Chilean side of the frontier with Bolivia, but runs towards Bolivian territory. In contrast, other smaller streams, like the Vilama, Silala, and Zapalieri, originate in Bolivia and flow into Chile. Given the extraordinary aridity of the Chilean lowlands and the increasing water demands, it is easy to see the growing value of water sources and transboundary rivers in this region.

Since the end of the 1970s, Chile has been regarded as one of the South American pioneers in moving towards a private, open, and outward-oriented globalized economy. As a consequence, between 1983 and 1996 this country received more than US$60 billion of direct foreign investment (Fazio 1997), rising from an annual amount of US$660 million in 1990, to US$5.4175 billion in 1997. The gross domestic product (GDP) increased 2.5 times between 1986 and 1998, at an average annual rate of growth of 7.5%. Direct investments have increased from 24% of the GDP in 1990 to 31% in 1997. Another US$27.4 billion in investments are expected between 2000 and 2004.

As indicators of economic success, at least in Latin American terms, Chile has also substantially reduced its annual inflation rate to less than 5% and the percentage of people living under the poverty threshold to less than 20%. Between 1990 and 1996, the latter percentage was reduced by 6%, but between 1997 and 1999 a decrease of only 1.5% was measured. This suggests that the fast rates of economic growth seen in Chile throughout much of the 1990s may be coming to an end.

Based on these figures, Chile is frequently presented as an example of the benefits of economic globalization and of the success of "neoliberal" principles, particularly due to its efficient and export-oriented economy; the privatization and internationalization of its natural resources basis, formerly state-owned factories, facilities, and services; the opening and deregulation of its economy; and the reduction of the size and activities of the public sector. As Mertins (2000) notes, Chile is now considered to be a Latin American model of the transformation process oriented entirely on neoliberal premises.

However, to evaluate the Chilean experience more completely, it is necessary to consider environmental and socioeconomic perspectives on different geographical scales. Several reports (Sunkel 1996; Chernela 1997; Claude 1997; Romero 2000a; Universidad de Chile 2000) mention that Chile, as a consequence of its vigorous economic growth, has severely damaged its natural and social environments: water, air, and soil pollution have increased; its native forests and biodiversity have been reduced substantially; its native populations and their habitats are threatened; and regional disparities and urban social segregation have been maintained, and even accelerated.

In the following, the environmental impacts of one selected economic activity under the condition of globalization, namely mining, will be analyzed at regional and local scales in the context of its contribution to the sustainability of the Chilean "model." The extraction of raw materials, such as copper, from mines is still the main basis of the Chilean economy. Mining developed in the arid regions of the north of Chile, in and around the Atacama Desert, is examined here, in consideration of its effects on the natural and social components of the environment. Major scientific concerns are focused on the availability of water to sustain economic and social needs, given the extreme aridity and the effects of climate change and fluctuations on water sources. Furthermore, most of the water can be considered to be a nonrenewable resource since it was formed under paleoenvironments that no longer exist.

## 2. CLIMATE CHANGE AND CLIMATE FLUCTUATIONS

### 2.1. Main Climate Characteristics

Climate is one of the most conspicuous environmental components of the Chilean Norte Grande regions, since at these latitudes (18°–24°S), the

Desierto de Atacama, one of the driest areas in the world, is found. In fact, it almost never rains at coastal areas of Norte Grande. Meteorological stations situated in the harbors of Arica (18°S), Iquique (19°S), and Antofagasta (23°S) record annual means less than 3 mm. These figures are, in turn, little more than statistical fictions, based on only one or two rainfall events, of about 10 mm, occurring in a few hours on perhaps 1 or 2 days over 10 years.

The extreme aridity of Norte Grande is the outcome of the spatial interaction of several factors: higher Andes mountains above 6,000 m act as a real barrier to humid air masses coming from the Atlantic Ocean; the coastal range also acts as a barrier to the inland penetration of marine fog ("camanchaca") and cloudiness. The region is under the permanent action of southeastern Pacific anticyclones, which prevent the arrival of polar fronts and any formation of local rains. High amounts of solar radiation and high evaporation rates are also important climate factors. Inversion layers are re-inforced near the ocean surface, by atmospheric stabilization produced by upwelling and the cold sub-Antarctic waters transported northwards by the Humboldt Current.

Due to the spatial convergence of all these factors, precipitation along the coast is practically nil. Scarce and irregular rainfalls are recorded at about 2,000 m altitude on the first western Andean slopes, but it is only in the Altiplano or highlands, above 3,000 m, that "genuine" summer rainfall occurs regularly. Therefore, two very distinctive climate regions could be differentiated in Norte Grande: the absolute desert along the lowlands (coastal areas and intermediate depressions or pampas) and the relaxation of desert conditions in the highlands (altiplano or puna).

### 2.1.1. ENSO and Climate Fluctuations

Annual precipitation varies dramatically in the altiplano of Norte Grande. These variations partially depend on the performance of the El Niño/Southern Oscillation (ENSO) phenomena. During the warm phase of ENSO (El Niño), South Pacific anticyclones are weaker, and warm waters reach the Norte Grande coast. In contrast, with the ENSO cool phase (La Niña), anticyclones are stronger, and cold seawaters accentuate the north-ward flowing Humboldt Current and coastal upwelling processes.

El Niño phenomena produce different climate patterns that accentu-ate the general environmental differences between lowlands and highlands. They have had great impacts on human settlements, severely damaging the regional infrastructure in recent times. During an El Niño event, in May 1992, for example, severe flooding in the coastal cities of Antofagasta and

Taltal resulted in the deaths of several people. In contrast, El Niño appears to be a major cause of persistent droughts in the Altiplano, as occurred between 1987 and 1991 (Aceituno 1992; Martin et al. 1992). These drought events have profoundly affected the desert oases, where native populations are concentrated (Romero and Rivera 1997).

According to Aceituno and Montecinos (1992) and Aceituno (1992), only recently has a more clear and meaningful relationship between ENSO and rainfall in the Altiplano been recognized. During 1965–66, 1982–83, and 1991–92, El Niño-related droughts were recorded in the Altiplano, but during the 1972–73 event, many stations in the Altiplano recorded positive anomalies. The strongest El Niño events, such as those in 1972–73, 1982–83, and 1991–92, have been followed by rainy summers, whereas moderate events such as those in 1976 and 1986–87 were followed by relatively dry summers in the region.

Romero and Rivera (1997) analyzed spatial variations in annual and seasonal precipitation data recorded between 1975 and 1990, specifically in Puna de Atacama, and classified according to ENSO categories (Quinn 1993). An enhancement of this relationship for the whole Altiplano is illustrated in Figure 1. The figure shows annual precipitation variations in the past 50 years in different regions of the Norte Grande region. While individual station records in the region is sufficient to estimate annual means (Table 1), the data were deemed insufficient to illustrate the high level of interannual variability in this region. Accordingly, we used a high-resolution gridded data set (Willmott and Matsura 1998) with monthly precipitation values available at half-degree intervals of latitude and longitude. We have assigned appropriate ensembles of the gridded values to the different river basins and developed precipitation indices (Fig. 1). During the moderate 1976 El Niño (El Niño years are depicted by narrow vertical lines with horizontal hatching), rainfalls were above the annual mean, but the previous and the following years had even larger amounts of precipitation. For 1982–83, considered to be the strongest El Niño, lower precipitation in Puna de Atacama appeared to be only part of a negative pluviometric phase that began in 1978. However, at the end of 1983 intense precipitation began, and 1984 was one of the wettest years in the observed period. The 1986–87 El Niño event coincided with normal precipitation, and in 1988 a new drought event began. This drought lasted until the winter of 1993, when a normalization of the pluviometric regime resumed.

On a seasonal scale, the El Niño events of 1976 and 1982–83 were characterized by the exceptional occurrence of winter rainfall. During the great El Niño (1982–83), almost no summer rainfall was recorded in Puna de Atacama, but winter precipitation was three times the seasonal mean. In 1984, heavy rainfall took place primarily in January (3.7 times greater than the mean), June (6.1 times), and October (13.3 times). Further complicating this pattern, during the 1986–87 El Niño, precipitation was greater than the mean in both summer and winter (June 1987).

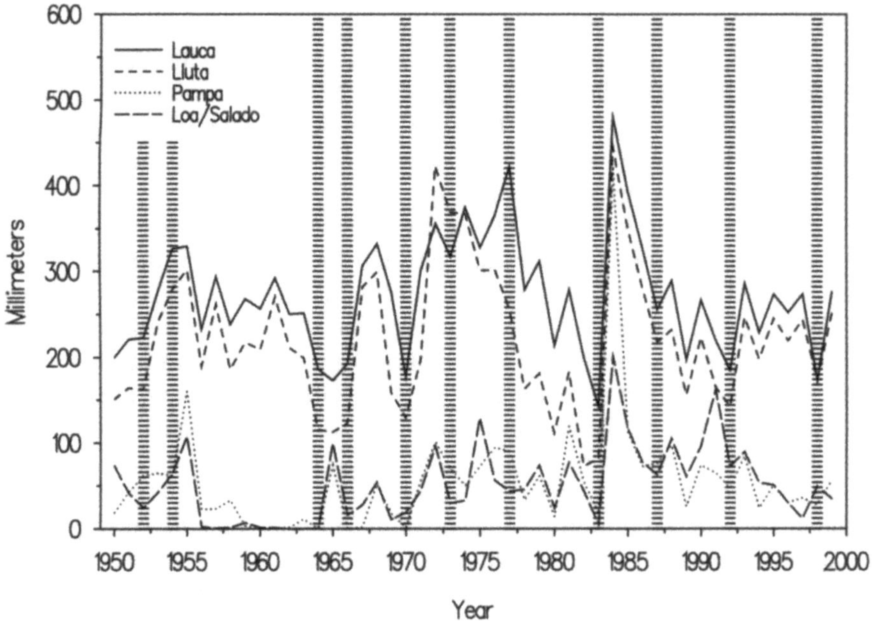


*Figure 1.* Annual precipitation (in mm) and El Niño years (vertical hatched lines) in the Chilean Altiplano, 1950–1999. Curves show the values for different watersheds in the Norte Grande region.

Precipitation patterns also present spatial variations. For example, in the upper and middle sections of the Loa River, rainfall during warm ENSO events has been more abundant during the spring (August–November) and almost nonexistent during the summer. At the bottom of the Puna (Fosa de Atacama), rainfalls were even more erratic than in the Loa Valley.

Analysis of the data leads to the conclusion that there is not a unique and strong correlation between El Niño events and precipitation in the Norte Grande region as a whole. This conclusion differs from those for other Chilean regions such as the semiarid Norte Chico and Mediterranean Central Chile (Romero and Garrido 1985; Romero and González 1989). An empirical generalization of data from Puna de Atacama indicates a reduction in summer convective rains and an increase in spring and winter season frontal rainfalls during warm ENSO events. The exceptional arrival of polar fronts in the northern regions of Chile indicates a weakness in the Pacific anticyclone as a consequence of the presence of warm waters offshore. Particularly interesting is the occurrence of heavier rainfall during the summer season immediately after warm ENSO years, as recorded in 1977, 1984, 1987, and 1993.

## 2.2. Spatial and Temporal Variations

The current climate of the Norte Grande highlands shows important spatial variations. Table 1 shows that mean annual rainfall varies between 21 mm at Embalse Conchi on the Loa River, the lowest (at 3,010 m asl) and westernmost meteorological station, and 392 mm in Cotacotani, in the Lauca River Basin, one of the highest stations (at 4,500 m asl), located in the northeast section. A linear regression model between annual rainfall, and latitude, longitude, and altitude as explanatory variables, produced a 0.7 correlation. Altitude is the primary factor influencing the increase in rainfall towards the north and to the east. The residuals in the linear regression could be associated with local orographic components related mainly to the generation of thermal convection for positive values, and to orographic rain-shadow effects for negative values (Romero and Rivera 1997).

Seasonal distribution of rainfall also presents notable spatial variations. At the highest and easternmost locations, summer rainfall (December through March) accounts for over 90% of the annual average. Conversely, the lowest and westernmost locations (Calama and Peine, for example), record only 71%. Puna de Atacama is the only Chilean region where climate is controlled by tropical influences and rainfall takes place mainly in summer. These rainfalls are generally called "Invierno Boliviano" because precipitation originates from the neighboring Bolivian Altiplano.

The spatial distribution of rainfall in Puna de Atacama is directly related to elevation and the geographic location of mountain passes. Other orographic effects also modify the local distribution of rainfall. Localized heat sources form where solar radiation is absorbed more readily due to specific local features related to the topography, exposure, land cover, slope, and lithology (Romero and Rivera 1997). They may generate super-adiabatic lapse rates and convection columns able to cause very intense rainfall (Fuenzalida and Rutlland 1986).

During the winter season, the Pacific anticyclone strengthens enough to deflect the traveling low-pressure cells to the south (Romero 1985). Sometimes, the anticyclones weaken and divide into two cells, allowing the arrival of high-altitude isolated low-pressure systems. These disturbances of polar maritime origin compress the tropical maritime air mass to below 700 millibars (mb), and are able to produce rainfall and snow at elevations greater than 5,000 m; i.e., above the higher Altiplano. The penetration of this cold air generates a large drop in the height of the tropopause, a lowering of temperature, and an intense pressure gradient, which produces strong zonal winds (northern and western jet streams).

*Table 1.* Rainfall stations in the Chilean Altiplano.

| Name | Watershed | Code | Rainfall Annual Average (mm) | Altitude (m) | Latitude (S) | Longitude (W) |
|---|---|---|---|---|---|---|
| Chungará Reten | Lauca | 01010050 | 277 | 4570 | 18 13 | 69 07 |
| Chungará Ajata | Lauca | 01010053 | 353 | 4570 | 18 14 | 69 07 |
| Isla Blanca | Lauca | 01020050 | 345 | 4500 | 18 11 | 69 13 |
| Cotacotani | Lauca | 01020051 | 392 | 4500 | 18 11 | 69 14 |
| Parinacota CONAF DGA | Lauca | 01020052 | 337 | 4390 | 18 12 | 69 16 |
| Chucuyo Carabineros | Lauca | 01020053 | 339 | 4200 | 18 13 | 69 20 |
| Parinacota Ex Endesa | Lauca | 01020054 | 339 | 4390 | 18 12 | 69 16 |
| Guallatire | Lauca | 01021050 | 280 | 4280 | 18 30 | 69 10 |
| Villa Industrial | Lluta | 01200050 | 346 | 4060 | 17 47 | 69 43 |
| Humapalca | Lluta | 01200051 | 291 | 3970 | 17 50 | 69 42 |
| Alcerrega | Lluta | 01201050 | 230 | 3990 | 18 00 | 69 40 |
| Pacollo | Lluta | 01202050 | 223 | 4050 | 18 10 | 69 29 |
| Putre | Lluta | 01202051 | 186 | 3530 | 18 12 | 69 35 |
| Coyacagua | Pampa | 01050050 | 134 | 3990 | 20 03 | 68 50 |
| Collahuasi | Pampa | 01080051 | 162 | 4250 | 20 59 | 68 44 |
| Huaytani | Pampa | 01700051 | 110 | 3720 | 19 33 | 68 37 |
| Pampa Lirima (Pueblo Nuevo) | Pampa | 01730053 | 131 | 3940 | 19 51 | 68 54 |
| T. Isluga (Col-chane) | Pampa | 01730054 | 110 | 3965 | 19 17 | 68 39 |
| Copaquire | Pampa | 01770050 | 46 | 3490 | 20 57 | 68 54 |
| San Pedro de Conchi | Loa | 02103050 | 38 | 3217 | 21 56 | 68 32 |
| Parshall N. 2 | Loa | 02103051 | 30 | 3318 | 21 57 | 68 31 |
| Ojos San Pedro | Loa | 02103052 | 64 | 3700 | 21 58 | 68 19 |
| Inacaliri | Loa | 02103053 | 135 | 4100 | 22 02 | 68 04 |

| | | | | | | |
|---|---|---|---|---|---|---|
| Conchi Viejo | Loa | 02104050 | 46 | 3491 | 21 57 | 68 44 |
| Conchi Embalse | Loa | 02104051 | 21 | 3010 | 22 03 | 68 38 |
| Muro Embalse Conchi | Loa | 02104052 | 20 | 3010 | 22 03 | 68 38 |
| Cupo | Loa | 02105050 | 56 | 3400 | 22 07 | 68 19 |
| Turi | Loa | 02105051 | 37 | 3053 | 22 14 | 68 18 |
| Linzor | Loa/Salado | 02105052 | 179 | 4096 | 22 12 | 67 59 |
| Toconce | Loa/Salado | 02105053 | 101 | 3350 | 22 15 | 68 10 |
| Salado Embalse | Loa/Salado | 02105056 | 63 | 3200 | 22 18 | 68 12 |
| Caspana | Loa/Salado | 02105057 | 87 | 3260 | 22 19 | 68 13 |
| El Tatio | Loa/Salado | 02105058 | 138 | 4320 | 22 22 | 68 00 |

Source: Dirección General de Aguas de Chile [DGA]

Daily concentration of rainfall is another typical climate characteristic of arid regions. The occurrence of only 1 day of rainfall can reach and even surpass the long-term annual mean. One example is San Pedro de Conchi, where 40 mm of rain fell in 24 hours, and where the long-term average for the year is only 39 mm! Similarly, in Ayquina, 44.5 mm has been recorded in 24 hours, compared with an annual mean of 42.7 mm. The average ratio of maximum daily rains and the annual mean varies between 21.5% in Inacaliri and 40.7% in San Pedro de Conchi. The average rate between the observed 72-hour rainfall over the annual mean varies between 28% in Linzor and 59.8% in Toconao.

## 2.3. Surface Temperature Distribution

There is a lack of systematic observations about other climate elements besides precipitation; for example, of the spatial and temporal distribution of temperature in the north of Chile. This fact is even more restrictive given the complex topography of Norte Grande highlands. Using such data as are available, including infrared images from National Oceanic and Atmospheric Administration (NOAA) satellites, daily winter temperature estimates ranged from –15° to 4°C in Puna de Atacama, on July 5, 1987. The lowest, swampy area located at the southern border of the Salar de Atacama had the highest temperatures.

Elevation determines temperature distribution. The 1.5°C isotherm follows Llano de la Paciencia and Peine at the eastern and southern borders of the Salar de Atacama and the 3,000 m contour line. The –2.5°C isotherm follows the 4,000 m contour line, and the –13°C isoline extends above 5,000 m elevation. In addition to this standard pattern of decreasing temperatures as a function of elevation, notable heat islands exist in winter along the Loa and Salado River Valleys, and in oases, such as Caspana and Ayquina. The Tara, Ascotán, and Carcote salars are other features that show positive thermal signatures. In January 1990, NOAA satellite images showed summer temperatures varying between –4° and 19.9°C. Maximum temperatures (>19°C) were found in the lower parts of Salar de Atacama and in Llano de la Paciencia. Elevation controls the distribution of the 12.6°C and 6.4°C isotherms, which follow the 4,000 and 5,000 m contour lines, respectively. Heat islands are also observed in the summer in several other places along the Loa and Salado Rivers, whereas all of the altiplano salars (Tara, Ascotan, Pujsa) and lagoons (Miscanti-Miñiquez, Lejia) always show more elevated temperatures than the surrounding landscapes.

## 2.4. Climate Change in Norte Grande

Arid conditions have been the dominant feature in Desierto de Atacama since the Miocene (Ortlieb 1995a,b; Abele 1991), tens of millions of years before the present (B.P.). In the Altiplano, climate has evolved as sequences of extremely arid and relatively wet phases. Large amounts of precipitation occurred during early and late Miocene times and during the Pliocene. During the Pleistocene, aridity was a predominant feature but higher precipitation is inferred to have taken place between 28 and 22 thousand years before present (ky B.P.) in the Altiplano. During the late maximum glaciation (around 18 ky B.P.), aridity was predominant in the Altiplano, but was followed by a period of greater precipitation during the late glaciation, between 17 and 11 ky B.P. (Vargas 1996).

This latter wet event has been called the Tauca phase, and is central to our understanding of climate evolution since the end of the late glaciation in Norte Grande. Climate characteristics were different during the Holocene. Several times, when Desierto de Atacama confronted severe droughts, the Altiplano was wet, or vice versa.

Knowledge about Altiplano paleoclimates is fundamental for understanding the problems of water resources in Norte Grande, because these paleoclimates have produced the aquifers located throughout the whole region. Study of sediments in Laguna Lejía, in the Altiplano of the Second Region of Antofagasta (Grosjean 1994), has indicated that the climate at the highlands was much wetter during the late glacial and in the early Holocene, with maximum precipitation occurring between 13.5 and 10.4 ky B.P. Precipitation at that time was more than double the present values, supporting the conclusion that much of the underground resources of the region originated during this period. These wet conditions were found, however, only at the Altiplano, and above 4,000 m, since in the lowlands, below 3,500 m, arid conditions have always been predominant (Grosjean and Nuñez 1994).

Grosjean et al. (1991) have reconstructed the formation of lakes and extensive underground water bodies between 14,000 and 12,000 B.P. These water bodies formed synchronously with Bolivian paleolakes (Tauca). Rainfall produced by summer easterlies was at least 120% above the present average. This period represented a climate transition from the drier and colder conditions of the late glacial maximum, towards warmer, more humid conditions. Paleohydrological evidence in the Bolivian-Peruvian Altiplano indicates that between 12,500 and 11,000 years B.P., paleolakes covered an area 4 times larger than that covered at present, indicating an increase in annual rainfall of 30–50%. The greater quantity of precipitation is inferred to have been caused by a southward shift of the Intertropical Convergence Zone (ITCZ).

In today's climate, as discussed previously, the ENSO phenomenon has a significant impact on climatic variations on interannual to interdecadal time scales. The question of how long the present mode of operation of ENSO has been in effect is still a matter of scientific debate. However, most researchers believe that for at least the past 5,000–6,000 years, ENSO has been a regular part of the climate of the Altiplano (Diaz and Markgraf 2000).

## 2.5. Water Availability

Hydrologically, a small amount of water accumulated on the highlands remains after strong evaporative processes, and descends towards lowland areas as surface or underground runoff. Only rarely does it reach the sea. In the northernmost Chilean region, distances between the sea and the mountains are shorter, and runoff basins extend as parallel systems towards the ocean. However, farther south, only a large basin, Pampa del Tamarugal, acts like a perennial stream, and other streams are not able to reach the sea. In the region of Antofagasta, the Loa watershed dominates.

Basins located in the highlands are orographically closed, and their drainage systems are endorheic. Waters flow mostly eastward, since they become part of the large altiplanic basin located in Bolivia. In contrast, infiltrated waters form underground flows westward, and become part of the lowland aquifers. The behavior of underground waters is still mostly unknown for many areas of Desierto de Atacama. Water flows are formed by infiltration and runoff of precipitation falling over varying periods of time. However, much of the precipitation that falls today evaporates before becoming an important part of the recharge system in the region. Recharge zones developed in a north-south direction along the Andean foothills, where a well-developed permeable strata of andesitic laves, rhyolites, ignimbrites, and tuffs (Harza Engineering Co. 1978) allow water infiltration.

Another recharge zone is found in the Altiplano proper, where rainfalls are heavier. In these places, ground is formed by Quaternary and Tertiary volcanic debris and composed of open fissures, tufas, andesitic breccias and ignimbrites. These strata are less permeable, mainly associated with faults that facilitate the infiltration of altiplanic precipitation. Below most of the Altiplano and Desierto de Atacama, underground water can be found in confined aquifers, lying between sand and gravel strata, mixed with ignimbrites, clay, and volcanic ashes. Aquifers here could reach hundreds of meters in depth (Harza Engineering Co. 1978). These strata are westward oriented, a consequence of the general inclination of the topography.

Hydrologeologic studies provide information about physical recharge processes without emphasizing the ages and the sources of underground waters. Only a few studies have been found for Norte Grande, and many of them are considered as confidential reports by the private sector, especially mining companies. Table 2 presents a summary of the aquifer characteristics and the estimated recharges for each of the basins. Compiled information is fragmentary, and there is little agreement between published recharge values and values estimated from theories about the origin of underground waters. In conclusion, most of the studies indicate that modern recharge of subsurface waters exists in almost all the zonal aquifers, and that they originate from altiplano rainfall occurring during the Bolivian winter.

However, the only quantitative study about the recharge of underground water was conducted for Sierra Gorda. The Sociedad Química y Minera de Chile (SQM 1998) has found one aquifer in the alluvial deposit of Pampa Blanca, for which recharge has been estimated by applying the model of a seasonal hydrological simulation called SEAMOD. This model incorporates hydrological variables to demonstrate quantitatively that underground recharge exhibits variations that are similar to those in annual precipitation, and that this recharge occurs mainly following the Bolivian winter, between March and July. Higher recharges have been estimated for the years 1968, 1972, 1981, 1987, and 1989.

*Table 2.* El Niño events, highland and lowland rainfall, and water recharge year for selected aquifers, 1962 through 1998.

| Sierra Gorda Recharge[1] | Rio San Jose Recharge | **Year** | EN | Atacama Desert Rain-fall[2] | Rainy | | | | Dry | | | |
|---|---|---|---|---|---|---|---|---|---|---|---|---|
| | | | | | Lauca | Lluta | Pampa | Loa | Lauca | Lluta | Pampa | Loa |
| | | **1962** | | | | | | | | | X | |
| | | **1963** | | | X | | | | | | | |
| | X | **1964** | | | | | | | | | X | |
| | | **1965** | X | | | | | | X | | X | |
| | | **1966** | X | | | | | | X | | X | |
| | | **1967** | | | | | | | | | | |
| X | X | **1968** | X | | | | X | | | | | |
| | | **1969** | X | | | | | | X | | | |
| | | **1970** | | | | | | | X | | X | |
| | | **1971** | | | | | | | | | | |
| X | X | **1972** | XX | X | X | | X | X | | | | |
| | | **1973** | XX | X | | | X | | | | | |
| | | **1974** | | | | | X | | | | | |
| | | **1975** | | | X | X | X | X | | | | |
| | | **1976** | X | X | | | | | | | | |
| | | **1977** | | | | | | | | | | |
| X | | **1978** | | | | | | | | X | | X |
| | | **1979** | | | | | | | | X | | |
| | | **1980** | | | | | | | X | X | X | X |
| X | | **1981** | | | | | | | | | | |
| | | **1982** | XX | X | | | | | | | | X |
| | | **1983** | XX | X | | | | | X | X | X | X |
| | X | **1984** | | | X | X | X | X | | | | |
| | X | **1985** | | | X | | | | | | | |
| | | **1986** | X | | X | | | | | | | |
| X | | **1987** | X | X | | | | | | | | |
| | | **1988** | | | | | | | | X | X | X |
| X | | **1989** | | | | | | | | | | |
| | | **1990** | | | | | | | X | | | |
| | | **1991** | XX | X | | | | | | | | X |
| | | **1992** | XX | X | | | | | X | x | | |
| | | **1993** | | | X | | | X | | | | |
| | | **1994** | | | | | | | | | | |
| | | **1995** | | | | | | | X | X | | |
| | | **1996** | | | | | | | | X | | X |
| | | **1997** | | | X | X | X | X | | | | |
| | | **1998** | X | | | | | | X | | X | |

EN = El Niño; X = moderate intensity; XX= strong intensity; Rainy Year: precipitation >25% higher than the annual average in all the stations; Dry Year: precipitation >25% less than the annual average in all the stations. [1]SQM 1998; [2]Ortlieb 1995b.

Isotopic and hydrological studies have been used as valid approaches to understanding the problem of the recharge of underground water in Norte Grande. However, one of the challenges in such studies is how to integrate their results. One set of studies indicates that some modern recharge should be taking place, based on hydrological models that show some relationship between altiplano rainfall and the aquifers' behavior. On the other hand, there is no evidence of tritium (from the period of atomic bomb testing in the atmosphere) in these waters. Different conclusions reached by both types of analyses could be based on several factors, like the season of sampling, rapid discharge of present rainfall, the speculative character of hydrological models, etc.

Both isotopic and hydrologic studies support, or at least do not reject, the idea that most of the underground water in Desierto de Atacama is fossil, and that the scarce precipitation in the Altiplano could constitute the only source of modern recharge for aquifers. However, there are still many uncertainties about the actual amount of available water and about the importance and mechanisms of altiplano recharge.

## 2.6. Precipitation and Current Water Table Levels

Precipitation data have been analyzed for various basins. We have utilized a high-resolution gridded data set (Willmott and Matsura 1998) with precipitation values available at half-degree of latitude and longitude, as noted in Section 2.1. The precipitation indices for the Lauca and Lluta Rivers shown in Figure 1, are representative of the higher altiplano (above ~4,500 m). The other indices are representative of areas that are lower in elevation and farther south (but still generally above ~3000 m). Precipitation is highly variable and the annual totals range from near 0 to ~800 mm. We note that there is reasonably good temporal coherence in the annual totals for the higher regional indices, and among the lower indices, but lower association between the upper and lower zones.

The index that represents the Pampa del Tamarugal (dotted line in Fig. 1), with stations located above 3,500 m altitude, shows that annual precipitation varies between ~0 and ~350 mm. This area exhibits somewhat different behavior in time, but the peaks and troughs coincide, particularly since the mid-1970s.

Stations located in the Loa River and in the Salado River subbasin of the Loa River make up the other precipitation indices shown in Figure 1, and comprise the southernmost areas of the Norte Grande region examined here. There is significant relief in these basins. Table 1 shows that observing stations in the Loa Basin are located between ~3,000 and 4,100 m altitude and mean annual precipitation varies between ~30 and 135 mm, whereas in the Salado Basin, stations are located between ~3,200 and 4,300 m altitude and average rainfall ranges between 60 and 180 mm. Both subgroups of stations demonstrate concordance in the annual trends, but precipitation is highly variable (Fig. 1).

# 3. MINING AND WATER RESOURCES

## 3.1 Underground Water

As was noted in the preceding sections, the Norte Grande of Chile is a region with little renewable water resources; it is a land of widespread aridity and extreme variability in precipitation. It does, however, possess extensive mining resources that are supported primarily by the mining of fossil water from underground aquifers and other groundwater resources. In the following discussion, we consider the record of ground water table changes in the region.

In the case of the San José River at Saucache and Riveras de Madrid (Fig. 2), a stream located immediately beyond and south of Río Lluta, there are major differences in the peaks and troughs in the depth of the water table. A pronounced lowering of the water table is observed, which began in the 1980s and accelerated after 1987. There is a considerable deepening of static levels, because during the 1970s the water table fluctuated between 0 and 40 m depth, but in the 1990s, depths ranged between 10 and 70 m.

In Pampa del Tamarugal at Pintados (Fig. 2), the totality of the wells exhibit relatively small variations in water table depth over time, but there is a lot of variation among the stations in the area. Places located in salars like Pintados and Bellavista (not shown) have the water table very near the surface. Most of the stations have static levels between 0 and 30 m, but occasionally water table levels below 70 m are found, probably as a consequence of their locations far away from the salar bottom. In Salar de Ascotán (Fig. 2), small variations in static levels over time are also found, but regionally, water table levels are between 5 and 85 m in depth. Similarly, there are small temporal variations in the Loa River wells, but there are substantial differences in water table depths among the stations, with fluctuations between 0 and 90 m.

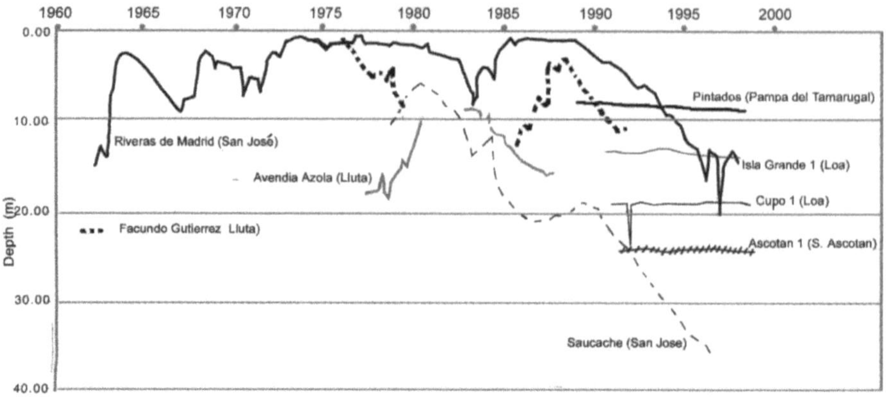


*Figure 2.* Temporal changes in water table depths for selected areas in Northern Chile

Unfortunately, the general lack of spatial coherence in the data does not allow precise knowledge of the interrelationships between meteorological and underground water data. As has been mentioned, only the San José River presents some fluctuations in the depth of the water table, which could be correlated with precipitation. Water table levels for Rivera de Madrid and Saucache could be considered good representatives of all the San José underground water stations.

In the rest of the region, there are large variations in the depths of water tables, but not in the time series. Such is the case for Pampa del Tamarugal, Ascotán Salar, and the Loa River. This means that, according to the available data, it is not possible to know the correlation between current rainfall amounts and variations in the depth of the water table. This conclusion, however, should be taken as preliminary. The period over which observations have been made is not long enough to support definite conclusions, and there is little spatial correspondence between the water table and precipitation records.

In terms of rainfall performance, it has been mentioned in previous sections that a complex relationship should be recognized between ENSO and precipitation, but that droughts and wet years can be found at the same time during El Niño occurrences. In Table 2, ENSO events are compared with annual rainfall. Wet years have been arbitrarily classified as those when precipitation exceeds 25% above annual averages.

In terms of modern recharge of the aquifers, the San José River demonstrates a close relationship between lower water depths and heavier altiplano rains (Table 2). The same table demonstrates an important aquifer recharge in Sierra Gorda, at the southeastern side of the Loa River (SQM

1998). These recharges coincide with rainfall records for the San José River in 1968 and 1972, but in other high-recharge years, such as 1978, 1981, 1987, and 1989, rainfalls in the Loa River altiplano were normal or less than the average. This fact indicates that recharges in the Loa Valley are not directly dependent on altiplano rainfall, and that many other factors likely influence small modern recharges in this area of Norte Grande. In addition, Messerli et al. (1993) suggest that El Niño rains could contribute to modern recharge of underground waters near the coast rather than in the altiplano.

Variations in static levels at other basins are not enough to clearly identify modern recharge amounts. Despite the fact that static levels are not the unique variable used to detect recharge variations, the observed lack of fluctuation indicates that the Bolivian winter does not contribute very much to the aquifer recharge, or rather that evaporation minimizes its contribution. Wells located south of Arica, in Pampa del Tamarugal, are located mainly in salars and therefore not much variation could be expected. Some tendencies, like an increase in water levels after altiplano rains, have been searched for without success.

Hydrogeologic studies indicate that there are significant modern recharges in Salar de Ascotán, in Vegas de Turi, and in Ojos del Salado in the upper Loa Valley. Given the high evaporation recorded at these places, it is assumed that some source of modern recharge, besides annual precipitation, must exist to maintain the constancy of water table levels. However, it is not known if this water is supplied by modern or old recharge; i.e., by old storage water or by modern precipitation. The stability of the water levels near the surface is perhaps the best indicator that they are provided by old waters rather than by variable modern precipitation. This fact could also explain why most of the isotopic studies done at the end of the dry season lower in the altiplano do not contain any modern component. Modern waters are normally lost by evaporation, runoff, or rapid discharge.

Another interesting result is the existence of regional differences in the capture of altiplano rainfalls, and the transformation of them into modern recharges between the northern and southern sections of Norte Grande. It seems that it depends also on the interconnectivity of the aquifers and on the watershed morphology. In the basins located farther north, where streams directly connect the Andes with the coast, and where waters are directly infiltrated to the aquifers, a remarkable impact of altiplano rainfall on the static levels could be observed.

In contrast, in such places as Pampa del Tamarugal, where basins are closed and water streams do not reach the sea, neither variations in water depth nor influences of altiplano rains are observed. Extreme dryness in this area and the presence of salars also contribute to this result. Static levels have been measured in salars, where, except in areas of extreme water withdrawals, they normally are equilibrated between high rates of evaporation and recharge from old water storage.

It is really very difficult to compare different studies about the performance of superficial and underground water resources in Norte Grande. Samples and aims of the studies are completely different. This could be why their conclusions are also very different.

Another serious limitation is the disagreement observed in data on water table depth presented by numerous studies. For example, it seems that current measurements in Salar de Pintados are now 20 m lower than those obtained in 1978 (Harza Engineering Co. 1978). However, Chilean Water Authority (DGA) measurements made between 1987 and 1998 show that in these 11 years water levels have changed very little. This makes it difficult to accept the large decreases published for 1978–87. Only a data harmonization process can validate existing information to support scientific conclusions, which are required today to make properly informed decisions. Most of these decisions are related to the amount and quality of available water for mining purposes.

## 3.2. Mining in Norte Grande

Since colonial times (sixteenth century), mining has been the main factor in economic growth for the Chilean economy. Today the country is highly specialized in copper exploitation (86.9% of the total national mineral exports in 1998), producing US$5.284 million; this makes Chile the world's largest producer, accounting for 35% of the international market with 4.382 million tons of minerals in 1999.

The production of copper in Chile accounts for 49% of total foreign direct investments, coming from a variety of industrialized nations: Australia (Broken Hill Proprietary), the United Kingdom (Rio Tinto), The Netherlands (Shell), Luxembourg (Corminco), South Africa (Anglo-American), Canada (Falcombridge, Cominco, Teck Corporation, Lac Minerals, and Río Algon), the United States (Phelps Dodge, Cyprus, and Exxon), Finland (Outokumpu), and Japan (Mitsui, Sumitomo, and JECO). Only a few domestic investments are worth mentioning, executed by two main national companies: the state company Corporación del Cobre (CODELCO) and Anaconda (a mine associated with the Luksic economic group, one of the three most relevant national holdings). While in 1988, 77% of the mineral was still produced by the public sector, by 1998 this percentage had already decreased to 49% as a consequence of the privatization process.

Mining takes place mainly in the center of the Atacama Desert and needs large amounts of water for production of the mineral and for associated economic processees and infrastructure, like industrial processing, transportation, facilities, and urban services. Currently, mines such as El Abra, Zaldivar, and Escondida require 2,000, 500, and 1,400 liters per second (l/sec) of water, respectively, and many of them need to increase their supplies to satisfy proposed expansions. For the mines located in the Region of Antofagasta, it has been estimated that future demand will be about 6,000 l/sec. Another 1,055 l/sec are required by urban users (Romero and Rivera 1996).

Demand is rapidly reaching the available water capacity of the north of Chile (Brown 1996; Pizarro 2000). This means that inter- and intrasectoral redistribution of the resources—e.g., from abandoned to new mines and from water conservation or agriculture to mining and urban uses—emerges as the only possible solution to satisfy increasing demand. The latest figures indicate that in recent times enormous amounts of water have been transferred from agricultural and from indigenous communities to mineral users (Dourejanni and Jouravlev 1999). To facilitate this transfer, the 1981 Chilean Water Codex allowed the existence of water markets—separated from land ownership—where, even if the resource is designated as a public good, the rights to use it are considered as private property, with rights of perpetual ownership that can be freely sold at market prices.

Some documentation of the present competition for water resources for natural conservation, the agropastoral indigenous communities, and mining can be found these days in articles such as in the Chilean newspaper *El Mercurio* (e.g., November 28, 2000). It was reported that the Australian Broken Hill Proprietary (BHP), which owned 57.5% of the largest private Chilean mine, La Escondida (the other owners were Rio Tinto with 30%, Japan Escondida Corporation [JECO] with 10%, and the International Finance Corporation with 2.5%), announced the implementation of a new investment (phase IV) for its mine, planning to invest US$1.045 billion to increase production from 800,000 to 1.2 million tons of copper by 2004. From the total capital, US$748 million would be directly spent in national goods and services: US$339 million in construction and US$409 million to purchase Chilean-made equipment. This was expected to create 6,000 direct and 20,000 indirect new jobs at a time when the national unemployment rate was nearly 10%, and has become a national priority. During the operational step, 300 permanent direct and 1,000 indirect jobs would be created. With this expansion, the mine would increase the annual purchase of national goods and services from US$400 million to US$500 million and tax payments to the Chilean government would be US$25 million for each additional 100,000 tons of copper. According to the newspaper, the total amount of taxes paid to the state since 1990 was US$1.550 billion.

All these apparent benefits depended intrinsically on the availability of water resources: 700 l/sec would be required to implement phase IV, and the company was expecting to buy 630.9 l/sec for another mine (Zaldivar), owned by Placer Dome, at annual costs of US$9 million over 15 years.

Two days after the above report on mining development plans appeared, the same newspaper (*El Mercurio*, November 30, 2000) published a new report from BHP, announcing that the proposed expansion of La Escondida could not be implemented because of difficulties in obtaining the required water, as negotiations with the Zaldivar mine were not successful.

This example illustrates how dramatically water scarcity may influence or even interrupt regional economic development. But it also can be interpreted purely as a sensitivity-building campaign by the mining company to request from the government water rights from sources located on the Altiplano (i.e., from salars or underground water) or for it to subsidize water imports from Bolivia, where privatization of the Silala River now allows the sale of previously free water at very high prices (Romero 2000b).

For purposes of planning economic growth in the north of Chile, water cannot be considered a renewable resource. Modern recharge of water is extremely limited there, and modern economic development (mines, tourism, urban growth), as well as preserving the cultural heritage and natural habitats in lower regions, depend largely on fossil water reserves that are barely renewable or perhaps even nonrenewable (Messerli et al. 1997). Accordingly, the water demand crisis urgently requires the development of a strategic policy at a regional scale in order to conserve the remaining resources, to prioritize the uses of waters, and to search for alternative sources.

Nevertheless, a general and dramatic change in the allocation and use of water at regional levels is taking place. To satisfy short-term economic priorities, water has been assigned to mining and urban needs. Areas where agriculture, ecological and environmental services, and traditional human uses have taken place for thousands of years are rapidly becoming seriously threatened. Mining has already caused negative direct in situ impacts on the areas where ores are located. Enormous amounts of water have been withdrawn from rivers, salars, lagoons, and wetlands for mineral concentration processes. About 2 l/sec of water are necessary to extract 1 ton of mineral, and by the year 2000, total annual production had grown above 4.3 million tons per year, most of this from Atacama Desert mines.

At the same time, many thousands of tons of pollutants have been discharged into air, waters, and soils. The air pollution around the melting chimneys at copper refineries like Chuquicamata and Paipote, both located in the arid north, have been recognized as major sources of arsenic and sulfur dioxide ($SO_2$.) Higher levels of water pollution could also be found in each of the rivers around the mines. Sometimes, the discharge of pollutants in the rivers has seriously damaged even the coastal environments, situated 100 km away. Water and airflows have transported the pollutants along the river basins, affecting agricultural soils and urban populations. Only since the approval of environmental laws during the 1990s have mineral companies begun to control their emissions.

The main indirect and cumulative impacts of mining and related economic activities on the regional environments in northern Chile are demonstrated in Figures 3 and 4. In the case of the First Tarapaca Region (Fig. 5), most of the sources of water and upper river catchments are occupied by indigenous communities and natural conservation zones: national parks, reserves, and sanctuaries. Many of these conservation areas have loosely defined boundaries. These boundaries were established not only to protect unique ecosystems that exist in these highlands (Rundel and Palma 2000), but also because of their locations in publicly owned areas since the end of the "Pacific War" more than a century ago. However, the map also shows that local indigenous communities claim relevant portions of this land. In fact, 700,000 hectares (ha) of highland areas should be considered the property of Aymara communities according to a recent inventory provided by the Ministerio de Bienes Nacionales in the year 2000. As is indicated in the map, there is an important overlapping between lands declared as natural reserve and land claimed by indigenous communities.

Land competition is by itself an important conflict challenging the Chilean land management public policy. It also has further outstanding relevance because it is not legal to request water rights in those areas where tacitly or explicitly land and water form a unique and inseparable ecosystem belonging to local peoples. This is based on the modified Chilean Water Codex (June 1992), and the implementation of the Indigenous Law (declared in October 1993) in order to protect the resources of the indigenous communities (see Peña 1996; Dourejanni and Jouravlev 1999).

Under these circumstances, pressure from mining interests and related demands on highland water resources means that even some national parks (like Lauca National Park, located above 4,000 m altitude, near Arica city) would be in danger of desertification of 25% of its surface area should its hydrologic resources be withdrawn. Paradoxically, scientific research has indicated that natural conservation areas should be *increased* to safeguard these unique Puna ecosystems, which are an important part of the South American arid landscape (Messerli et al. 1997; Rundel and Palma 2000).

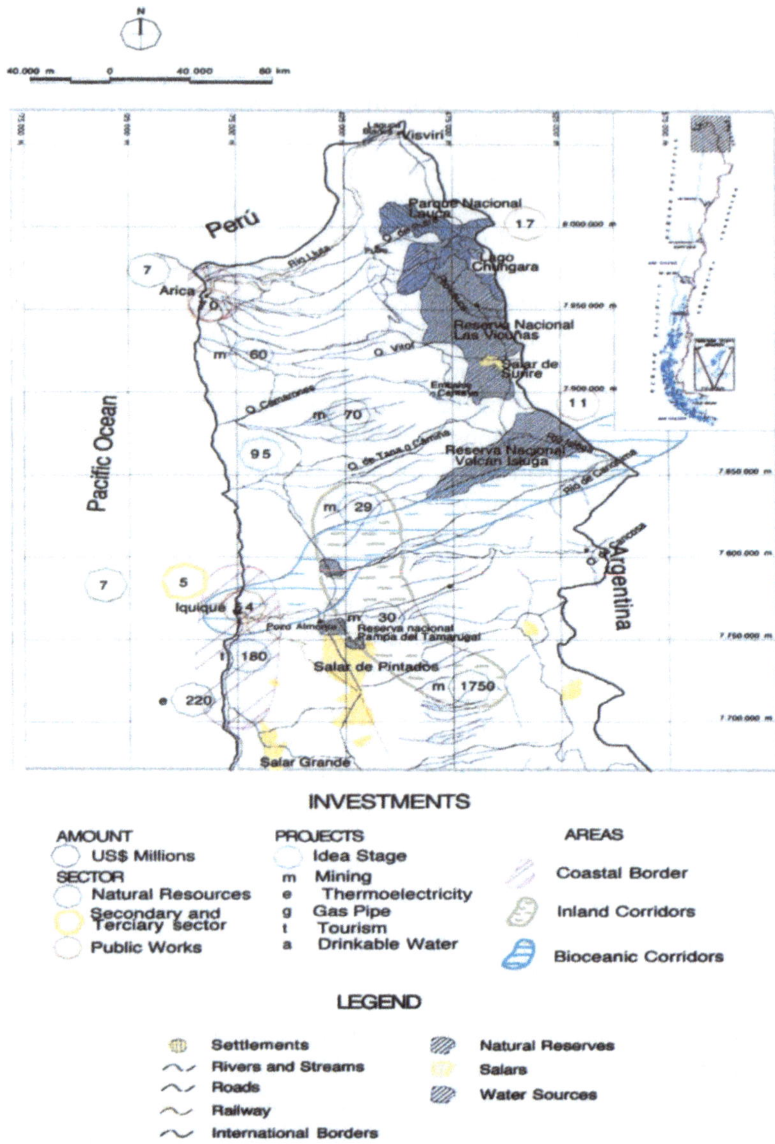


*Figure 3.* Areas, points of investment, the drainage system, and natural conservancy reserves in the Tarapaca I Region.

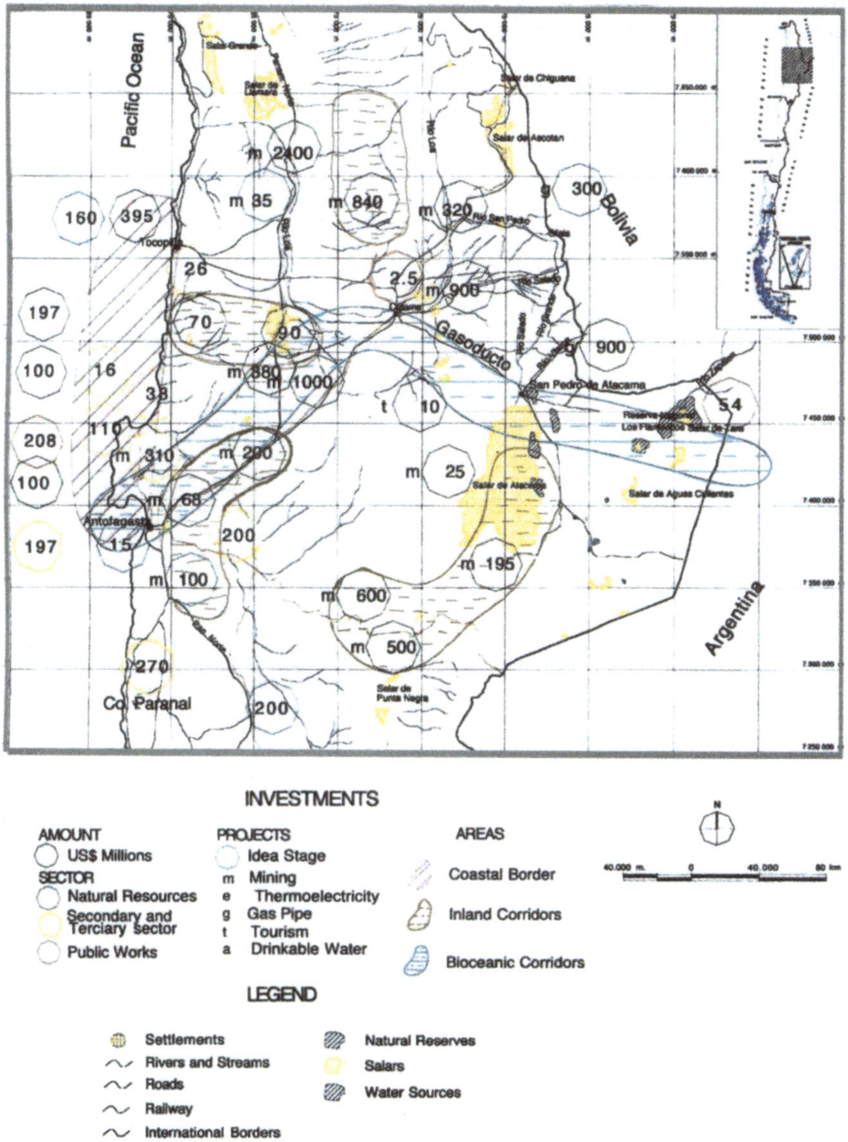


*Figure 4.* Areas, points of investment, the drainage system, and natural conservancy reserves in the Antofagasta II Region.

In the case of the Antofagasta Region (Fig. 6), there exist only a few small natural reserves, representing exclusively salar ecosystems, but not necessarily the complex arid landscapes of the high Andes. Given the extreme aridity and water scarcity, these saline water bodies represent an invaluable environmental resource.

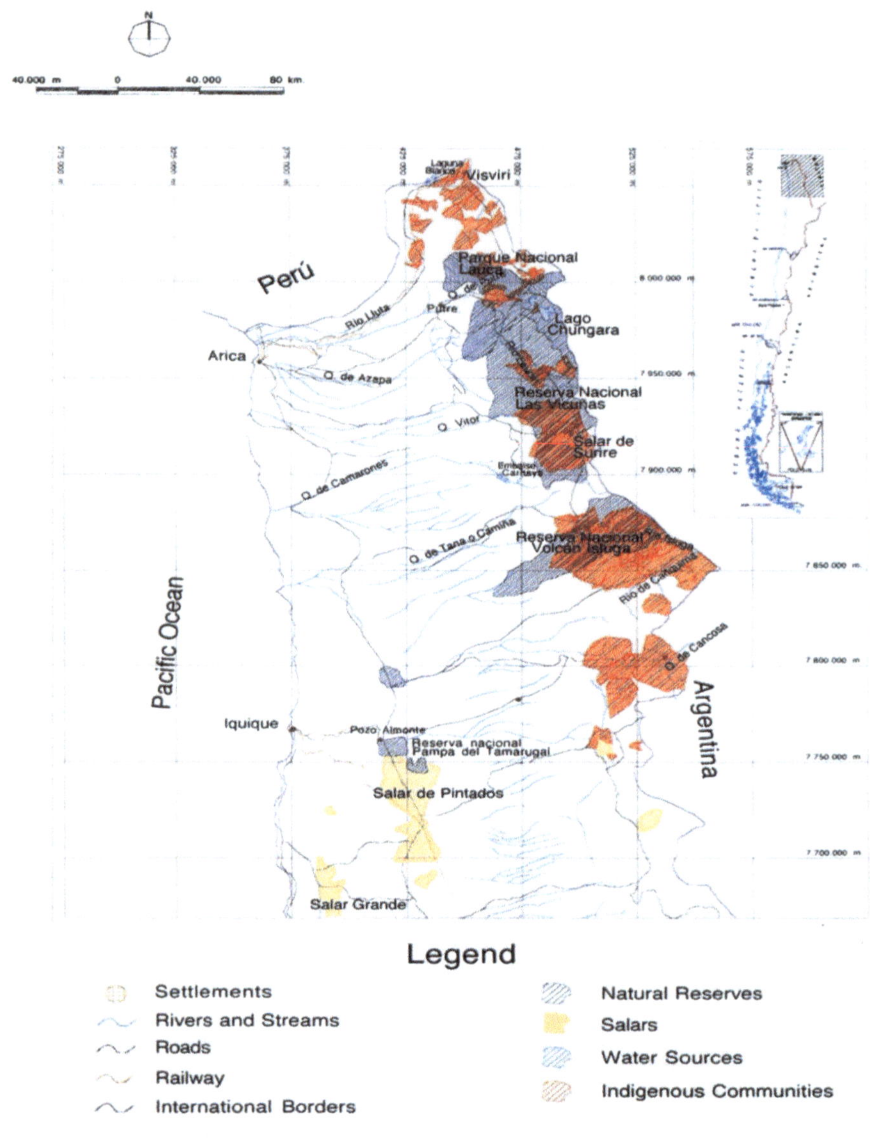


*Figure 5.* Indigenous communities, natural conservancy areas, and drainage systems in the Tarapaca I Region.

Land claimed by the "Atacameños," the regional ethnic peoples, corresponds mainly to small elongated areas, strategically located along the main river valleys. Consequently, conflicts over land here for control of the rivers and streams and their upper catchments occur primarily among agricultural communities, mining interests, and drinking water–supply companies supporting the major cities, and among the agencies and organizations defending the integrity of natural reserves located in lakes and salars.

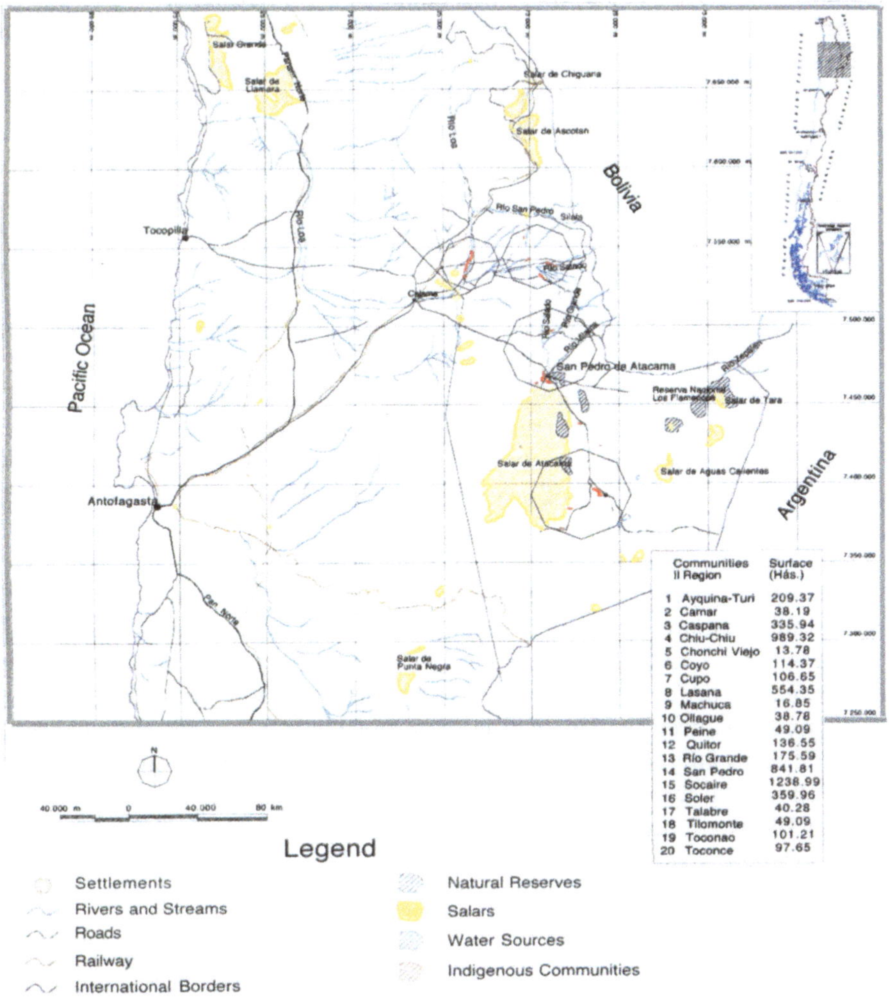


*Figure 6.* Indigenous communities, natural conservancy areas, and drainage systems in the Antofagasta II Region.

In terms of environmental services, water resources and associated biological species that live permanently and seasonally in salars and lagoons, creating oases on the boundary of the desert, are seriously threatened by water withdrawals. Preliminary observations made in 1999 in the Ascotán and Caracote salars, which act as water sources for the El Abra mine (more than 1,000 l/sec of surface and underground water are extracted), have demonstrated that as a consequence of the withdrawals, strong evaporation, and resulting salinization, the composition of the local vegetation is rapidly changing towards more halophytic species. However, it has not yet been possible to separate this process from the lowering of water levels produced by droughts, which have recently affected the region.

Furthermore, traditional herders of llamas, guanacos, and vicuñas have witnessed the drying of their grasslands and the loss of their agricultural irrigated floodplains. Mining companies and water speculators have been purchasing or acquiring water rights, which have belonged tacitly or legally to local people. Indigenous people in fact owned the water for centuries, but often do not possess legal titles. As an example, in 1995 a price of US$250,000 was paid for one water right of the Loa River.

Many of the traditional settlements have already disappeared, and most of the traditional agricultural and livestock lands have been abandoned (Romero and Rivera 1997). Economically, traditional land uses cannot compete against the rapid economic growth in the north of Chile. This is unfortunate because fields and pastures have been the main support of the local populations for thousands of years, while the average time for the life cycle of a mining project could be less than 20 years.

On the other hand, the crises caused by lack of available water and increasing competition among users have increased the necessity to access the water resources of the altiplano. One of the most important water sources is the Lauca River, which originates in Laguna de Cotacotani in the Chilean side of the altiplano and runs towards the Bolivian Coipasa Basin. This river has been fundamental in the provision of water for hydroelectricity and especially irrigation water for the rich agricultural zone of the San José River and for urban drinkable water more recently. For these purposes, in the 1960s a long pipe was built to transport water from the highlands to the lowlands. This construction project caused a severe dispute with Bolivia, and since then, political relationships between the countries have been strained.

Diplomatic disputes over water resources between both countries have threatened the necessary integration between the neighboring countries. Chile and Bolivia have found a new reason to work together in the need to solve water issues. Joint efforts should allow increased knowledge of the hydrologic features that control the flow of both surface and underground resources. There likely is a common source for the many aquifers feeding salars and the upper section of the Loa River.

In the case of rivers, there are many transboundary sources where joint exploitation and conservation are needed. Some of the most interesting flows of water crossing the Andes and reaching the Chilean altiplano are the Vilama, Silala, and Zapalieri Rivers.

The Silala River has provided water for the Antofagasta-Calama Railway Company since the end of the nineteenth century through a license given by Bolivia to the Chilean government as a way to connect the Pacific coast and inland territories. This water not only feeds the mechanization along the streams, but also the human settlements that arose around the railway stations. Silala River flow was also used for potable water for the city of Antofagasta at the beginning of the twentieth century. Silala River water now serves agricultural, mining, and urban users.

In the year 2000, and as a result of the privatization process, water from the Silala River was acquired by the Bolivian company DUCTEC; this company purchased the water rights from the Bolivian government, paying a substantial price of US$3 million for a discharge of 400 l/sec. This private company must also pay annually US$1.3 million for the allowance of the water rights (a kind of resource severance payment). To support these expenditures, the company has planned to sell the water to Chilean mining companies. For this purpose they have invested US$38 million to increase the water supply by 800–1,200 l/sec and presented the first bill, of several hundred thousand dollars, to the companies located in Chile. While companies have refused to pay higher prices, the Chilean government is negotiating on the basis that the Silala River is an international transboundary river, and that according to the law, both countries should share the discharge in equal proportions.

Taking into consideration the increasing demand for water from Puna de Atacama, it is quite certain that new conflicts are going to emerge when resources from other rivers, streams, and wells become necessary. On the other hand, the current frontier between Chile and Bolivia is growing considerably in terms of the transportation and geographical integration of people, goods, and services. Both countries have natural conservation areas on the highlands and traditional altiplano communities sharing their cultures. It appears that the time is right to replace the idea of a separated frontier with an integrated concept of areas and corridors. This is an opportunity to advance the creation of common resource areas, as has been agreed upon already in the recently signed mining treaty between Chile and Argentina. Along the border of Bolivia and Chile, water exchange treaties and agreements for the conservation of common natural resources will serve the best interests of both countries and in particular those of the Altiplano region, and they must be completed as soon as possible.

## 4. CONCLUSION

Direct foreign investments in capital and technology in Chile can be considered as key components of the country's globalization and economic success; these economic processes are driven by very specific comparative advantages: namely, an abundance of high-quality mineral resources, and arid climatic conditions and high evaporation that facilitate the easier concentration of minerals. In addition, lower labor costs, good infrastructure (harbors, roads, and telecommunications), and governmental institutional arrangements that guarantee the repatriation of capital revenues have created an attractive investment climate.

However, environmental accountability should be considered to determine if the Chilean process is really a sustainable development process instead of a narrowly economic and ephemeral growth process. The absence of strategic environmental planning to evaluate economic efficiency, social equity, and environmental protection—together with the absence of cost and benefit analyses of growth policies, plans, and programs, and the weakness of regional institutions and urban planning systems—have been mentioned as indicators of a lack of sustainable development policies.

The rather late approval of environmental (1994) and indigenous peoples (1995) laws, the incipient application of appropriated environmental standards and regulations, and the lack of enforcement caused by the small scientific, technical, and professional capacity of the public sector may have been attractive factors for foreign investments, at least in the first stage (1975–83). According to many analyses, the Chilean economy's rapid growth has been implemented mainly at the cost of reducing natural resources, reducing diversity (biodiversity, sociodiversity, and landscape heterogeneity), and the fragmentation of habitats.

## 5 REFERENCES

Abele, G. 1991. The influence of age, climate, and relief on the preservation of volcanic landforms in the North Chilean Andes. *Bamberger Geographische Schriften* 11: 45–57.

Aceituno, P. 1992. Estudio del régimen pluviométrico en el altiplano: variabilidad interanual y teleconexiones asociadas. Proyecto Fondecyt 1245-90, Santiago, Abril.

Aceituno, P., and A. Montecinos. 1992. Precipitación en el altiplano sudamericano: variabilidad interanual e intraestacional y mecanismos asociados. Congreso Iberoamericano de Meteorología, España, Octubre.

Brown, E. 1997. Disponibilidad de recursos hídricos en Chile en una perspectiva de largo plazo. Sustenabilidad Ambiental del Crecimiento Económico Chileno, Osvaldo Sunkel, ed. Programa de Desarrollo Sustentable; Centro de Análisis de Políticas Publicas, Universidad de Chile.

Chernela, J. 1997. The Wealth of Nations. *Hemisphere* 8 N°1, Fall): 43–49.

Claude, M. 1997. *Una Vez Más la Miseria ¿Es Chile un País Sustentable?* Santiago de Chile: Lom Ediciones, 216 pp.

Diaz, H.F., and V. Markgraf (eds.). 2000. *El Niño and the Southern Oscillation: Multiscale Variability and Global and Regional Impacts*. Cambridge: Cambridge University Press, 496 pp.

Dourejanni, A., and A. Jouravlev. 1999. El Código de Aguas de Chile: entre la ideología y la realidad. División de Recursos Naturales e Infraestructura, Comisión Económica para América Latina y El Caribe (CEPAL), Santiago de Chile, Octubre.

Fazio, H. 1997. *El Mapa de la Extrema Riqueza en Chile*. Lom Ediciones, Santiago de Chile.

Fuenzalida, H., and J. Rutlland. 1986. Estudio sobre el origen del vapor de agua que precipita en el invierno boliviano. Facultad de Ciencias Físicas y Matemáticas de la Universidad de Chile.

Grosjean, M. 1994. Paleohydrology of the Laguna Legía (north Chilean Altiplano) and climatic implications for late-glacial times. *Paleogeography, Paleoclimatology, Paleoecology* 109: 89–100.

Grosjean, M., and Nuñez, L. 1994. Late glacial, early and middle Holocene environments, human occupation, and resource use in the Atacama (northern Chile). *Geoarchaeology* 9: 271–286.

Grosjean, M., B. Messerli, and H. Schreier, 1991. Seenhochstände, Bodenbildung und Vergletchscherung im Altiplano Nordchiles: Ein interdisziplinärer Beitrag zur Klimageschichte der Atacama. Erste Resultate. *Bamberger Geographische Schriften*. 11: 99–108.

Grosjean, M., L. Nuñez, I. Cartajena, and B. Messerli. 1997a. Mid-Holocene climate and culture change in the Atacama Desert, northern Chile. *Quaternary Research* 48: 239–246.

Grosjean, M., B.L. Valero-Garces, M.A. Geyh, B. Messerli, U. Schotterer, H. Schreier, and K. Kelts. 1997b. Mid- and late-Holocene limnogeology of Laguna del Negro Francisco, northern Chile, and its palaeoclimatic implications. *The Holocene* 7: 151–159.

Harza Engineering Company International, S.A. 1978. Desarrollo de los recursos de agua en el Norte Grande. Project Chi-69/535 CORFO-DGA-CCC-PNUD.

Martin, L., M. Fournier, Ph. Mourguiart, A. Sifedine, and B. Turcq. 1992. Some climatic alterations recorded in South America during the last 7000 years may be expounded by long-term El Niño-like conditions. Paleo ENSO Records International Symposium, ORSTOM-CONCYTEC, Lima.

Mertins, G. 2000. Transformation research in Latin America: Approaches, results so far, and some of the future aims of research. In, Borsdorf, A. (ed.), *Perspectives of Geographical Research on Latin America for the 21st Century*, Herausgegeben Vom Institut für Stadt-Und Regionalforschung, Verlag der Österreichischen Academie der Wissenschaften, Heft 23:21–28, Wien 2000.

Messerli, B., M. Grosjean, G. Bonani, A. Bürgi, M. Geyh, K. Graft, K. Ramseyer, H. Romero, U. Schotterer, H. Schreier, and M. Vuille. 1993. Climate change and natural resource dynamics of the Altiplano during the last 18.000 years: A preliminary synthesis. *Mountain Research and Development* 13(2): 117–127.

Messerli, B., M. Grosjean, and M. Vuille. 1997. Water availability, protected areas, and natural resources in the Andean desert Altiplano. *Mountain Research and Development* 17: 229–238.

National Research Council (NRC). 1999. *Our Common Journey: A Transition Toward Sustainability*. Washington, D.C.: National Academy Press, 363 pp.

Ortlieb, L. 1995a. Paleoclimas Cuaternarios en el Norte Grande de Chile. In, Argollo, J., and Ph. Mourguiart (eds.), *Climas Cuaternarios en America del Sur*, ORSTOM-La Paz, Bolivia, pp. 225–246.

Ortlieb, L. 1995b. Eventos El Niño y episodios lluviosos en el Desierto de Atacama: El registro de los últimos dos siglos. *Bulletin de l'Institute Français d'Etudes Andines* 24: 519–537.

Peña, H. 1996. Oral presentation at Foro del Sector Saneamiento sobre el Proyecto de Ley General de Aguas, Lima, Perú, January.

Pizarro, R. 2000. Recursos Hídricos, Estado del Medio Ambiente en Chile-1999. Informe País, Universidad de Chile, Centro de Análisis de Políticas Públicas, pp. 75–130.

Quinn, W.H. 1993. The large-scale ENSO event, the El Niño and other important regional features. *Bulletin de l'Institute Français d'Etudes Andines* 22: 13–34.

Romero, H. 1985. Geografía de los Climas de Chile. Tomo (Vol.) IX, *Geografía de Chile*, Instituto Geográfico Militar, Santiago, Chile, 250 pp.

Romero, H. 2000a. Environment, regional and urban planning in Latin America. In, Borsdorf, A. (ed.), *Perspectives of Geographical Research on Latin America for the 21th Century*. Herausgegeben Vom Institut für Stadt- Und Regionalforschung, Verlag der Österreichischen Academie der Wissenschaften, Heft 23, 29–58, Wien.

Romero, H. 2000b. Efectos de las fluctuaciones climáticas sobre la sustentabilidad del desarrollo de las regiones áridas del Norte de Chile. III Simposio Brasilero de Climatología Geográfica, Río de Janeiro, January.

Romero, H., and A.M. Garrido. 1985. Influencias genéticas del Fenómeno El Niño sobre los patrones climáticos de Chile. *Investigaciones Pesqueras* 32: 19–35.

Romero, H., and P. González. 1989. La variabilidad climática interanual en Chile en el período 1982–88 y su relación con el fenómeno El Niño. *Pacifico Sur* (Número Especial): 105–112.

Romero, H., and A. Rivera. 1995. Régimen pluviométrico actual del altiplano de Antofagasta: Antecedentes para una evaluación paleoclimática. *Cambios Cuaternarios en America del Sur*, pp. 215–223.

Romero, H., and A. Rivera. 1996. Global Changes and Unsustainable Development in the Andes of Northern Chile. Umwelt Mensch Gebirge. Beitrage zur Dynamik von Naturund Lebensraum. Jahrbuch der Geographischen Gesellschaft Bern, Band 59.

Romero, H., and A. Rivera. 1997. Desarrollo Sostenible de Ecosistemas de Montaña. Manejo de Areas Frágiles en los Andes. In, Liberman, M., and C. Baied (eds.), *La Fragilidad de los Ecosistemas de Montañas del Altiplano de Antofagasta Frente a la Modernización Económica de Chile*. La Paz, Bolivia: Editorial Instituto de Ecología, 343–355.

Rundel, P., and B. Palma. 2000. Preserving a unique Puna ecosystem of the Andean Altiplano: A descriptive account of Lauca National Park, Chile. *Mountain Research and Development* 20: 262–271.

Sociedad Química y Minera de Chile (SQM). 1998. Pontificia Universidad Católica de Chile, Escuela de Ingeniería, Departamento de Ingeniería Hidráulica y Ambiental. Informes hidrogeologicos: Salar Sur Viejo, I Región y Sector Sierra Gorda, II Región.

Sunkel, O. (ed.). 1996. *Sustentabilidad Ambiental del Crecimiento Económico Chileno*. Programa de Desarrollo Sustentable, Centro de Análisis de Políticas Públicas, Universidad de Chile, 380 pp.

Universidad de Chile, Centro de Análisis de Políticas Públicas. 2000. *Estado del Medio Ambiente en Chile-1999*. Santiago de Chile: Informe País, 409 pp.

Vargas, G. 1996. Evidencias de cambios climáticos ocurridos durante el cuaternario en la zona de Antofagasta, IIa región. Masters Thesis, Departamento de Geología, Faculatad Ciencias Físicas y Matemáticas. Santiago de Chile: Universidad de Chile, 174 pp.

Willmott, C.J., and K. Matsuura. 1998. Global air temperature and precipitation: Regridded monthly and annual climatologies (version 2.01). Newark, Delaware: Center for Climatic Research, Department of Geography, The University of Delaware. (http://www.lba-hydronet.sr.unh.edu/tsgrid/precip/WWstd/).

6

# WATER RESOURCE MANAGEMENT IN RESPONSE TO EL NIÑO/SOUTHERN OSCILLATION (ENSO) DROUGHTS AND FLOODS

*The Case of Ambos Nogales*

Terry W. Sprouse* and Lisa Farrow Vaughan**
**Water Resources Research Center, University of Arizona, Tucson, Arizona 85719*
***National Oceanic and Atmospheric Administration (NOAA), Office of Global Programs (OGP), Silver Spring, Maryland 20910*

**Abstract** Ensuring the continued availability of freshwater will present major social, economic, political, and environmental challenges in the twenty-first century. Water is a most fundamental resource, essential to basic human survival and development. A large number of transnational river basins depend upon this precious commodity, and the shared utilization of this water can serve as a point of conflict, or cooperation, between the countries that jointly exploit them. One such transnational river basin is the Santa Cruz River Basin, shared by the state of Arizona in the United States and the Mexican state of Sonora.

The cities of Nogales, Arizona, and Nogales, Sonora (together known as Ambos Nogales), which both utilize Santa Cruz River water, are located in a semiarid region with high climate variability characterized by frequent drought and occasional serious flooding. The sharing of surface and groundwater resources has been a source of both cooperation and conflict between the two cities. Population pressures, which place added demand on the region's resources, have caused the communities to search for additional water sources to supplement traditionally shared supplies. Because of the limited nature of water supplies in the area, the impacts of climate variations on water resources can be severe. Recent developments in understanding and predicting El Niño/Southern Oscillation (ENSO) events, as well as important lessons learned during the 1997–98 El Niño episode, present new challenges and opportunities for averting disaster and improving water management in border areas such as Ambos Nogales.

This chapter examines water resources and management strategies in Arizona and Sonora. The authors describe the regional climate and how it is affected by El Niño events. Local management responses to cope with drought are investigated, as well as future scenarios for improving border water management in the area.



*Henry F. Diaz and Barbara J. Morehouse (eds.), Climate and Water*, 117-143.

# 1. INTRODUCTION

The predictability and effective management of freshwater resources looms on the horizon as a leading scientific, institutional, political, social, economic, and environmental challenge of the twenty-first century. Water is a fundamental resource, essential to basic human survival and development. Over 300 river basins span the territories of more than one nation. Given the high number of transnational river basins that yield this precious commodity, water can serve as a point of conflict, or concordance, among countries that exploit and are affected by shared resources. Hundreds of international agreements, formal and informal, have been negotiated to address the allocation, management, and utility of shared water systems. However, these arrangements are generally based on assumptions of static systems (both human and natural), and thus they were not designed to reflect variations in climate, and the associated impacts on natural resources, and public safety and well-being.

For example, these arrangements were not designed to give cognizant and explicit consideration to El Niño/Southern Oscillation (ENSO), a phenomenon that influences water resources and climate-related natural disasters throughout the world (e.g., floods and droughts). The impacts of ENSO on natural resources, and the potential use of climate data and forecasts to prepare for these impacts, present unique challenges for border regions. Climate variability can affect border regions in several ways, including: (1) the impact of temperature and precipitation fluctuations on shared natural resource systems; (2) the transfer of impact from one side of the border to another;[1] and (3) the migration of individuals from one side of the border to another, due to unfavorable environmental or economic conditions influenced by climatic changes. Recent developments in understanding and predicting ENSO, as well as important lessons learned during the 1997–98 El Niño episode, present new challenges and opportunities for averting disaster and improving water management in transboundary areas.

The cities of Nogales, Arizona, U.S.A., and Nogales, Sonora, Mexico (often referred to jointly as Ambos Nogales), are linked by many cultural and historic bonds. In addition, the sister cities are linked by a physical environment that includes the surface and groundwaters of the Santa Cruz River. Wastewater from both cities is treated 9 miles north of the border in Rio Rico, Arizona, at the Nogales International Wastewater Treatment Plant (NIWTP). These resources have been the source of both cooperation and

[1] The transfer can be an artifact of physical topography, or socioeconomic and environmental conditions. An example of this type of impact is the transfer of ENSO-induced floodwaters from one side of a border to another due to a lack of flood control.

dissension over the years. Ambos Nogales is located in a semiarid region, which is subject to substantial climate variability and is characterized by frequent drought and occasional flooding. Due to the tenuous nature of the region's water supplies, climate variability, such as that associated with ENSO, can have critical implications for natural resource management and public safety.

Population pressures and economic development of the border region, most dramatically in the last decade, have generated increasing demand on these common resources. The communities thus have been motivated to search for additional water sources to supplement traditional supplies and to develop improved management mechanisms for utilizing these and existing resources (Sprouse 2001; Liverman et al. 1999).

To examine these issues, this chapter proceeds in the following way. First, water resources on both sides of the border are identified, and management strategies in Arizona and Sonora are examined. Second, the authors look at regional climate, including ENSO impacts on the region. Third, several recent management responses to cope with climate and water resources are examined, including the development of a groundwater model for the Santa Cruz River, the proposed establishment of a regional water district, and binational cooperative efforts during the 1997–98 El Niño event. The conclusion explores scenarios for improving water management in the Ambos Nogales area in the future.

## 2. REGIONAL WATER RESOURCES

The major water resource that Nogales, Arizona, and Nogales, Sonora, share is the Santa Cruz River. From its headwaters in the San Rafael Valley of southern Arizona, the river flows southward into Mexico, travels 25 miles through Mexico, and reenters Arizona again 5 miles east of Nogales, Arizona (Map 1) (Arizona Department of Water Resources [ADWR] 1999). Water flows in the Santa Cruz River can be characterized as ephemeral (with flows dependent upon rains) or intermittent (a spatially interrupted flow) with some perennial (year-round flow) reaches. A perennial reach of the river exists downstream of the NIWTP, created by Mexican and U.S. effluent discharged into the river from the plant. On both sides of the border, the Santa Cruz River aquifer system is generally shallow with limited storage capacity, and is extremely sensitive to drought. The system also rebounds quickly when recharge is present. Population pressures on both sides of the border have put increasing demands on area water resources (see Table 1). Use of the Santa Cruz River and other regional water resources, by Arizona and Sonora, is described in the following section.

*Table 1.* Historical population and water use in Ambos Nogales.

| Year | 1950 | 1960 | 1970 | 1980 | 1990 | 1995 |
|---|---|---|---|---|---|---|
| Nogales, Arizona, population | 6,153 | 7,286 | 8,946 | 15,683 | 19,489 | 20,500 |
| Nogales, Arizona, potable water use (acre-feet) | 845 | 1,218 | 1,881 | 2,552 | 4,529 | 4,290 |
| Nogales, Sonora, population | 26,016 | 39,812 | 53,494 | 68,076 | 107,936 | 180,000 |
| Nogales, Sonora, potable water use (acre-feet) | 857 | 2,229 | 3,084 | 3,923 | 14,113 | 18,535 |

ADWR 1999; Anderson 1956; Cervera 1997; Comisión Nacional de Agua (CNA) 1996; International Boundary and Water Commission (IBWC) 1999; Putman et al. 1983; Rodriguez and Cervera 1999.

## 3. WATER RESOURCES AND MANAGEMENT IN NOGALES, ARIZONA

The Santa Cruz Active Management Area (AMA) manages water resources in the Nogales, Arizona, area. The Santa Cruz AMA is one of five areas (along with Prescott, Phoenix, Pinal, and Tucson) established by the ADWR where competition for groundwater is most severe, to provide more intensive regulation of groundwater use. Groundwater regulations stem from the Arizona Groundwater Management Code, enacted in 1980. The Code requires conservation of water and promotes use of renewable supplies.

The stated management goal of the Santa Cruz AMA is, "to maintain a safe yield condition and to prevent local water tables from experiencing long-term declines" (ADWR 1999). The AMA attempts to achieve its goal through state code provisions that address well drilling, well registration, and construction requirements; water supply adequacy requirements for new subdivisions; limitations on transportation of groundwater across watershed boundaries; prohibition of development of new farmland; and enforcement of well spacing and impact criteria.

Water management goals of the Santa Cruz and Assured Water Supply Rules promulgated by ADWR place significant importance on renewable water supplies to meet existing consumption and future demands (ADWR 1999). The ADWR requires that persons proposing to offer subdivided lands for sale or lease must demonstrate that sufficient water is con-

**MAP 1**
**Regional Water Resources**
**Ambos Nogales**

*Map1.* Regional Water Resourses Ambos Nogales. Fry and Cerverax1995.

tinuously available to meet the water needs of the proposed use for at least 100 years.

The AMA is required to produce a series of five management plans, each covering the succeeding 10-year period, in which mandatory conservation requirements for industrial, municipal, and agricultural water users are stated. Other programs in the AMA include grants for water conservation and augmentation projects, and monitoring of water supplies.

Surface and groundwater resources in the Nogales, Arizona, area include the Santa Cruz River aquifer; the Nogales Wash; the Potrero Creek aquifer; and two lakes, Patagonia Lake and Peña Blanca Lake (see Map 1). Drinking water supplies for the city of Nogales, Arizona, presently come from the Santa Cruz River and the Potrero Creek well fields (see Table 2).

The Santa Cruz River provides half of the needed 4,386 acre-feet per year[2] required by residents (ADWR 1999). Potrero Creek, located to the northwest of the city of Nogales, Arizona, provides the other half of Nogales' demand for water. The Santa Cruz River aquifer is more susceptible to drought than the Potrero Creek aquifer, as the latter is much deeper. However, hydrogeologic analysis by ADWR indicates that a dual aquifer exists in the Potrero Creek area. A shallow aquifer rests on top of a semipermeable clay layer, which separates it from the lower aquifer. The deeper, underlying aquifer supplies the city's well field. The degree of hydraulic connection between the shallow aquifer and the deeper aquifer has not been scientifically established, so it is not known how quickly the aquifer can be naturally recharged (ADWR 1999); this means that the aquifer could be subject to depletion. The city is also developing a new well field in the Guevavi Ranch area, 3 miles north of the Santa Cruz River well field, to supplement supplies.

Agricultural water use in the AMA is concentrated on the Santa Cruz River and averages 11,713 acre-feet/year (1990–97) (ADWR 1999). Other water demands include industrial water use of 1,300 acre-feet/year, and riparian demand of 25,000 acre-feet/year. Between 6,600 and 10,600 acre-feet of groundwater leaves the AMA each year. By comparison, water consumption in Nogales, Sonora, is twice that of Nogales, Arizona, for potable water use and half that of Nogales, Arizona, for agricultural use (ADWR 1999; Cervera 1997; International Boundary and Water Commission [IBWC] 1999). Industrial water use is roughly 1,000 acre-feet more in Arizona (most of which is used for irrigating golf courses).

*Table 2.* Potable water sources for Ambos Nogales, 1995.

| | Nogales, Sonora | | Nogales, Arizona | |
|---|---|---|---|---|
| Source | Quantity (acre-feet) | Percentage (%) | Quantity (acre-feet) | Percentage (%) |
| Santa Cruz River | 8,999 | 49 | 2,145 | 50 |
| Los Alisos, Sonora Creek<br>Potrero Creek, Arizona | 7,670 | 41 | | |
| Potrero Creek, Arizona | | | 2,145 | 50 |
| City wells | 1,866 | 10 | 0 | 0 |
| Total potable use | 18,535 | 100 | 4,290 | 100 |

Rodriquez and Cervera 1999; ADWR 1999.

[2] Based on data for the years 1995 through 1997.

The Santa Cruz AMA is strongly dependent upon the 13,000 acre-feet of effluent, two-thirds of which is generated in Mexico, that is discharged annually from the NIWTP into the Santa Cruz River below the plant (ADWR 1999). The effluent recharges the Santa Cruz River aquifer, which provides irrigation and drinking water for the downstream communities of Rio Rico, Tumacacori, and Tubac (Scott et al. 1997). The effluent also sustains a lush riparian area for a reach of about 12 miles from Rio Rico to the town of Tubac. Interruption in the flow of Mexican effluent could seriously affect water resources in the Santa Cruz AMA (Morehouse et al. 2000). Mexican participation in the Border Environmental Cooperation Commission (BECC) wastewater project, certified in June 2000, would appear to provide some guarantee of continued effluent flows for the 30-year life of the project.[3] However, because Mexico retains the right to recapture its share of the effluent under the provisions of IBWC Minute 227 (IBWC 1967), it cannot be used to establish an assured water supply designation. One way to assure continued flows of effluent to Arizona could be through the establishment of a regional water district, which is discussed later in this chapter.

## 4. WATER RESOURCES AND MANAGEMENT IN NOGALES, SONORA

Drinking water for the city of Nogales, Sonora, is provided by the Comisión de Agua Potable y Alcantarillado del Estado de Sonora (COAPAES), a state agency with headquarters in Hermosillo, Sonora. The COAPAES operates solely as a water delivery and wastewater service provider, and does not have the same regulatory mandate to conserve water supplies as does the Santa Cruz AMA. The Nogales office of COAPAES supplies water to 85 % of the area within the municipio (municipality), and to 64 % of the total population (IBWC 1999). Some of the areas within the municipio are difficult to supply because they lie in steep, hilly areas on the outskirts of town. Only 38 % of the population receives water 24 hours a day. The 36 % of the population not connected to the drinking water system, as well as the 62 % of the population who do not receive 24-hour water de-

[3] The project upgrades and expands the NIWTP, while creating a new wastewater treatment plant 10 miles south of the border in Mexico. The Mexican plant will treat wastewater generated from new growth in the area and allow the present flow volumes to the NIWTP to continue at 9.9 million gallons/day. (See IBWC 1999.)

livery, must rely on illegal connections or delivery by large "pipa" trucks[4] to meet their water needs (Ingram et al. 1995). Per capita water consumption is estimated at a very austere 150 liters/capita/day (39.6 gallons/capita/day) (IBWC 1999).

The city of Nogales, Sonora, has three sources that supply its residents with drinking water: the Santa Cruz River to the east, the Los Alisos Basin to the south, and wells located within the city itself (Table 2). The three sources provide about 18,500 acre-feet/year of water to Nogales, Sonora (Rodriquez and Cervera 1999).

The Sonoran section of the Santa Cruz River is much more rural and unpopulated than the Arizona section. In its south-flowing stretch, the Santa Cruz River crosses the border into Mexico (at Lochiel, Arizona) 5 miles east of Ambos Nogales. Five miles south of the border, the river passes through the town of Santa Cruz,[5] a farming and ranching area with a population of 741 (Cervera 1997). Eleven miles south of the border, the river turns west at Miguel Hidalgo (also known as San Lázaro), travels 6 miles, then turns north at Agua Zarca. Most agricultural and livestock water use takes place between Lochiel and Miguel Hidalgo. Major crops grown along the river are forage grains for livestock, corn, and beans (Cervera 1997). From Agua Zarca the river travels 15 miles in a northwesterly direction as it makes its way back up to the U.S. border, where it crosses near Buena Vista, Arizona. It is in this last section of the river between Miguel Hidalgo and the border where the city of Nogales, Sonora, has its well fields. Water production capacity for the two Santa Cruz River well fields is 19,700 acre-feet/year; however, the average flow from this area is 9,000 acre-feet/year (IBWC 1999; Rodriguez and Cervera 1999).

The Mexican portion of the Santa Cruz River, which historically was a perennial river, is today an ephemeral stream (Solis-Garza 1998; Fry and Cervera 1995). In addition to drinking water use, agricultural activities consume an average of 8,693 acre-feet/year (Cervera 1997). Of the three sources of water for Nogales, Sonora, the Santa Cruz River is the most sensitive to drought. As in Arizona, the shallow depth of aquifers and the nature

---

[4] Those who cannot receive water delivery by legal means often make illegal connections to the public water system. Ingram et al. (1995) report that at least 3,000 such connections exist.

[5] The city of Santa Cruz is also the seat of government for the municipio of Santa Cruz, or political district, which includes the community of Miguel Hidalgo and several other small communities. The total population of the Municipio of Santa Cruz is 1,580. The Municipio of Nogales also extends to include communities on the Santa Cruz River. The communities of Mascareñas, Centauro del Norte, López Mateos, Cárdenas Valdez, and Alvaro Obregón all belong to the Municipio of Nogales. The total population of these communities is 274. The vast majority of water use for agriculture, over 70%, takes place in the Municipio of Santa Cruz.

of the alluvium soil, which transmits surface water quickly to the aquifer, make Santa Cruz River aquifers in Sonora responsive to precipitation events and susceptible to droughts (Ingram et al. 1995; ADWR 1999).[6]

The Los Alisos aquifer, which lies in a separate watershed from the other two aquifers that supply the city,[7] is considered the most important future source of water for the city (IBWC 1999). Located approximately 10 miles south of the border at Ambos Nogales, the aquifer supplies about 7,670 acre-feet/year to the city's water supply from its six wells (Rodriguez and Cervera 1999). As is the case for Arizona, the Santa Cruz River well fields are pumping water, which is derived from rainwater runoff, a renewable source, but the same may not be true of the Los Alisos Basin. A 1998 hydrogeologic study indicated that the Los Alisos aquifer may not be receiving renewable recharge water from surface flows (CDM 1998). Soil borings point toward the existence of two aquifers, similar to the Potrero Creek aquifer in Arizona: a shallow alluvial aquifer and a deeper aquifer, which are separated by a confining layer. Radioisotope dating of the age of the water at two water supply wells suggests that the aquifer is not receiving significant amounts of new recharge water from the Los Alisos River.[8] Increased pumping of the Los Alisos aquifer means tapping into a potentially nonrenewable source of water, possibly resulting in depletion of that source.

A third source of water for Nogales, Sonora, is the 18 urban wells located within the city itself, which tap into the Nogales Wash aquifer. The Nogales Wash moves through a valley about 15 miles long and half a mile wide and flows though the downtown sections of Ambos Nogales (Ingram et al. 1995). Water flows in Nogales Wash are composed primarily of fugitive flows from breaks in sewage and potable water pipes, although when heavy rains occur, it can also carry substantial flood flows. The confluence of the Nogales Wash and the Santa Cruz River takes place near the NIWTP, 9 miles north of the border. The well field in the city of Nogales is the smallest source of drinking water for the city, providing about 1,866 acre-feet per year to the water budget (Rodriguez and Cervera 1999). One significant

---

[6] Yield can decrease greatly during dry seasons.

[7] The Los Alisos Arroyo is a tributary of the Magdalena River, which lies in an adjacent watershed.

[8] Most of the water in the aquifer was deposited there before 1952, and in the case of one well the age of the water was estimated to be at least 1,700 years old. If the aquifer were being recharged by streamflow from the Los Alisos River streambed, the age of the water (identified by tritium and carbon-14 dating) would be expected to be very young. (See CDM 1998.)

complication with this source is that it has experienced water contamination problems (IBWC 1998; Ingram et al. 1995).[9]

The supply infrastructure in Nogales, Sonora, poses a chronic problem in that the delivery system is old and large sections have outlived their useful life, thought to be 30 years (IBWC 1999). Water losses from leaks and breaks are estimated to be at 51.2 %, and CNA reports 7,000 leaks a year in the system (IBWC 1999). Much of the water that leaks from the system runs into the wastewater collection system, or the Nogales Wash, both of which flow into Arizona.

## 5. OVERVIEW OF THE REGIONAL CLIMATE

The Arizona/Sonora region of the U.S.-Mexico border experiences a high degree of climatic variability, punctuated throughout history by periods of extended drought and heavy precipitation.[10] Such variability has implications for numerous socioeconomic sectors in the region, including municipal and industrial water resources, disaster management, agriculture, ranching, and energy consumption (Merideth et al. 1998)

Precipitation in the region is highly influenced by the North American monsoon. The North American (NA) monsoon, sometimes referred to as the Mexican monsoon, is a spatially extensive phenomenon, extending through much of the western United States and northwestern Mexico (Adams and Comrie 1997). This process is characterized in the southern Arizona/northern Mexico region by a summer rainfall peak occurring from July through early September (Table 3). This peak is preceded and followed by drier conditions in June and mid-September (Adams and Comrie 1997). The region typically receives up to 50 % of its annual rainfall during this period (Sheppard et al. 1999).

A secondary rainfall peak occurs in this region during the winter months; this rainfall is essential to the recharge of the Santa Cruz River valley's aquifers. Winter rainfall is influenced by cyclonic storms and the

[9] Most recently, the IBWC reported that two monitor wells on the Nogales Wash in Mexico were found to have elevated levels of the volatile organic compound tetrachloroethylene (TCE). TCE is a probable human carcinogen with an EPA Maximum Contaminant Level (MCL) of 5.0 parts per billion (ppb).

[10] An excellent overview of the region's climate is presented in *The Climate of the Southwest* (Sheppard et al. 1999). This review was developed as part of the Climate Assessment for the Southwest (CLIMAS) Project; see http://www.ispe.arizona.edu/climas/index.html for further information about the CLIMAS. *Los Impactos de El Niño en México* (Magaña 1999a) provides a description of El Niño–related climate variability in Mexico.

southward shift of storm tracks from the eastern North Pacific (Sellers and Hill 1974; Woodhouse 1997). While greater amounts of rain typically fall in summer, the monsoon season is less conducive than the winter rains to the recharge of increasingly stressed aquifers. Summer rainfall typically evaporates prior to distribution in the reservoirs, and occurs in greater intensities, while winter rains tend to last for several days and ultimately result in replenishing spring runoff from snowpack (Sheppard et al. 1999).

*Table 3.* Average precipitation Nogales, Arizona, 1952–2000 (inches)

| Jan | Feb | Mar | April | May | June | July | Aug | Sept | Oct | Nov | Dec |
|---|---|---|---|---|---|---|---|---|---|---|---|
| 1.15 | .90 | .92 | .36 | .24 | .50 | 4.46 | 4.20 | 1.59 | 1.42 | .68 | 1.46 |
| Annual 17.87 | | | | | | | | | | | |

Western Regional Climate Center 2002.

Temperatures in the broader southeastern Arizona/northern Mexico region reflect a seasonal pattern, hitting their peak in midsummer and their minimum in the middle of winter. In the southwest United States and northern Mexico, ENSO has the most significant impact on precipitation during the months of October through March (Western Regional Climate Center 1997; Sheppard et al. 1999; Magaña et al. 1999b). The linkages between the North American monsoon and larger-scale processes such as ENSO are not well understood by scientists. This uncertainty makes forecasting summer rainfall in this region an even more difficult challenge (Adams and Comrie 1997; Sheppard et al. 1999). However, an international effort, the North American Monsoon Experiment (NAME), is beginning to explore more fully the evolution and variability of the North American monsoon system. The NAME initiative seeks to provide the fundamental understanding necessary for improved prediction of climatic variability in the region.[11]

El Niño events, or the warm phase of ENSO, tend to be associated with cooler, wetter winters in this region. Precipitation amounts were 4 inches above normal in southern Arizona during the 1982–83 ENSO warm phase (Climate Prediction Center [CPC] 1997a). Over the 102-year period between 1895 and 1996, the average precipitation during El Niño events in southeastern Arizona (Climate Division Seven, which includes Nogales, Arizona) shows an excess of 180 % of normal from November through December and 170 % of normal from January through March (CPC 2001).

In general, El Niño events are associated with enhanced streamflow in the southwestern United States (Cayan and Webb 1992). Increases in

[11] Additional information about NAME and its Science and Implementation plans can be found at www.clivar.org.

rainfall and the associated impacts on streamflow can have significant societal impacts in semiarid terrains such as the southwestern United States and northern Mexico. The dramatic impacts of flooding in this region were demonstrated by the events of 1983 and 1993. From September 27 through October 3, 1983, intense rainfall affected much of southeastern Arizona, with the most significant rains occurring less than 10 miles north of the U.S.-Mexico border (McHugh 1995). The greatest damage from flooding and erosion was located along the Santa Cruz River. The event resulted in direct losses of an estimated $227.8 million, and eight deaths throughout southeastern Arizona. Rainfall amounts during the January 1993 floods were lower than those that affected the region in 1983; however, flooding still occurred, as this storm involved warm rain falling on a substantial snowpack in the surrounding mountains. Losses due to this event were estimated at over $100 million, and seven deaths occurred (McHugh 1995). A disaster declaration was issued for the state on January 19, 1993 (U.S. Federal Emergency Management Agency [FEMA] 1998).

While there is a tendency for cooler and wetter winters in this region during El Niño events, a global forecast of a warm ENSO phase does not guarantee such conditions. However, during an El Niño event, the probability increases sharply, particularly during strong events, such as the one in 1997–98 (Water Resources Research Center [WRRC] 1997).

In addition to El Niño, southern Arizona and northern Mexico are impacted by the cool phase of the ENSO cycle, known as La Niña events. Reduced precipitation during the winter of 1995–96 associated with cooler sea surface temperatures (SSTs) in the tropical Pacific affected the Southwest's ranchers, forcing some to implement emergency measures and others to abandon ranching altogether (Conley et al. 1999). By the summer of 1997, water levels in the Santa Cruz aquifers had reached their lowest levels ever recorded, and conservation measures had to be implemented (Horton 1998).

## 6. MANAGEMENT RESPONSES TO DROUGHT AND FLOODING

Located on the floor of a narrow valley, Ambos Nogales has a history of damaging floods. Individuals living in the colonias (new urban settlements) of Nogales, Sonora (some of which are squatter settlements), and in the Arizona neighborhoods located near the border are perhaps the most vulnerable to flooding, including that associated with El Niño. Many of the colonias are built on hilly terrain, and some of them are located in natural arroyos (channels). When significant flooding occurs in Nogales, Sonora, these waters can rush across the international border into the city of Nogales, Arizona. These waters have been documented to contain contaminants such as ammonia, heavy metals, and high fecal coliform bacterial levels, as well as *Cryptosporidium* (U.S. Environmental Protection Agency [EPA] 1996). As a result, the subject of flood control at the Nogales border has been an issue of contention between the two countries. There is currently no formal agreement between the two cities that provides for the elimination of this critical problem.

Although droughts clearly pose challenges for the socioeconomic sectors and institutions affected by them, there are some benefits that can be derived from this phenomenon. For example, a drought-induced crisis can create the political will necessary for actions that might otherwise be deemed infeasible (Ingram 1999). Drought can also create an awareness of water problems and a willingness to consider the kinds of conservation measures that under normal conditions would not have been possible (Utton 1999). The severe effects of droughts can be attributed partly to poor planning and misallocation of water as much as it can be attributed to weather (Rincón 1999). A key feature of drought planning must include the sensible use of water even when it is plentiful. This section describes some recent efforts to anticipate and plan for future droughts and floods in the Ambos Nogales region.

## 6.1. Groundwater Flow Model

A groundwater flow model is one tool being developed by ADWR for the Santa Cruz AMA to predict how the Santa Cruz River will respond to drought, flooding, and other potential stresses and changes. As a planning and management tool, this numerical computer model will assist ADWR to assist regional planning and operational efforts in the area (ADWR 1999). In addition to drought, other conditions that may affect the Santa Cruz River that will be addressed by the model include the effects of various levels of municipal and nonagricultural growth in the basin; the effects of retirement of agricultural lands or increased agricultural activity; the effects of municipal and agricultural conservation measures; the results of increased Mexican use of groundwater; and the effects of discharge of contaminants into the surface water systems and into the floodplain aquifer on the groundwater system of the Santa Cruz AMA (ADWR 1995). The model will include data being collected by the University of Sonora in the upper Santa Cruz River in Sonora. Upon completion, the model will be shared with the University of Sonora and Mexican planning agencies, so that it can be used for regional management studies in Sonora.

As a tool to prepare for drought, the model allows identification of specific reaches of the Santa Cruz River that would be most affected. This could allow preparations to ease the effects of drought. For example, if the model projected that farmlands were going to be strongly affected by drought, farmers might consider the option of planting more drought-resistant species, or of planting fewer acres than normal. If the model indicated that city well fields were going to be effected by drought, the city might be able to anticipate the effects by beginning conservation measures long before the consequences of a drought took place. Identification, by the model, of locations on the Santa Cruz River where the strongest effects of the drought will be felt will allow these locations more time for anticipation and preparation to help soften the blow. The model could work in conjunction with the proposal for establishing an ENSO Task Force for Ambos Nogales, which is discussed in the Future Directions section of this chapter.

## 6.2. Water Augmentation and River Restoration

An initiative has been introduced by the Santa Cruz AMA to augment local water supplies through its grants program that funds studies to examine alternative sources of water for the area. A portion of the withdrawal fees collected on water withdrawn within the AMA is used for funding augmentation and recharge projects to supplement the water supply, and for conservation assistance projects to assist water users in achieving conservation requirements.[12] The 1999–2000 funding cycle financed studies specifically aimed at identifying potential sites where ground and surface water can be stored, and to identify new sources of water for the AMA.[13] Potential storage sites include Patagonia Lake and Peña Blanca Lake for surface water, and older alluvium basins for groundwater.[14] If potential sites

---

[12] Groundwater withdrawal fees in the Santa Cruz AMA are $2.00/acre-foot. A total of 20,777 acre-feet of water was withdrawn in the AMA in 1997 for a total of $40,346 in fees collected. (See ADWR 1999.)

[13] The Santa Cruz AMA allocated $140,000 to be spent on these studies for 1999–2000. (See ADWR 1999.)

[14] The older alluvium is located in the mountainous areas outside of the floodplain of the Santa Cruz River and its major tributaries. The older alluvium consists of locally stratified layers of boulders, gravel, sand, silt, and clays with cemented zones, or caliche. Investigation of older alluvium areas outside the Santa Cruz River floodplain, by assessing the hydrogeologic characteristics of various potential U.S. sites, could provide additional groundwater storage reservoirs for the AMA.

are physically and financially feasible, then water that presently passes through the AMA unused could be captured and stored as insurance against droughts. These waters, carried by the Santa Cruz River, include storm water flows, effluent from the NIWTP, and mountain front runoff.

In the area of river restoration, the Coronado Resource Conservation and Development Service[15] is coordinating an erosion control project on the Santa Cruz River 5 miles northeast of Nogales, Arizona (Larkin 2000). Flood flows in that reach of the river have undermined and eroded away thousands of tons of topsoil and property. Raw banks exposed to further erosion and subsequent floods have increased flood flow velocity, and reduced opportunities for aquifer recharge, since there is no longer any vegetation to slow waters down. The goal of the project is to restore this reach of the river to as near a natural state as possible by reestablishing a healthy riparian corridor. As vegetation increases in the degraded area, floodwaters will be slowed and sediment will settle out. A riparian corridor that reduces erosion, both on site and downstream, will provide habitat and improve the health of the river system. If successful, this project could serve as a model for use on other reaches of the Santa Cruz River.

## 6.3. Water Management District

A grassroots approach to water management, currently being examined in southern Arizona, would involve establishment of a local water district (ADWR 1999). While the Santa Cruz AMA is responsible for managing water resources in southern Arizona, it may need more tools to meet the stated AMA goal (ADWR 1996). Some urgent issues that adversely affect management of the AMA include the facts that more legal claims to Santa Cruz River water exist than there is actual water; the AMA has authority to manage only groundwater, not surface waters; and the AMA has no powers to move water from sources of supply to sources of demand. In order to more comprehensively manage water resources in southern Arizona, the prospect of establishing a water district has been proposed by some local water interests (ADWR 1999). Such a district could provide water to water right–holders, regulate all well drilling, manage both surface and groundwater, and build infrastructure to move water around.

While the idea is presently still in the discussion stage, the district could also be the vehicle to establish a dialogue with Mexico about continuing wastewater flows to Arizona (Sprouse 2000). If the idea were carried to

[15] The Coronado Resource Conservation and Development Service is a branch of the Natural Resources Conservation Service of the U.S. Department of Agriculture.

the next logical step, the district could be extended across the border to include Nogales, Sonora, as a full member. As a member of the district, the city of Nogales, Sonora, could both deliver water to, and receive water from, the district, just like any other member of the district. The district, while admittedly facing political obstacles, might be an entity that could comprehensively manage wastewater, potable water, and effluent for the entire watershed. The key to making a binational water district work would be to assure that both Nogales, Sonora, and Nogales, Arizona, have their respective water needs met. For Arizona this would mean some guarantee that it would continue to receive wastewater flows from Mexico, and for Sonora it would mean some type of help with potable water supplies, possibly including the return of some water to Mexico (Barcenas 2000).[16]

The advantage of having a water district in times of drought is that it could provide more detailed management of the limited waters that exist. If one entity manages all the water in southern Arizona, or possibly on both sides of the border, then in times of shortage priority of water use could be established prior to a drought, and the water would go where it was most needed. In addition, planning would be facilitated by the fact that a water district would probably have an infrastructure budget so that structures could be in place to store water in anticipation of drought.

## 6.4. Flood Control

In 1977, as much as 14 inches of rain fell in 3 days in southern Arizona and northern Mexico. Areas as great as 4 miles wide lay under water, at least 16,000 acres of farmland were covered, several lives and over 40 homes were lost, the NIWTP was negatively impacted, and damage in the region was estimated to be $15 million (Aldridge and Eychaner 1984; Ingram et al. 1995). Similar flooding occurred in 1983 and 1993 (see the above discussion of regional climate).

In the hopes of promoting improved forecasting and management of floodwaters in Ambos Nogales, several U.S. entities (including the Army Corps of Engineers, the IBWC, ADWR, and Santa Cruz County) are considering an initiative to develop a joint Flood Warning System. (As envi-

[16] Perhaps the most efficient way for Arizona to return water to Mexico would be to transfer a portion of the effluent from the NIWTP to the Mexican well fields on the Santa Cruz River, located just south of the U.S.-Mexican border. This would allow the effluent to be recharged into the aquifer, taking advantage of the soil to remove any impurities still in the water, and then pumped back for use as drinking water in Nogales, Sonora. This plan has the advantage of creating a semiclosed water loop in which the water would continuously cycle from Mexico to the NIWTP and back to Mexico.

sioned, the system would be composed of a series of rain gauges located in critical areas throughout both cities. The data would be conveyed in real time to a central location in Nogales, Arizona, where it would then be forwarded to the weather service in Tucson for analysis, and finally returned to Nogales for application in emergency response systems. As of December 2001, the project has not moved beyond the planning stage due to several pivotal issues that remain unresolved, including the role that Mexico would play in this system, potential cost sharing, and the potential for the Army Corps of Engineers to operate and maintain equipment on Mexican land. The path and challenges of this initiative highlight the difficulties of binational cooperation, even in the case of two cities that have a long and intertwined history.

## 6.5. The 1997–98 El Niño Event and Binational Cooperation in Ambos Nogales

The 1997–98 El Niño episode was at least as strong as the powerful 1982–83 event (Barnston et al. 1999), which caused heavy flooding throughout parts of southern Arizona and northern Mexico (Guttman et al. 1993; Magaña 1999b)

In January 1997, the Climate Prediction Center of the NOAA National Centers for Environmental Prediction issued an ENSO Advisory that called for Pacific SSTs to be slightly above normal by the end of 1997 (CPC 1997b). By March 1997, the predicted warming in the tropical Pacific began to occur. The print, Internet, and television media around the world picked up this information and were quick to relay it to their diverse constituencies. Major news organizations such as CNN, MSNBC, *USA Today*, ABC News, and the *Los Angeles Times*, and more than 100 private and public sector groups established Web pages dedicated to the ENSO phenomenon (World Wide Web search, spring 1998).

On a global scale, and within some regions, El Niño 1997–98 met and exceeded expectations. Estimates of direct dollar losses from the event exceeded $35 billion worldwide. Lives lost exceeded 21,000, while the number of displaced persons was estimated to exceed 4,000,000. In excess of 129 million people were affected by El Niño throughout the world (Sponberg 1999). The global population experienced primary and secondary impacts through a variety of disasters triggered by the El Niño–driven shifts in rainfall and temperature patterns, including droughts, floods, mudslides, tropical cyclones, wildfires, and damage of transportation (e.g., bridges) and public services infrastructure (Sponberg 1999). The degree of impact varied across regions and sectors. Forest fires raged out of control in Mexico, Central America, Indonesia, and Brazil, and severe flooding affected parts of Peru, California, Ecuador, the Sudan, and Kenya. Outbreaks of diseases such as malaria and cholera were reported in Kenya and Uganda.

In the region encompassing southern Arizona and northern Mexico, the description of El Niño's potential impact was often presented to the public in terms of comparisons with the 1982–83 warm event (FEMA 1997). The loss of life and property damage due to heavy rains and severe flooding frequently were cited as the primary potential effects (Erickson 1997a,b, 1998). In October 1997, the Water Resources Research Center of the University of Arizona issued the following advice to decision makers. The paragraph below immediately followed a discussion regarding the uncertainty of the forecast's implications for Arizona:

> Water resource officials, however, cannot afford a wait-and-see attitude, especially since many experts are predicting that this El Niño event will be the biggest, most potent in recorded history and will deliver extremely heavy rains to Arizona. El Niño might be the big water event of the year, and possibly also of next year. Water officials in various capacities, whether flood control, dam and/or river management, fish and wildlife protection, or delivery of water supplies, not to mention those concerned with disaster relief, need to take stock of the current situation and possibly plan for a worst-case scenario.
>
> *El Niño News*, Water Resources Research Center, October 1997

*El Imparcial*, a newspaper in Sonora, Mexico, developed a series of articles beginning in the second half of 1997 regarding the incipient El Niño event and the potential impacts on the state. Experts in the fields of climate, economics, agriculture, fisheries, and public works were interviewed regarding the potential social and economic impacts of the El Niño event. In many articles, the pending climatic event was discussed in the context of a serious water shortage that had been affecting the region for several years:

> The problem of a water shortage could disappear sooner than anyone could imagine. According to experts at the National Water Commission, the water shortage could disappear completely with the arrival of the natural phenomenon known as "El Niño". On the other hand, another problem will emerge—that of a surplus of water; in other words, Sonora will have too much water on its hands.
>
> Translation from Ponce de León in *El Imparcial*, November 7, 1997

While many of the projections of El Niño-related flooding in southeast Arizona and northern Mexico proved to be significantly exaggerated, rainfall in the region of Ambos Nogales exceeded normal levels during the winter months, most notably during February (see Table 4 for the 1997–98 departures from normal precipitation amounts in Nogales, Arizona). There were several cases of flash flooding in Ambos Nogales (National Climatic Data Center 1999); however, the damage did not approach that which occurred in the southwest United States and northern Mexico during the 1983 and 1993 floods.

*Table 4.* Precipitation departure from normal for Nogales, Arizona during the winter of 1997–98. Source: National Climatic Data Center, Asheville, NC USA

| Month | Precipitation (inches) |
|---|---|
| October | −1.48 |
| November | 0.22 |
| December | 1.93 |
| January | −0.97 |
| February | 1.62 |
| March | 0.65 |

Unable to foresee the final outcome of the forecasts and given the increased likelihood of heavy rains during an El Niño event, governments on both sides of the border initiated preparations for the event in the fall of 1997. Reported actions included removing debris and other physical obstacles from drainage systems; alerting citizens living in flood-prone areas; and practicing water rescue operations (Horton 1998; Tencza 1998; H. Ayuntamiento de Nogales, Unidad Muncipal de Proteccion Civil 1997; Yedra 1997). In Nogales, Sonora, the Unidad Municipal de Protección Civil issued a *Contingency Plan for Flooding and Lower Temperatures* that provided an overview of El Niño and identified areas of the city at risk for floods; the locations of temporary refuge sites for people driven out of their homes by floodwaters; the managers of these relief sites; the institutions responsible for responding to certain types of disasters (e.g., floods, health crises); and recommendations for individual citizens about coping with hazardous situations (H. Ayuntamiento de Nogales, Unidad Muncipal de Proteccion Civil 1997). For the first time in the city's history, flood-warning notices related the risk to the ENSO phenomenon, an important step in educating the general public about the impacts of climate variability.

Both Nogales, Arizona, and Nogales, Sonora, recognized the potential impact of El Niño on transboundary water issues that have historically shaped relations between the two cities, including the movement of floodwaters across the international border, water availability and quality, and the effect on the NIWTP. In order to avert potential disaster, the two cities took some initial steps toward working together in planning for El Niño's potential impacts. These joint activities were facilitated in large part by existing mechanisms for collaboration.

The two border cities have long enjoyed a cordial relationship that set the stage for cooperation on cross-border ENSO events. Parades commonly wound through the cities on both sides of the border, and baseball clubs frequently crossed the border to play games (Ready 1973). In past emergencies, the two cities have generously shared their water supplies with one another during droughts, jointly dealt with public health issues, and allowed their fire departments to fight fires side by side (Ready 1973). Collaboration has been a common thread that ties the two communities together.

In early November of 1997 the mayor of Nogales, Sonora, hosted a meeting of the Border Liaison Mechanism to discuss preparations for the 1997–98 El Niño event.[17] The special session of the Border Liaison Mechanism focused entirely on the subject of El Niño, including potential impacts, the current condition of common rivers and streams, the identification of obstructions to the flow of streams and washes that could cause water to pool and spill over, and coordination among parallel organizations and agencies regarding response to and communication about public emergencies. Institutions invited to the meeting by the Consulado de México (Mexican Consulate) fell into two main categories: (1) agencies with operations that potentially could be affected by the El Niño event and (2) agencies that could participate in preparations for the event. More than 20 agencies were invited.

The Nogales, Sonora, office of the IBWC created a risk map for the border area to help focus discussions during the meeting. The map identified seven areas along the border that had the potential to flood, and delineated the potential flow of water along the border. A primary concern was the potentially exacerbating impact of the border fence in the case of heavy rainfall and flooding. Following the November meeting, road grading activities occurred on both sides of the border in an attempt to mitigate the impact. The IBWC concluded in a March 1998 follow-up reconnaissance visit to the border that surface flooding had been abated through the joint effort (Tencza 1998).

While the most significant degree of activity was unilateral in nature, the transboundary aspect of the threat was recognized and discussed by both parties within the framework of existing international coordination mechanisms (e.g., IBWC, Border Liaison Mechanism). The November 1997 bilateral meeting was intended to facilitate a common understanding between the two cities of the science of the ENSO phenomenon, potential impacts, and finally, response options.

---

[17] The Border Liaison Mechanism allows the Consuls General of border twin cities to facilitate dialogue on regional issues. (See Brown 2001.)

This meeting demonstrated the recognition on the part of the involved parties of the cross-boundary effects of ENSO events, and also a willingness to work together to reduce disruption and damage. While many meeting participants were already familiar with each other through their cooperation in the areas of flood control, wastewater management, and water quality and supply, new institutional connections were made at the gathering (Horton 1998). These institutional arrangements and the experience gained from the 1997–98 event can serve as the basis for continued development of a new focus of collaboration between the sister cities.

## 7. FUTURE DIRECTIONS

This final section offers a brief overview of some steps that could be taken to address future drought and flood events in the Ambos Nogales area. While the region has a history of positive relations and cooperation, it stands to benefit from increased, long-range binational planning (Mumme and Sprouse 1999). Transboundary cooperation in the face of climate variability and its potential impacts is one opportunity for the communities of the U.S.-Mexico border region to build on this existing foundation. Advances in climate science and prediction offer unprecedented opportunities for the private and public sectors to mitigate the potential negative impacts, and to take advantage of the opportunities associated with seasonal to interannual fluctuations in climate. Present bilateral efforts to respond to climate fluctuations have been positive steps forward and should be continued, including the construction of the physical flood control infrastructure needed to deal with potential impacts. A bilateral ENSO task force could provide a forum for education, analysis, and dissemination of climate information in the border region, including forecasts issued by the respective authorities that could give communities the precious time they need to make preparations for imminent fluctuations in climate.

### 7.1. Infrastructure Initiatives

The Ambos Nogales border region could benefit from more systematic planning for droughts and floods, followed by actual construction of appropriate infrastructure. Useful initiatives would include small check dams on the Santa Cruz River and its tributaries to slow down flood flows and to promote aquifer recharge. Low-cost alternative building materials, such as old tires, could make check dams readily available for the area. Water catchment basins on the sides of the Santa Cruz River to capture and store flood flows aboveground could be particularly applicable in Sonora, which has more open space upon which to store water. It is not too soon to begin taking the legal steps necessary to commence building the infrastructure required to deliver lake water for municipal use. The same actions are necessary to increase the size of reservoirs as necessary to adequately prepare for drought. [18]

Because local economic resources are limited and mayors in Nogales, Arizona, are restricted to a 2-year term of office, only short-term pressing issues can normally be addressed by the local government. One way to overcome this short-term political time frame would through increased reliance on institutions such as the Border Liaison Mechanism. This could provide a longer-term tool for addressing water management and infrastructure issues. The Border Liaison Mechanism could be a means to set priorities and to support needed projects such as river restoration and a joint flood warning system structure. It could also help to encourage the use of new management tools, such as binational hydrologic modeling and regional water district planning. Another way to help prepare for weather-related emergencies would be through the establishment of a binational border Climate Prediction Center.

## 7.2. Binational ENSO Task Force

The 1997–98 experience provided insight regarding several important dimensions of transboundary collaboration in dealing with ENSO events. First, the two cities benefited from existing formal and informal binational mechanisms, including the IBWC and the Border Liaison Mechanism. The 1999 National Research Council report *Making Climate Forecasts Matter* identifies the need for "superorganizational" entities that operate within the context of a broader network to prepare for natural hazards (Stern and Easterling 1999). The region could invest its resources and time

[18] On a larger scale, a Bureau of Reclamation study in 1976 reported that a reservoir storage facility built on the Santa Cruz River at the international border could provide 12,000 acre-feet of floodwater annually for municipal and industrial uses for both Nogales, Arizona, and Nogales, Sonora. The projected construction of an earth-fill dam to be constructed about 2.5 miles downstream from the international border would have a 50,000 acre-foot reservoir that would extend into Sonora. Environmental implications aside, this type of project could seriously address the problem of drought emergencies in the area. (See Bureau of Reclamation. 1976. *San Pedro–Santa Cruz Project, Arizona, Concluding Report*, United States Bureau of Reclamation.)

in equipping and empowering existing mechanisms to cope with ENSO. For example, a standing task force within the context of the Border Liaison Mechanism or the IBWC could be established and tasked with the collection, analysis, and dissemination of climate information in the Ambos Nogales region, including early warnings when such information is available. A binational ENSO task force could work in a non-crisis mode to establish a network of institutions involved in transboundary natural resource management and early warning and preparedness. This network will be more effective if the roles, linkages, and authority of participants are well defined prior to an ENSO event (Stern and Easterling 1999).

In addition, this task force could foster a common perception of ENSO through local participation in transboundary education and awareness efforts, and the collection, analysis, and dissemination of climate information. In order to make joint and mutually beneficial decisions regarding ENSO, parties need to be working from a common understanding and information base. Historically, weather-related forecasts have stopped at the international boundary, as they are issued by institutions with national mandates. In recent years, however, new cooperative activities and institutions have emerged to address the transboundary nature of climate variability and its impacts. For example, the International Research Institute (IRI) for Climate Prediction, a joint venture of Columbia University and the U.S. National Oceanic and Atmospheric Administration, is an innovative science institution dedicated to helping countries adapt to climate fluctuations through the use of predictive information. Important elements of this endeavor include the study of impacts, the collaborative development of applications methodologies with decision makers in various regions, and the advancement of capacity building activities (International Research Institute for Climate Prediction 2001). The International Research Institute partners with NOAA and other institutions around the world in a suite of regional scale efforts designed to increase adaptive capacity through the production and application of climate information. For example, hundreds of institutions have participated in regional Climate Outlook Forums over the last few years. These forums provide opportunities for scientists and decision makers from neighboring countries to gather on a regular basis to examine the state of the climate system, its implications for their region, and potential implications for applications in key social and economic sectors (Buizer et al. 2000; NOAA Office of Global Programs [OGP], NOAA 1999). While there is not currently a Climate Outlook Forum for North America that would facilitate cooperation between the United States and Mexico, a binational task force could collect forecast information from multiple sources, including national institutions such as the NOAA Climate Prediction Center and the Mexican Meteorological Service, as well as from international sources such as the

IRI. This information could then be analyzed and distributed to participants in natural resource management and early warning and preparedness efforts.[19]

The experience of Ambos Nogales offers critical lessons for other regions, given the widespread need for effective water and disaster management practices. Local construction of the infrastructure necessary to anticipate floods and droughts, federal cooperation, and the formation of local task forces are actions that could help to reduce the impact of future ENSO events.

## 8. REFERENCES

Adams, D., and A. Comrie. 1997. The North American Monsoon. *Bulletin of the American Meteorological Society* 78(10): 2197–2213.

Aldrige, B., and J.H. Eychaner. 1984. *Floods of October 1977 in Southern Arizona and March 1978 in Central Arizona*, U.S. Geological Service Water Supply Paper No. 2223.

Anderson, C. 1956. *Potential Development of Water Resources of the Upper Santa Cruz River Basin in Santa Cruz County, Arizona and in Sonora, Mexico*. Phoenix: State Land Department.

Arizona Department of Water Resources (ADWR). 1995. Workplan for Santa Cruz AMA Hydrologic Assessment, Groundwater Model, and Hydrologic Database. Hydrology Division, Arizona Department of Water Resources.

Arizona Department of Water Resources (ADWR). 1996. *State of the AMA: Santa Cruz Active Management Area*. Phoenix: Arizona Department of Water Resources.

Arizona Department of Water Resources (ADWR). 1999. *Third Management Plan for the Santa Cruz Active Management Area*. Phoenix: Arizona Department of Water Resources.

Barcenas, A. 2000. Director of the Santa Cruz Active Management Area in Nogales, Arizona. Personal communication. January.

Barnston, A.G., M.H. Glantz, and Y. He. 1999. Predictive skill of statistical and dynamical climate models during the 1997–1998 El Niño episode and the 1998 La Niña onset. *Bulletin of the American Meteorological Society* 80(2): 217–244.

Brown, C. 2001. Exploration of the Border Liaison Mechanism as an option for bi-national water resource management in the U.S.-Mexico borderlands. Paper presented at the Annual Meeting of the Association of Borderland Studies, Reno, Nevada, April 18–21.

Buizer, J., J. Foster, and D. Lund. 2000. Global impacts and regional actions: Preparing for the 1997–1998 El Niño. *Bulletin of the American Meteorological Society* 81(9): 2121–2139.

[19] The development of a transboundary prediction system may require legal analysis focused on liability issues associated with distribution and application of climate forecasts. A regionally based climate services function could provide a structure for developing and disseminating such information if the structure incorporated a binational border area focus.

Camp, Dresser & McKee, Inc. (CDM). 1998. *Hydrogeological Assessment for the Candidate Discharge Site at Rio Los Alisos, Sonora.* Camp Dresser & McKee, Inc.

Cayan, D., and R. Webb. 1992. El Niño–Southern Oscillation and streamflow in the western United States. In, Diaz, H.F., and V. Markgraf (eds.), *El Niño: Historical and Paleoclimatic Aspects of the Southern Oscillation.* Cambridge: Cambridge University Press, pp. 29–68.

Cervera Goméz, L.E. 1997. Planeación de la demanda de agua en Nogales, Sonora: La Sustentabilidad de su Utilización en una Región. *Desarrollo Fronterizo y Globalización.* Associación Nacional de Universidades e Instituciones de Educación Superior.

Climate Prediction Center (CPC). 1997a. The 1997–1998 El Niño: Potential Effects in Arizona. National Oceanic and Atmospheric Administration. http://www.cpc.ncep.noaa.gov/products/analysis_monitoring/ensostuff/states/az.disc.html

Climate Prediction Center (CPC). 1997b. *ENSO Advisory.* National Oceanic and Atmospheric Administration (NOAA).

Climate Prediction Center (CPC). 2001. National Oceanic and Atmospheric Administration. http://www.cpc.ncep.noaa.gov/products/analysis_monitoring/ensostuff/states/ndpstat_az.gif.

Comisión Nacional de Agua (CNA). 1996. *Actualización del Plan Maestro de los Servicios de Agua Potable, Alcantarillado y Saneamiento de la Ciudad de Nogales.* Mexico City: Comisión Nacional de Agua.

Conley, J., H. Eakin, T. Sheridan, and D. Hadley. 1999. *CLIMAS Ranching Case Study, Year 1.* CL3-99 CLIMAS Report Series. Tucson: Institute for the Study of Planet Earth, University of Arizona.

Erickson, J. 1997a. Winter floods possible, say El Niño experts. *Arizona Daily Star*, June 18.

Erickson, J. 1997b. Robust El Niño could triple winter rainfall, scientist says. *Arizona Daily Star*, September 4.

Erickson, J. 1998. Shy El Niño still likely to get us, experts say. *Arizona Daily Star*, January 29.

Federal Emergency Management Agency (FEMA). 1997. El Niño of 1997–1998 Could Resemble the Destructive 1982–1983 Event. Federal Emergency Management Administration. http://www/fema/gov/nwz97.nino905.html

Federal Emergency Management Agency (FEMA). 1998. Major Disaster Declarations, January 1–December 31, 1993. Federal Emergency Management Administration, http://www.fema.gov/library/cy93.htm

Fry, J., and L.E. Cervera Goméz. 1995. *Recursos Hidraulicos en la Cuenca Alta Del Rio Santa Cruz.* Arizona State University and El Colegio de la Frontera Norte.

Guttman, N.B., J.J. Lee, and J.R. Wallis. 1993. *1992–1993 Winter Precipitation in Southwest Arizona.* Climatic Data Products, NOAA National Data Centers, Maryland.

Horton, K. 1998. Superintendent, Water Department, City of Nogales, Arizona. Personal communication (spring).

H. Ayuntamiento de Nogales, Unidad Muncipal de Proteccion Civil. 1997. *Plan de Contingencias Ante Inundaciones y Bajas Temperaturas.* Nogales, Sonora, Mexico.

Ingram, H. 1999. Lessons learned and recommendations from coping with future scarcity. *Natural Resources Journal* 39(1): 179–188.

Ingram, H., N. Laney, and D. Gillilan. 1995. *Divided Waters: Bridging the U.S.–Mexico Border.* Tucson: University of Arizona Press.

International Boundary and Water Commission (IBWC). 1967. *Minute 227. Enlargement of the International Facilities for the Treatment of Nogales, Arizona and Nogales, Sonora Sewage.* El Paso: International Boundary and Water Commission.

International Boundary and Water Commission (IBWC). 1998. *Binational Nogales Wash United States/Mexico Groundwater Monitoring Program, Interim Report*. El Paso: International Boundary and Water Commission.

International Boundary and Water Commission (IBWC). 1999. *Final Report. Ambos Nogales Wastewater Facilities Plan*. El Paso: International Boundary and Water Commission.

International Research Institute for Climate Prediction (IRI). 2001. The *International Research Institute for Climate Prediction: Linking Science to Society. A Look Forward*. IRI Annual Report, Columbia University, Palisades, NY.

Larkin, Mark. 2000. Water protection on the Santa Cruz River. *Santa Cruz River Watershed Update*, July.

Liverman, D.M., R.G. Varady, O. Chávez, and R. Sánchez. 1999. Environmental issues along the United States–Mexico border: Drivers of change and responses of citizens and institutions. *Annual Review of Energy and the Environment* 24: 607–643.

Magaña, V. (ed.). 1999a. *Los Impactos de El Niño en México*. Mexico City: Centro de Ciencias de la Atmósfera, Universidad Nacional Autónoma de México.

Magaña, V., J.L. Pérez, J.L. Vázquez, E. Carrisoza, and J. Pérez. 1999b. El Niño y el Clima. In, Magaña, V. (ed.), *Los Impactos de El Niño en México*, Mexico City: Centro de Ciencias de la Atmósfera, Universidad Nacional Autónoma de México, pp. 23–66.

McHugh, C.P. 1995. Preparing public safety organizations for disaster response: A study of Tucson, Arizona's response to flooding. *Disaster Prevention and Management* 4: 25–36.

Merideth, R., D. Liverman, R. Bales, and M. Patterson (eds.). 1998. *Climate Variability and Change in the Southwest: Impacts, Information Needs and Issues for Policy Making*. Final Report of the Southwest Regional Climate Change Symposium and Workshop, September 3–5, 1997. Tucson: Udall Center for Studies in Public Policy, University of Arizona, Tucson.

Morehouse, B.J., R.H. Carter, and T.W. Sprouse. 2000. Assessing transboundary sensitivity to drought: The importance of effluent in Nogales, Arizona and Nogales, Sonora. *Natural Resources Journal* 40(4): 783–817.

Mumme, S.P., and T.W. Sprouse. 1999. Beyond BECC: Envisioning needed institutional reforms for environmental protection on the U.S.-Mexico border. Chapter 38 in Soden, D.L., and B. Steel (eds.), *The Handbook of Global Environmental Policy and Administration*, New York: Marcel Dekker, Inc.

National Climatic Data Center. 2002. Data collected from Nogales 6 N (025924). http://lwf.ncdc.noaa.gov/servlets/ACS.

National Climatic Data Center. 1999. http://www4.ncdc.noaa.gov/cgi-win/wwcgi.dll?wwevent~storms.

National Research Council. 1999. *Making Climate Forecasts Matter*. Panel on the Human Dimensions of Seasonal-to-Interannual Climate Variability (K. Miller, member), P.C. Stern and W. Easterling (eds.), National Academy Press, Washington, D.C.

Office of Global Programs (OGP), National Oceanic and Atmospheric Administration (NOAA). 1999. *An Experiment in the Application of Climate Forecasts: NOAA-OGP Activities Related to the 1997–1998 El Niño Event*. Silver Spring, MD.

Ponce de León, G. 1997. Definen estrategias por El Niño. *El Imparcial*, November 1. Sonora, Mexico.

Putman, F., T. Turner, A. Hellerud, and V.O. Chatupron. 1983. *Report on the Hydrology of the Buena Vista Area, Santa Cruz County, Arizona*. Phoenix: Arizona Department of Water Resources.

Ready, A. 1973. *Open Range and Hidden Silver: Arizona's Santa Cruz County*. Nogales: Alto Press.

Rincón, C. 1999. Comments on the 1994–1995 drought: What did we learn from it?: The Mexican perspective. *Natural Resources Journal* 39(1): 61–63.

Rodríguez Esteves, J.M., and L.E. Cervera Goméz. 1999. Aspectos de la relación sociedad-ambiente natural en la cuenca binacional del río Santa Cruz, Sonora. *Frontera Norte* 11(22): 81–112.

Scott, P.S., R.D. MacNish, and T. Maddock. 1997. *Effluent Recharge to the Upper Santa Cruz River Floodplain Aquifer, Santa Cruz County, Arizona.* Tucson: University of Arizona Press.

Sellers, W.D., and R.H. Hill (eds.). 1974. *Arizona Climate 1931–1972*, 2nd ed. Tucson: University of Arizona Press.

Sheppard, P.H., A.C. Comrie, G.D. Packin, K. Angersbach, and M.K. Hughes. 1999. *The Climate of the Southwest.* CL1-99 CLIMAS Report Series. Tucson: Institute for the Study of Planet Earth, University of Arizona.

Solis-Garza, G. 1998. Estudio de la Calidad de Agua y Sedimento del Rio Santa Cruz y Arroyo Los Nogales, Sonora, Mexico. Universidad de Sonora.

Sponberg, K. 1999. Compendium of Climate Variability. Draft Report to the NOAA Office of Global Programs.

Sprouse, T.W. 2000. Resolving border wastewater problems in Ambos Nogales: An evaluation of issues. Paper presented at the Annual Meeting of the Association of Borderland Studies, San Diego, California, April 26–29.

Sprouse, T.W. 2001. Preserving the rural nature of land on the upper Santa Cruz River. *Arizona Water Resources* 9(4): 6.

Stern, P.C., and W. Easterling (eds.). 1999. *Making Climate Forecasts Matter.* A report of the National Research Council's Panel on the Human Dimensions of Seasonal-to-Interannual Climate Variability. Washington, D.C.: National Academy Press.

Tencza, S. 1998. Nogales Project Manager, IBWC-US. Personal communication. January.

United States Bureau of Reclamation. 1976. *San Pedro-Santa Cruz Project, Arizona, Concluding Report.* Washington D.C.: United States Bureau of Reclamation.

United States Environmental Protection Agency (EPA). 1996. *U.S.-Mexico Border XXI Program, Framework Document.* Washington, D.C.: United States Environmental Protection Agency.

Utton, A. 1999. Coping with drought on an international river under stress: The case of the Río Grande/Río Bravo. *Natural Resources Journal* 39(1): 27–33.

Water Resources Research Center (WRRC). 1997. Publication focuses on Arizona El Niño happenings. *Arizona Water Resource, El Niño News 1.* Water Resources Research Center, University of Arizona, October.

Western Regional Climate Center Webpage. 2002. Arizona Climate Summaries http://www.wrcc.dri.edu/summary/climsmaz.html

Western Regional Climate Center. 1997, 2001. Frequently Asked Questions About El Niño. http://www.wrcc.sage.dri.edu/enso

Woodhouse, C.A. 1997. Winter climate and atmospheric circulation patterns in the Sonoran Desert region, U.S.A. *International Journal of Climate* 17: 859–873.

World Wide Web. Spring 1998. Internet search conducted by L. Farrow Vaughan.

Yedra, G. 1997. Previenen contra El Niño. *El Imparcial.* November 3, Sonora, Mexico.

7

# TRANSBOUNDARY WATER PROBLEMS AS "JURISDICTIONAL EXTERNALITIES"

*Increased Economic Efficiency Through Institutional Reforms*

Charles W. Howe

*Environment and Behavior Program, Institute of Behavioral Science and Professor Emeritus of Economics, University of Colorado, Boulder, 80309 Colorado*

**Abstract** Problems of interjurisdictional water allocation and management are found worldwide. Negative externalities are likely to be created when the boundaries of physical water systems fail to correspond with the boundaries of the institutions that have power to govern or otherwise affect them. Transboundary situations create these "jurisdictional externalities" almost by definition and range from international and interstate scope to local issues. In the water sector, jurisdictional externalities are especially difficult to modify because of the unidirectional nature of the resource flow. While the well-known "Coase theorem" states that solutions can be bargained out, in reality only in situations where "interconnected games" are being played is bargaining likely to lead to efficient and equitable solutions. Many transboundary problems would be solved by a return to the river basin as the unit of management, while the expansion of water markets within the river basin context would serve to provide for the internalization of many current externalities. Flexible allocation rules that reflect real-time hydrologic conditions can greatly improve on the fixed rules that are incorporated in most interstate and international river compacts.



*Henry F. Diaz and Barbara J. Morehouse (eds.), Climate and Water*, 145-161.

# 1. BACKGROUND ON TRANSBOUNDARY WATER CONFLICT AND AGREEMENT

The Food and Agriculture Organization of the United Nations has listed (1978, 1984) more than 3,000 international river agreements. A large literature has developed on transboundary water allocation, associated conflicts, conflict resolution, and instruments of agreement. In this context, transboundary can refer to international, interregional, interstate (in the U.S. sense), or intrastate relationships. As an example of this literature, the book *Water Quantity/Quality Management and Conflict Resolution* (edited by Dinar and Loehman 1995) contains papers dealing with interstate water management in Australia, multinational Nile River issues, conflict and agreement on the Jordan River, conflicts over Native American claims to water, intraregional conflicts in California, and water allocation in the Great Lakes region of North America.

An important subset of this literature relates to issues of water scarcity and related conflicts in the Middle East. The Jordan and Yarmuk Rivers are overappropriated, while coastal and mountain aquifers are being mined (Zeitouni et al. 1994). Various studies have suggested the establishment of water markets or auctions to effect economically efficient trades of water among the countries of that region and to defuse potential conflicts (Dinar and Wolf 1994). Wolf's analysis (1993) suggests that water, being a critical factor in regional sustainability, may actually lead to peaceful, successful negotiations rather than serious conflict. *International Waters of the Middle East: From Euphrates-Tigris to Nile* (edited by Biswas 1994) also concludes on an optimistic note regarding the possibility of broader negotiations motivated and initiated by agreement on water issues.

Many international river negotiations have been bilateral in structure, even where the river systems have included numerous riparian countries. The Nile Water Agreement was negotiated by Egypt and Sudan alone, although the Nile and its tributaries are shared by nine countries. The 1951 agreement on the Mekong River excluded Burma and China. A 1991 agreement by India and Nepal on management of the Ganges River excluded Bangladesh. Future problems will certainly occur in these cases.

An example of successful balanced bargaining over a single shared river is that of the Columbia River Compact of 1961 negotiated between Canada and the United States (Krutilla 1967). In that case, the river is unidirectional in a strict flow sense but it and its tributaries cross the international border several times, making each country the downstream party in at least part of the basin. Many of the best storage sites were in Canada, while the demand for power was primarily in the United States. The benefits of coordinated system development were shared through electric power sharing, construction cost sharing, and cash side payments. More recently, however, Canada has expressed dissatisfaction with the terms of the compact.

In the United States, many transboundary conflicts over the sharing of water exist in both the eastern and western parts of the country. Some of these conflicts are interstate and others are intrastate. Interstate conflicts are usually resolved through the negotiation of interstate compacts. Table 1 (Bennett 1995) lists and describes the interstate compacts in effect in the western United States.

*Table 1.* Interstate river compacts in the western United States.[a,b]

| Compact | Ratification Date | States | Quantity Apportionment |
|---|---|---|---|
| Arkansas River Compact | 1948 | Colorado<br>Kansas | Fixed: by releases from John Martin Dam (winter and summer rates in cubic feet per second-[cfs]). |
| Arkansas River Basin Compact | 1965 | Kansas<br>Oklahoma | Fixed: by reservoir storage capacity within each state. Unrestricted use. Kansas gets 100% of average natural flow. |
| Arkansas River Basin Compact of 1970 | 1973 | Arkansas<br>Oklahoma | Percentage: percent of annual yield (agreement to not deplete annual yield by more than x% - % maximum). |
| Bear River Compact | 1955 amended 1980 | Idaho, Utah<br>Wyoming | Combination: under declared "water emergencies" percent allocations for upper and central divisions; lower division based on combination (storage capacity is allocated). |

| Compact | Ratification Date | States | Quantity Apportionment |
|---|---|---|---|
| Belle Fourche River Compact | 1943 | South Dakota Wyoming | Percentage: percent of unappropriated waters (South Dakota, 90%; Wyoming, 10%); some on storage capacity (fixed). |
| Big Blue River Compact | 1971 | Kansas Nebraska | Fixed: by mean daily flow; must maintain a given mean daily flow (varies by month). |
| Costilla Creek Compact | 1945 amended 1963 | Colorado New Mexico | Combination: fixed by cfs for diversion, percentage for reservoir releases and surplus waters. |
| Colorado River Compact | 1922 | Arizona Colorado California Nevada New Mexico, Utah Wyoming | Fixed: upper Upper Basin must deliver 75 maf /10 yr to Lower Basin and Lower Basin states division of 7.5 maf/yr fixed by acre-feet/year. maf/yr: million acre-feet per year. |
| Canadian River Compact | 1950 | Oklahoma Texas New Mexico | Misc.: allocates reservoir storage; flow decided by commission (New Mexico and Texas have free and unrestricted use of waters subject to certain storage limitations). |

| Compact | Ratification Date | States | Quantity Apportionment |
|---|---|---|---|
| Klamath River Basin Compact | 1957 | California Oregon | Misc.: fixed by irrigated acreage limitation, use and vested right; no release or diversion requirement. |
| La Plata River Compact | 1922 | Colorado New Mexico | Percentage: if mean daily flow < 100 cfs Colorado must deliver 50% (unrestricted 12/1–2/15 or when flow > 100 cfs ); apportionment comes out of each state's share of Upper Colorado River Compact. |
| Pecos River Compact | 1948 | New Mexico Texas | Combination: beneficial consumptive use of salvaged and unappropriated floodwaters is by percent; 50% of floodwaters to each state, 43% of salvage water from New Mexico to Texas. |
| Red River Compact | 1978 | Arkansas Louisiana Oklahoma Texas | Combination: by percentage of annual flow or by acre-feet (by subbasin), (for low flow, by weekly runoff), and cfs. |

| | | | |
|---|---|---|---|
| Republican River Compact | 1942 | Colorado<br>Kansas<br>Nebraska | Percentage: specific allocations by total acre-feet per year calculated from average annual virgin water supply (includes both surface and groundwater; if supply varies >10%, allocations change to new specific allocations of the same percentage). |
| Rio Grande Compact | 1938 | Colorado,<br>Texas,<br>New Mexico | Percentage: variable percentage based on an index of upstream flow (smaller index, smaller percentage to be delivered). |
| Sabine River Compact | 1953, amended 1962 and 1977 | Texas<br>Louisiana | Percentage: 50% to each state of all free water after maintaining minimum flows. |
| Snake River Compact | 1949 | Idaho<br>Wyoming | Percentage: 96/4 for storage or direct diversion exclusive of established Wyoming rights (excludes Wyoming water for generation of electric power). |
| South Platte River Compact | 1923 | Colorado<br>Nebraska | Fixed: 4/1–10/15 Colorado may not divert beyond an amount that will diminish mean daily flow below 120 cfs measured at Julesburg, Colorado; unrestricted 10/15–4/1. |

| | | | |
|---|---|---|---|
| Upper Colorado River Basin Compact | 1948 | Arizona Colorado New Mexico, Utah Wyoming | Combination: percentage for Upper Basin states; fixed for Arizona (per annum) after required deliveries to Lower Basin[c]. |
| Upper Niobrara River Compact | 1962 | Wyoming Nebraska | Misc: mostly unrestricted; restrictions on storage quantities and groundwater development at Anchor Dam, and based on priority dates. |
| Yellowstone River Compact | 1950 | Montana North Dakota, Wyoming | Percentage: of unused and unappropriated waters (percentage of each tributary to given state based on annual water year). |

[a]Source: L. L. Bennett (1995).
[b]Source: U.S. Statutes: 52 Stat. 150, 65 Stat. 663, 83 Stat. 86, 58 Stat. 94, 44 Stat. 195, 86 Stat. 193, 57 Stat. 86, 63 Stat 145, 80 Stat. 1409, 87 Stat. 569, 66 Stat. 74, 68 Stat. 690, 76 Stat. 34, 77 Stat. 350, 53 Stat. 785, 63 Stat. 159, 63 Stat. 31, 45 Stat. 1057, 1064, 43 Stat. 796, 72 Stat. 38, 94 Stat. 4, 64 Stat. 29, 71 Stat. 497, 94 Stat. 3305, and U.S. Water Resources Council, *The Nation's Water Resources*, 1978, pp. 124–130 and Simms et al. (1989). Apportionment is based only on "water emergencies."
[c]Subject to Colorado Compact provision of providing 7.5 maf to Lower Basin.

The most famous interstate river compact in the western United States is the Colorado River Compact (Meyers 1966). The fixed delivery mechanism dictated by the compact allocates the risk imposed by variable river flows very unevenly between the Upper and Lower Basins, since the Upper Basin is required to deliver 7.5 million acre-feet annually to the Lower Basin regardless of total flow (although this risk has been offset in part by the construction of Glen Canyon Dam and its reservoir, Lake Powell, which can be used to offset low flows).

Booker and Young (1994) have shown that the current compact allocations are highly inefficient, largely because they ignore instream values from hydrogeneration, recreation, and water quality improvement. Booker and Young concluded that economically efficient administration would require a large reallocation from the Upper Basin to the Lower Basin to reflect the low marginal values of irrigation water in the Upper Basin and the high instream values generated "between" the two basins.

There are also important intrastate conflicts over water allocation in the western United States. Vaux and Howitt (1984) showed that reallocation of water from north to south (largely agricultural to urban uses) in California could generate $200 million of added benefits and cost savings annually. The main impediments to such reallocation are the inflexible contracts under which the water is delivered and the large subsidies for agricultural water that reduce the motivation for transferring water.

In the state of Colorado, there is continuing pressure to divert water from the western slope of the Rocky Mountains where the heaviest precipitation occurs to the eastern side where most of the population resides. The western slope Gunnison River Basin still has "unappropriated water" that is subject to claim under the state's priority system of water rights. The Union Park Project is a proposal to divert 60,000 acre-feet annually out of the upper Gunnison River Basin into rapidly expanding Arapahoe County on the eastern side of the mountains. Under Colorado water law and the Colorado River Compact, the downstream values that would be lost are in no way recognized by or imposed on the exporting party. Howe and Ahrens (1988) estimated that lost downstream values within the state of Colorado alone would amount to $45/acre-foot from power generation, while substantial recreational and ecosystem values would also be lost to the state. Total downstream system values lost, including hydropower and dilution benefits (but not recreation), were estimated to be $140/acre-foot (Howe and Ahrens 1988).

## 2. ECONOMIC EFFICIENCY AND TRANSBOUNDARY WATER ALLOCATION

Two economic issues related to transboundary water allocation warrant discussion: (1) the failure of political and decision-making boundaries to coincide with those of the river basin or watershed, leading to the creation of negative "jurisdictional externalities"; and (2) the unidirectional nature of river systems, which frequently places downstream parties at a bargaining disadvantage.

The lack of correspondence between the physical boundaries of river basins or watersheds and political or decision-making boundaries has long been noted as the cause of problems that I prefer to call "jurisdictional externalities." Typically, river basins are divided into several political and/or decision-making jurisdictions, each having some responsibilities for water management within its geographical area. The separation of management decision-making responsibilities combined with the hydrologic unity of the basin result in the imposition of externalities (usually negative but occasionally positive) on other participants. For example, the water diversions in the Upper Basin of the Colorado River impose negative externalities on the Lower Basin through the former's consumptive use and the deterioration of water quality resulting from return flows. Since the existing compact framework for water allocation between the two basins does not consider water quality issues and since downstream water use has intensified since the Compact was signed, current arrangements impose large economic losses compared to what could be achieved under basinwide optimization. These externalities are legally ignored by Upper Basin users, although the opportunity cost of the consumptive use of an acre-foot in the Upper Basin is about $140 in foregone hydroelectric and salinity dilution benefits (Howe and Ahrens 1988).[1] While these patterns of water use and water quality degradation may be considered "fair" in a historical and legal context, they have become increasingly damaging and inefficient from an economic point of view.

According to the famous "Coase theorem" (Coase 1960), unidirectional externalities can be expected to be efficiently resolved[2] through bargaining regardless of the direction of the externality (i.e., regardless of who is "upstream" and who is "downstream"). The conditions assumed for the validity of the theorem are quite stringent: the availability of complete information on damages created and the costs of mitigation; absence of transactions (bargaining and information) costs; and isolation of the issue from

---

[1] Until recently, water quality degradation from agricultural activities has been explicitly excluded from regulation. The Grand Valley Project in western Colorado historically was estimated to add 10 (short) tons of salts per year per acre irrigated to the Colorado River through its irrigation return flows (Skogerboe and Walker 1973). Studies have estimated that downstream damages to Lower Basin agriculture and urban users as a result of that historic salt input ranged from $150 to over $400/acre irrigated in the Grand Valley (Anderson and Kleinman 1978). The Grand Valley is located just upstream of the Utah border, so the state of Colorado had little concern over this externality. The 1948 Upper Basin Compact, which allocates water among the Upper Basin states, does not deal with water quality. In the last 20 years, the Bureau of Reclamation has had a program of water management aimed at reducing the saline return flows from the Grand Valley.

[2] "Efficiently resolved" doesn't mean eliminated, since it is generally inefficient to eliminate externalities totally. Damaged parties should be willing to pay for the reduction of damages as long as marginal damage reduction is greater than the cost of the reduction.

other issues of concern. Finding these conditions in the "real" world seems unlikely.

Barrett (1994) discusses the unique complexities associated with international water allocation. In these situations, there is usually no higher-level authority for enforcement of agreements, so agreements must be self-enforcing. Self-enforcement is often difficult to accomplish, since many shared water systems are unidirectional, making balanced bargaining difficult unless there are other issues that can be conegotiated.

Nonetheless, in spite of the unlikelihood of the Coase theorem and the difficulties just cited, successful single-issue upstream-downstream problem resolution does take place from time to time, as was shown by the U.S.-Mexican Colorado River salinity issue (1973). As consumptive uses and agricultural return flows increased in the United States, Mexico's Colorado River water that supported agriculture in the Mexicali Valley became saltier. The 1944 treaty between the United States and Mexico covered the sharing of water volume but did not address water quality. In 1960, a crisis was reached when a new drainage system for the Wellton-Mohawk irrigation project on the Arizona-California border began emptying heavily saline water into the river, just below the last U.S. diversion but above the diversion for Mexico. Total dissolved solids (TDS) increased from 785 mg/l in 1959 to 1,490 mg/l by 1962 (Jonish 1994), causing extensive damage to crops in Mexicali with large acreages taken out of production (Oyarzabal-Tamargo and Young 1976). Mexico protested strongly and the issue became important in the Mexican political scene. Negotiations ensued and finally, after Mexico suffered extensive losses, a new agreement was approved at a meeting of Presidents Echevaria of Mexico and Nixon of the United States (Minute 242 of the International Boundary and Water Commission [IBWC]). The United States guaranteed that water delivered to Mexico would have annual average TDS of no more than 115 mg/l over the TDS level at Imperial Dam (the point of last U.S. diversion).[3] The advantages of this agreement for the United States were better relations with Mexico and the waiving by Mexico of any compensatory payments for the damages done historically.

Resolution of unidirectional conflicts over externalities is more likely if two or more issues can be negotiated simultaneously, issues in

---

[3] To help in meeting this guarantee, the United States chose to build an enormously expensive reverse osmosis desalting plant that has proven both to be unnecessary (the salts from Wellton-Mohawk have mostly been leached out) and too costly to operate. The Bureau of Reclamation happily built the plant to offset the tremendous externality created by one of its own irrigation projects.

which the externalities (positive or negative) run in different directions. Such a situation is referred to in game theoretic terms as an "interconnected game." A substantial literature has developed concerning such "interconnected games" as an extension of traditional game theory. While the foundations are highly mathematical, it is argued that these techniques may provide an approach to international water and environmental problems that will enhance cooperation and provide an alternative to financial side payments, which traditionally have been used to induce cooperation (Folmer et al. 1993). The insights from this theory may also help explain the resolution of "unidirectional externalities" where water appears as the first object of bargaining but where other (possibly tacit) objectives are present; e.g., generation of goodwill that will aid in future bargaining. Such bilateral advantage is clearly necessary for the "self-enforcement" of international treaties referred to by Barrett.

The "interconnected game" is nicely illustrated by the simultaneous negotiations between Mexico and the United States on the division of the waters of the Colorado-Tijuana and the Rio Grande (Río Bravo) (Jonish 1994; Bennett and Ragland 1998). Although the United States and Mexico had negotiated for many years over these rivers, the severe drought of 1943 brought a resumption of serious negotiations. On the Colorado, the United States obviously held the trump cards, but Mexico held the trumps on the lower Rio Grande, the flow of which depended heavily on inflows from Mexican tributaries (the Conchos, San Diego, San Rodrigo, Escondido, and Salado Rivers). The resultant treaty (U.S. Government Printing Office: Treaty Series 994, 1946) guaranteed 1.5 million acre-feet annually to Mexico from the Colorado (although no guarantee on the quality of the water, as we have seen), while the United States was guaranteed 350,000 acre-feet annually from the Mexican tributaries. The recurrence of extreme drought in the mid-1950s stimulated further cooperation along the Rio Grande in the form of Minute 293 of the IBWC, under which the two countries agreed on the obligation of mutual assistance in loaning water during critical periods, the mutual enforcement of conservation measures, and the mutual sharing of data and operational information (Mumme 1998).

## 3. INCREASING TRANSBOUNDARY ALLOCATION EFFICIENCY THROUGH A RETURN TO RIVER BASIN COMMISSIONS, THE EXPANDED USE OF WATER MARKETS, AND FLEXIBLE ALLOCATION RULES

We have seen that the widespread existence of "jurisdictional externalities" and the tendency of legal compacts to become outdated result in large economic losses in transboundary water management. To improve the situation, we propose three steps, the first two of which may be considered "old hat," and one that is relatively new to the literature: a return to the use of river basin commissions; an expanded role for water markets within the context of river basins; and the substitution of flexible allocation rules for the fixed rules currently embodied in most compacts.

The establishment of river basin commissions, the role of which encompasses entire drainage basins, is obviously suggested by simple hydrology: What is done upstream will affect conditions downstream. There is a long history of such commissions in the United States, which helps it to better deal with basinwide issues of navigation and flooding. A Mississippi River Commission was established by the U.S. Corps of Engineers in 1918 to unify flood control management (Barry 1997). The Potomac River Basin Commission was established in 1940 to bring together four riparian states (Maryland, Virginia, West Virginia, and Pennsylvania) and Washington, D.C., to better deal with division of the water supply and control of water quality (Davis 1968). The Delaware River Basin Commission was established in 1967 to control water quality and to allocate water during droughts between the main source state of New York and Pennsylvania (Council on Environmental Quality 1975).

Under the Water Resources Planning Act of 1965, six river basin commissions (RBCs) were established to improve coordination among their riparian states: the New England RBC; the Great Lakes RBC; the Ohio RBC; the Upper Mississippi RBC; the Missouri RBC; and the Pacific-Northwest RBC (U.S. Water Resources Council 1980). Some of these commissions made headway in the coordinated planning of water resources and water quality in their basins, while others were stymied by a requirement of unanimity of all commission members on any substantial policies or projects. The potential for such commissions was intensively studied by the National Water Commission, which reported very favorably on their potential (Ingram 1971; Hart 1971). A thorough legal and institutional analysis of the river basin commission as an instrument of water management is provided by Muys (1971). Unfortunately, the Reagan administration eliminated all the commissions created under the 1965 act and essentially eliminated the powers of the Water Resources Council in 1981.

The establishment of intrastate, interstate, and international river commissions could overcome many of the "jurisdictional externalities" that have been cited above. The greatest problems will be the vested interests, which benefit from current inefficient arrangements; e.g., low-valued irrigated agriculture in the Upper Basin of the Colorado River. However, studies have shown that for the Colorado, improved basinwide management could create sufficient added benefits to make it a "win-win" situation for all parties if imaginative compensatory mechanisms were used (e.g., Booker and Young 1994).

Within the context of established river commissions, there would be broad opportunities for the use of water markets to further eliminate jurisdictional externalities to allow for broader "Coasian" bargaining to equalize marginal values of water in all its applications. Attempts to extend water markets across state lines have been heavily resisted by traditional water interests, partly on the grounds that compacts would have to be revised, opening the door to possible political redistribution of water. In 1985, the Galloway Group proposed the leasing of Colorado Upper Basin water to Lower Basin users (*U.S. Water News*, Dec. 1985). This proposal was strongly resisted by water officials and never moved ahead. In 1990, the Resource Conservation Group proposed the leasing of water that had been consumed in agriculture through an arrangement with Western Slope farmers whereby they would dry up part of their land each year, making their consumptive use available for lease (*Denver Post*, Jan. 29, 1990, p. 1B). This proposal was also defeated by water interests and officials. In 1994, California proposed the establishment of interstate water trading on the Colorado River (California, State of, 1991). This would have involved annual leases only, arranged through the offices of the state engineers and not by private bilateral arrangement. The proposal was quickly rejected by the Upper Basin governors. The fear seems to be that water once leased would be forever lost—if not through the courts then through politics. The development of interstate water markets is likely to proceed in spite of the resistance to date, but probably with a high degree of social oversight that will protect the long-term interests of the participating states and prevent the monopolization of major water resources.

Regarding the use of more flexible allocation rules between states and in compacts, Kilgour and Dinar (1995, pp. 3–4) state:

> Studies by the UN Food and Agriculture Organization (1978,1984) indicate that most water allocation agreements (said to be in excess of 3,000) include static descriptions of both the supply of and demand for water.... A water-sharing

> scheme that accounts for the stochastic nature of water supply and the dynamic nature of water demand will almost certainly produce more political stability. Destabilizing shortages are already a risk ...the ability to match allocations to current conditions may reduce regional water allocation transaction costs, allow states to plan water-related investments more effectively, and increase regional social welfare.

Review of Table 1 shows that the interstate compacts in the western United States are based on the assumption of a static hydrologic regime. The Colorado Compact was derived from the assumption that there were 15 million acre-feet of water to be distributed, giving the Upper and Lower Basins each 7.5 million acre-feet of consumptive use. The effects of the randomness of the streamflow apparently were overlooked when the requirement of a fixed delivery of 7.5 million acre-feet from the Upper Basin was imposed. With annual streamflows ranging from 4 million to 24 million acre-feet, tremendous hydrologic risk was placed on the Upper Basin (Jacoby 1975).

The stochastic nature of climate and related hydrologic events must be incorporated in the design of transboundary agreements. By incorporating continuous information about the current state of the system (flows, storage, soil moisture, etc.) and making use of better climate forecasts, the net payoff from transboundary allocations can be substantially increased relative to those allocations that result from the currently used inflexible allocation rules.

Kilgour and Dinar (1995) simulated allocation rules under which allocations were optimized from a systemwide point of view, based on assumed knowledge of actual contemporary flows and withdrawals, presumably from satellite observations. The resultant allocations did not follow any of the static rules; e.g., required fixed deliveries or percentage sharing. Bennett (1995) and Bennett et al. (2000) similarly used period-by-period optimization based on known total flow as a baseline for comparing the efficiencies of fixed delivery and percentage sharing compacts, each of which resulted in smaller net payoffs.

A study of the effects of severe sustained drought on the Colorado River system (Powell Consortium 1995) also brought out the importance of management flexibility in the face of extreme climate events. Tree-ring studies have long suggested that droughts of much greater magnitude and duration than those exhibited in modern records have taken place in the Colorado Basin. Taking the infrastructure of the Colorado system as given, the study estimated the hydrologic and, thence, the economic, social, and environmental impacts of a drought under present and alternative institutional arrangements for water allocation. Existing institutional arrangements were found to protect traditional consumptive uses, but the nonconsumptive instream uses were severely damaged. Introducing flexibility in the allocation of water both between the Upper and Lower Basins and within individual states exhibited "win-win" possibilities across all water uses, although decision-making games with players representing the various states and the federal government showed that agreement on efficient institutional changes would be very hard to accomplish.

In addition to adapting allocation arrangements to climate and hydrologic variability, river agencies face the possibility of significant climate change. Studies sponsored by the U.S. Environmental Protection Agency (EPA) of five major world river basins under climate change emphasize both the importance and complexities of projecting the effects of climate change on international rivers (Strzepek and Smith 1995). The Nile, Zambezi, Indus, Mekong, and Uruguay Rivers were modeled under various scenarios of climate change. The effects of those scenarios on streamflows, likely water management, hydropower production, irrigation, urban water supply, flood protection, and ecosystem impacts were then estimated. Management schemes based on "normal" flow conditions became inefficient under climate change. Few clear positive effects of climate change (other than increased power production in some basins) emerged from the analysis. The drier basins exhibited greater sensitivity to climate change. Again, flexibility in water management in the face of climate variability and change was seen to be important to both the efficiency and equity of water allocation schemes.

In sum, a return to the river basin as the basic unit for water management, an expanded role for water markets within those basins, and the use of more inventive flexible allocation rules can lead to significant improvement in transboundary water management.

## 4. REFERENCES

Anderson, J.C., and A.P. Kleinman. 1978. Salinity management options for the Colorado River. In, *Water Resources Planning Series Report P-78-003 (33-37)*, Utah Water Research Laboratory, June.

Barrett, S. 1994. Conflict and Cooperation in Managing International Water Resources. The World Bank, Country Economics Department, Washington, D.C.

Barry, J.M. 1997. *Rising Tide: The Great Mississippi Flood of 1927 and How It Changed America.* New York: Simon and Schuster Publishers.

Bennett, L.L. 1995. The interstate river compact as a water allocation mechanism. Ph.D. dissertation, Department of Economics, University of Colorado, Boulder.

Bennett, L.L., and S. Ragland. 1998. Facilitating international agreements through an interconnected game approach: The case of river basins. In, Just, R.E., and S.

Netanyahu (eds.), *Conflict and Cooperation in Transboundary Water Resources*. Dordrecht, The Netherlands: Kluwer Academic Publishers, Elgar.

Bennett, L.L., C.W. Howe, and J. Shope. 2000. The Interstate River Compact as a water allocation mechanism: Efficiency aspects. *American Journal of Agricultural Economics* 82(4)(November).

Biswas, A.K. (ed.). 1994. *International Water of the Middle East: From Euphrates-Tigris to Nile*. Bombay and Delhi: Oxford University Press.

Booker, J.F., and R.A. Young. 1994. Modeling intrastate and interstate markets for Colorado River water resources. *Journal of Environmental Economics and Management* 26: 66–87.

California, State of. 1991. Conceptual Approach for Reaching Basin States Agreement on Interim Operation of Colorado River System Reservoirs, California's Use of Colorado River Water Above Its Basic Apportionment, and Implementation of An Interstate Water Bank. Prepared for the Colorado River Basin States meeting, Colorado River Board of California, August 28, 1991.

Coase, R. 1960. The problem of social cost. *Journal of Law and Economics*, October.

Council on Environmental Quality. 1975. *The Delaware River Basin: An Environmental Assessment of Three Centuries of Change*. Washington, D.C.: U.S. Government Printing Office, August.

Davis, R.K. 1968. *The Range of Choice in Water Management: A Study of Dissolved Oxygen in the Potomac Estuary, Baltimore, Maryland*. Published for Resources for the Future, Inc., Johns Hopkins University Press, Baltimore, Maryland.

*Denver Post*. Jan. 29, 1990. Colorado River plan rekindles debate. Sec. B, p. 1.

Dinar, A., and E.T. Loehman. 1995. *Water Quantity/Quality Management and Conflict Resolution: Institutions, Processes and Economic Analyses*. Westport, Connecticut: Praeger.

Dinar, A., and A. Wolf. 1994. Potential for regional water transfer and cooperation: The case of the western Middle East. Paper presented at the International Workshop on Economic Aspects of International Water Resources Utilization in the Mediterranean Basin, Fondazione ENI Enrico Mattei, Milan.

Folmer, H., P.V. Mouche, and S. Ragland. 1993. Interconnected games and international environmental problems. *Environmental and Resource Economics* 3(4): 313–335.

Food and Agriculture Organization. 1978. *Systematic Index of International Water Resource Treaties, Declarations, Acts and Cases by Basin*, Legislative Study No. 15, Rome.

Food and Agriculture Organization. 1984. *Systematic Index of International Water Resource Treaties, Declarations, Acts and Cases by Basin*, Vol. II, Legislative Study No. 34, Rome.

Hart, G.W. 1971. *Institutions for Water Planning—Institutional Arrangements: River Basin Commissions, Inter-Agency Committees, and Ad Hoc Coordinating Committees*, National Water Commission, National Technical Information Service, September.

Howe, C.W., and W.A. Ahrens. 1988. Water resources of the Upper Colorado River Basin: Problems and policy alternatives. In, El-Ashry, M.T., and D.C. Gibbons, *Water and Arid Lands of the Western United States*. Cambridge and New York: Cambridge University Press.

Ingram, H. 1971. The New England River Basin Commission: A Case Study Looking Into the Possibilities and Disabilities of a River Basin Commission Established Under Title II of the Water Resources Planning Act of 1965. National Water Commission, National Technical Information Service, November.

Jacoby, G.C. 1975. Lake Powell Research Project Bulletin No. 14, November.

Jonish, J.E. 1994. Transboundary water resource management in arid lands: Lessons from the United States–Mexico. Paper presented at the annual Conference of the Universities' Council on Water Resources, San Antonio, Texas.

Kilgour, D.M., and A. Dinar. 1995. Are Stable Agreements for Sharing International River Waters Now Possible? Policy Research Working Paper 1474, Agriculture and Natural Resources Department, The World Bank, June.

Krutilla, J.V. 1967. *The Columbia River Treaty: The Economics of an International River Basin Development*. Published for Resources for the Future, Inc., by the Johns Hopkins University Press, Baltimore, Maryland.

Meyers, C.J. 1966. The Colorado River. *Stanford Law Review* 19(1): 1–75.

Mumme, S.P. 1998. Managing acute water scarcity on the U.S.-Mexican Border: Institutional issues raised by the 1990's drought. *Transboundary Resources Report* 11(1): 6–8.

Muys, J.C. 1971. Interstate Water Compacts: The Interstate Compact and Federal-Interstate Compact. Published by the National Water Commission through the National Technical Information Service.

Oyarzabal-Tamargo, F., and R.A. Young. 1976. The Colorado River salinity problem: Direct economic damages in Mexico. Paper presented at the annual conference of the Western Agricultural Economics Association, Fort Collins, Colorado, July 18–20.

Powell Consortium. 1995. Severe sustained drought: Managing the Colorado River System in times of water shortage. Reprinted from *Water Resources Bulletin* 31(5): 779–994.

Skogerboe, G.W., and W.R. Walker. 1973. Salt pickup from irrigated lands in the Grand Valley of Colorado. *Journal of Environmental Quality* 2: 377–382.

Strzepek, K.M., and J.B. Smith. 1995. *As Climate Changes: International Impacts and Implications*. Cambridge: Cambridge University Press.

United States Government Printing Office. 1946. Treaty Series 994, Utilization of Water of the Colorado and Tijuana Rivers and of the Rio Grande: Treaty Between the United States of America and Mexico (signed in Washington, D.C., Feb. 3, 1944).

*U.S. Water News*. December 1985. Can the Galloway Group lease the Colorado? p. 9.

U.S. Water Resources Council. 1980. State of the States: Water Resources Planning and Management. April.

U.S. Water Resources Council. 1978. The Nation's Water Resources, 1975–2000: Second National Assessment, U.S. Government Printing Office.

Vaux, H.J., Jr., and R.E. Howitt. 1984. Managing water scarcity: An evaluation of interregional transfers. *Water Resources Research* 20(7): 785–792.

Wolf, A. 1993. Water for peace in the Jordan River Watershed. *Natural Resources Journal* 33: 797–839.

Zeitouni, N., N. Becker, M. Shechter, and E. Luk-Zilberman. 1994. Two models of water market mechanisms with an illustrative application to the Middle East. Paper presented at the International Workshop on Economic Aspects of International Water Resources Utilization in the Mediterranean Basin, Fondazione ENI Enrico Mattei, Milan.

8

# IMPACTS IN MEXICO OF COLORADO RIVER MANAGEMENT IN THE UNITED STATES

## *A History of Neglect, A Future of Uncertainty*

David H. Getches
*Raphael J. Moses Professor of Natural Resources Law*
*University of Colorado School of Law, CB 401, Boulder, Colorado 80309-0401*

**Abstract** The Colorado River terminates in Mexico as a threaded delta above the Sea of Cortez (Gulf of California). Decades of water depletion, enabled largely by water projects in the United States, resulted in little or no water flowing in the channel. This lack of water has caused losses of wetlands, riparian vegetation, and estuarine habitat, reducing fish and wildlife populations, including some endangered species. Cucupá Indians who depended on fishing and gathering lost their traditional occupations. Recent years of high runoff revitalized the ecosystems of the Colorado River Delta and this animated nongovernmental organizations and scientists to call attention to the issue. Water users in the United States have objected in principle to addressing the issue because they resist the concept of ensuring the delivery of additional water to Mexico beyond current legal requirements, even in the small quantities needed for ecological maintenance. Mexico has not formulated a clear official position on the issue. Past transboundary water conflicts between the two countries over the quantity and quality of Colorado River water owed by the United States to Mexico have been characterized by neglect, followed by vigorous protest from Mexico, strong U.S. resistance from states and water users, and finally belated concessions at a high diplomatic level. There are several possible remedies to the delta problem that would help the countries avoid the conflictive approach of the past.



*Henry F. Diaz and Barbara J. Morehouse (eds.), Climate and Water*, 163-191.

## 1. INTRODUCTION

The Colorado River Delta in Mexico was once a wilderness, abundant in migratory birds, native fish, and wildlife. When Aldo Leopold and his brother explored in 1922, they found "green lagoons," plump deer, and "a wealth of fowl and fish." They feasted on quail and goose and were haunted at night by the specter of "el tigre" (Leopold 1949). In the same year as Leopold's visit, the United States began to lay the legal groundwork for an elaborate system of dams and diversions in the main stem of the Colorado River that was to sap the life of the delta. By the last quarter of the twentieth century, the delta was reduced to a ghastly moonscape. After visiting the delta around 1980, Philip Fradkin described the Colorado as "a river no more" (Fradkin 1995).

Several episodes of high precipitation and runoff in the Colorado River Basin began in 1983 and continued intermittently until 2000. This extraordinarily wet period filled reservoirs along the river and sent excess flows to the delta, which sprang again to life. Wetlands were recreated and riparian habitat was revitalized. The estuary at the Sea of Cortez (Gulf of California) was again nourished. The threat of extinction eased a bit for some endangered species (Pitt et al. 2000).

As flows of the Colorado River return to normal, depletion in the United States will again imperil the ecosystems of the Colorado River Delta unless there are legal or management changes. The legal framework that allocates rights to use water from the Colorado River plainly is designed to allow the consumption of all the water available in normal years without regard to the environmental consequences; the recurrence of high flows that have inadvertently restored the delta in recent years are less likely in the future. This gloomy prognosis for the delta is based on a number of factors, especially the historically wide variations in climate that will produce periods of low flows and growing demands for consumptive use of water from the river.

Controversy between the United States and Mexico over the ecological condition of the delta appears to be inevitable. The Colorado River has been a source of conflict between the two nations in the past. Two major transboundary disputes, first over Mexico's entitlement to a quantity of water, and later over the quality of the water left in the river when it passed into Mexico, erupted into international incidents (Ward 1999).

The delta issue has been ignored until recently. Nongovernmental organizations and scientists have put the issue before the public, and it has now begun to receive attention at an official level as well as in the literature.[1] The two nations now have an opportunity to avoid conflict, learning from the high-pitched controversies of the past. Ideally, the parties will act before the next episode of sustained low flows produces conditions that provoke, once again, international transboundary conflict over the Colorado River.

## 2. THE COLORADO RIVER—AN OVERVIEW

The Colorado River rises in the Rocky Mountains in Wyoming and Colorado. The 247,600-square-mile watershed drains high mountain snowmelt from the Rockies in Wyoming and Colorado, the Uintas in Utah, and the San Juans in Colorado and New Mexico. After flowing some 1,400 miles, mostly in the United States, the river ends at the Sea of Cortez, known in the United States as the Gulf of California. Beginning in the 1920s with the construction of Hoover Dam, the United States developed an impressive system of dams and reservoirs on the main stem of the Colorado. Today there are 10 major dams in the basin. These water projects helped to sustain major population growth and economic expansion in the southwestern United States, especially in Southern California (Meyers 1966; Getches 1985).

The Colorado River has been the object of tensions among states, with Indian tribes, and between the United States and Mexico ever since political boundaries were drawn. It is an internationally known example of a river instigating transboundary water issues. The river is prized by its users, being the main source of surface water for seven states in the United States, and sustaining a population of over 20 million people and the irrigation of 1.7 million acres. Although its annual flow of 13.5 million acre-feet is impressive, the river is overwhelmed by the demands of growing urban areas, agricultural uses in the arid region, and intensive hydroelectric power generation (Culp 2001).

The river is managed under a rigid legal regime known to specialists as "the law of the river." To understand the history of U.S.-Mexico relations concerning the Colorado River requires an understanding of this collection of legal instruments, which includes two interstate compacts, an international treaty, and numerous federal statutes and regulations. These laws

---

[1] This chapter is based on a paper presented at the "Climate, Water and Transboundary Challenges in the Americas" conference, held in Santa Barbara, California, July 16–19, 2000. Since that time, scholarly and official attention to the issue has multiplied. Two important journals have dedicated entire issues to the delta: *Journal of Arid Environments*, Vol. 49, No.1, September 1, 2001, and *Natural Resources Journal*, Vol. 40, No. 4, Fall 2000.

were designed to fulfill the United States' desires to develop resources for economic gain and growth and to forestall disagreements among the states that touch the river. Only recently—and awkwardly in light of the absence of such considerations by established institutions—has the natural environment been considered at all in river management.

The cornerstone of the law of the river is the 1922 Colorado River Compact, an agreement of the seven riparian states—Arizona, California, Nevada, Colorado, Wyoming, Utah, and New Mexico (*Colorado River Compact of 1922*; see Figure 1). The compact resulted in a rather simple allocative decision among the states. First, they divided the Colorado River Basin into an upper and a lower basin. Each of these subbasins was to have the right to use 7.5 million acre-feet each year, which was believed to be less than half of the total river flow. At the time, the assumption seemed reasonable because the period of record beginning in 1896 showed average flows in excess of 17 million acre-feet. Later, the assumed flows were shown to be grossly unrealistic. Indeed, studies of tree rings going back some 400 years show that the average flow is closer to 13.5 million acre-feet (Stockton and Jacoby 1976). This amount was inadequate for even the 15 million acre-feet allocation agreed to between the upper and lower subbasins (Getches 1985).

Since the 1922 compact, the four states of the upper basin have allocated their share of the river among themselves by another compact that gives each a percentage of the available water each year (*Upper Colorado River Basin Compact* 1949). A different approach was taken by the lower basin states; each of the three lower basin states was given a specific allocation (rather than a percentage) allocation of water under the Boulder Canyon Project Act, a federal enactment that authorized construction of Hoover Dam (*Boulder Canyon Project Act of 1928*).

Accordingly, California got the right to use 4.4 million acre-feet per year of the lower basin's 7.5 million acre-feet compact entitlement. Even this large quantity has not been enough to sate the demands of southern California's growth (Hundley 1992). Indeed, for many years California has used over 5 million acre-feet per year. This was possible because the other states generally did not use their full allocations. Demands in most of the other states have expanded, however, and overall depletions have consumed or stored nearly all the water except the amount committed to Mexico. As demands on the Colorado River grew, the volume of water delivered to Mexico by the river was reduced to the minimum required by a 1944 treaty between the two countries—1.5 million acre-feet. All of that water was committed to agriculture in northern Mexico. For most years in the twentieth century, the trickle of water that passed this last diversion on the Colorado River died in the sands of Mexico, well short of its natural destination in the Sea of Cortez (Getches 1985, 1997). The impacts of this heavy water use have been buffered, however, in high-flow years.

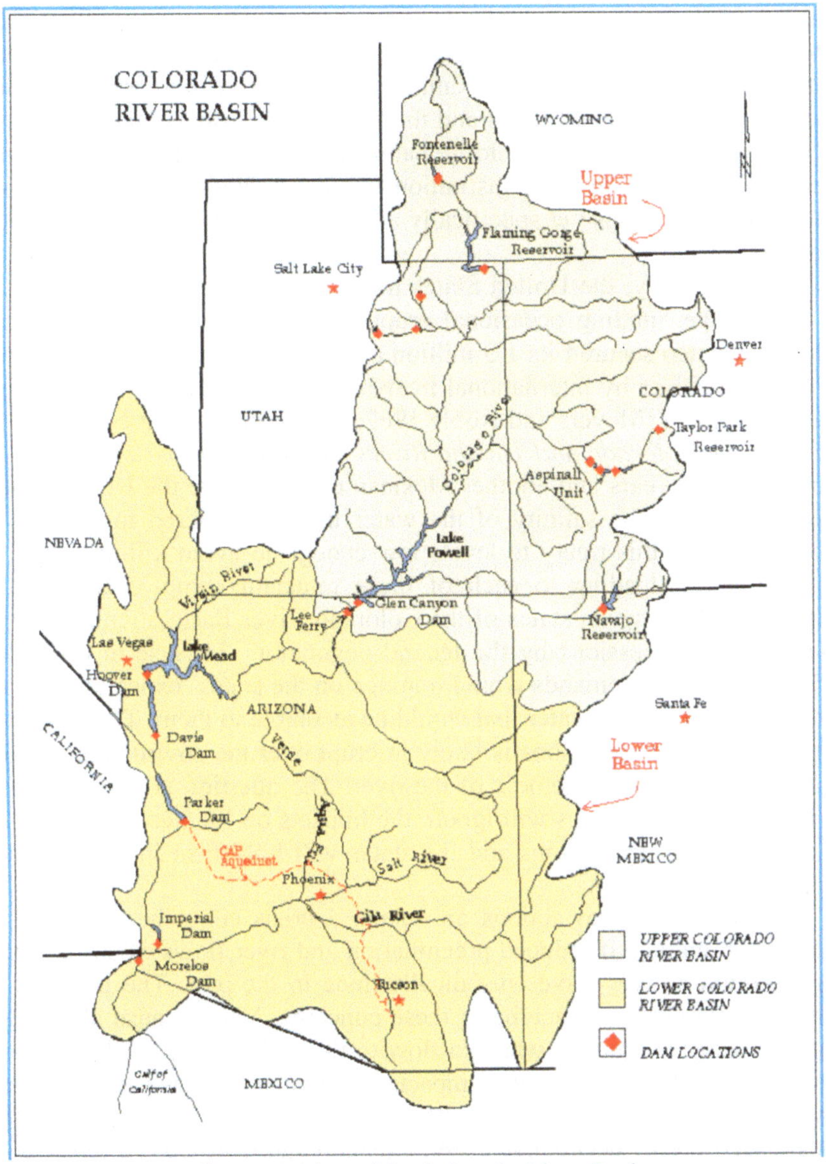


*Figure 1.* Map of the Colorado River Basin

## 3. U.S.-MEXICO RELATIONS

The United States and Mexico share an 1,800-mile border. It is the focal point for common interests, concerns, and, occasionally, conflicts. The principal concerns today are immigration, drug traffic, trade, and the use and protection of natural resources. Natural resource protection issues include allocation of the Colorado River and the Rio Grande, use of shared aquifers, and ensuring the integrity of wildlife habitats that transcend the international frontier. None of the existing institutions for borderlands environmental action or cooperation has yet specifically addressed the problems of the Colorado River Delta.

Historically, the United States has used its upstream position to its own advantage, inciting occasional eruptions by its disregard of Mexican interests. Mexico secured its 1.5 million acre-feet allocation of the water of the Colorado River by international treaty only after protesting against U.S. overdevelopment (Meyers and Noble 1967; *Treaty Respecting Utilization of Waters of the Colorado and Tijuana Rivers and of the Rio Grande* 1944). It then took many years and another Mexican outcry before the United States agreed to control the salinity of the water that is delivered to Mexico in compliance with this treaty to levels low enough to avoid killing crops. In the negotiations leading to each of these reluctant compromises by the United States, the seven states of the Colorado River Basin strenuously objected to any concessions by the federal negotiators, because in each case fulfilling Mexico's demands would impinge on the states' expectations concerning the quantity of water that would be available to them. The next conflict between the two nations is likely to erupt over the environmental consequences of U.S. management of the river. The question is whether, as in the past, the United States will ignore the impacts on Mexico until the matter reaches crisis proportions and the states will hold firm in resisting any compromise with Mexico.

Unless remedial actions are taken, a crisis could be triggered by variations in climate if reduced precipitation and river flows exist for a prolonged period as they have episodically done in the past. The prospect of global climate change exacerbates these concerns. Yet, because allowing a relatively small amount of water to flow to the Colorado River Delta could ameliorate most environmental impacts, it should be feasible to prevent grave environmental harm to the delta without either country making significant sacrifices (Luecke et al 1999; Glennon and Culp 2001). There is no reason to expect, however, that such accommodations will be reached quickly or easily in the absence of extraordinary statesmanship by U.S. leaders.

When the Colorado River Compact was entered into, Mexico was left out of the negotiations. The states recognized, however, that in the future it would be necessary to allow Mexico to use some of the river's water (Meyers and Noble 1967). This expectation was obvious at the time because Mexican farms were already using water in the region just south of the border. The compact provided only that:

> [i]f, as a matter of international comity, the United States of America shall hereafter recognize in the United States of Mexico any right to any use of any waters of the Colorado River System, such waters shall be supplied first from the waters which are surplus over and above the aggregate of the quantity specified in paragraphs (a) and (b); and if such surplus shall prove insufficient for this purpose, then, the burden of such deficiencies shall be equally borne by the Upper Basin and the Lower Basin, and whenever necessary, the States of the Upper Division shall deliver at Lee Ferry water to supply one-half of the deficiency so recognized.

The Colorado River Compact enabled Hoover Dam to be built (Hundley 1975). With the waters roughly allocated among the states, Congress was willing to fund its construction. Hoover Dam was largely for the benefit of California in that it would produce inexpensive and plentiful electric power for Los Angeles. In addition, it would provide water for farming in the Imperial Valley of California and deliver the water through the All-American Canal, replacing a canal that passed through Mexico. The new canal was believed to be more secure and free of the threat of Mexicans taking the water for use on their farms.

Once Hoover Dam was completed in 1935, Mexico began to object to the United States' monopoly on the river. The United States at first took a hard-line position, asserting the so-called "Harmon Doctrine." This doctrine essentially holds that an upstream state can do as it pleases with water before a waterway reaches a downstream state. Sensitized by the national security concerns connected with World War II, the United States eventually altered its position. The treaty that was consummated with Mexico in 1944 provides 1.5 million acre-feet of water to Mexico annually. When there is sufficient surplus, the treaty provides that Mexico can receive an additional 200,000 acre-feet (Meyers and Noble 1967).

The damming of the Colorado River began with Hoover Dam and continued into the 1960s (Reisner 1986; Martin 1989). The last major impoundment to be constructed on the main stem of the Colorado River was Glen Canyon Dam. In 1961, the United States began to fill Lake Powell, the enormous reservoir created by the dam. Lake Powell is approximately the size of Lake Mead, which was formed by Hoover Dam. The two reservoirs have a combined capacity almost equal to the mean annual flow of the Colorado River for 3 years.

While Lake Powell was filling, only the minimum deliveries of water required by the 1944 treaty were made to Mexico. During this period, salt concentrations in the river as it passed into Mexico spiked, reaching 2,700 parts per million (ppm) of dissolved solids (Miller et al. 1986). This far exceeded the maximum concentration that could be tolerated if the water was to be used for agriculture. The salinity in the Colorado River is naturally high because of saltwater springs and salt loading from natural runoff into the river. The problem is exacerbated by the addition of salts picked up by irrigating saline soils. With repeated uses and returns of water to the river, and with evaporation from the downstream reservoirs in which the already-saline water is stored, concentrations of dissolved solids reached the highest levels in history. Much of the water that reached Mexico was undiluted brackish return flows.

Mexico objected vigorously to these conditions. At first, the seven Colorado River Basin states that use the water in the United States took the position that there was no requirement that the United States deliver treaty water of a particular quality. The users feared, correctly, that ensuring a reduction in salinity would mean either changing consumption and use patterns in the United States or loss of credit under the treaty for the most saline water passing to Mexico. The United States' adamant resistance was eventually relaxed and negotiations resulted in a resolution, but only after federal diplomatic intervention in response to continued Mexican protests (Friedkin 1988). This change of position occurred amidst an oil crisis. U.S. supplies were in doubt and Mexican oil production was substantial.

Ultimately, the United States and Mexico negotiated agreements (called "minutes") supplementing the 1944 treaty. These agreements were manifested in Minutes 218, 241, and 242 to the 1944 treaty. The treaty did not address salinity, although it had been a cause for some concern among treaty negotiators. Minute 218 in 1965 provided for a 5-year program to manage discharges of return flows differently in order to reduce the salinity of deliveries to Mexico. This brought levels down to approximately 1,240 ppm of dissolved solids. Minute 241 in 1971 then agreed to bypass even more of the worst return flows so that the most saline waters would be put in the river channel below the Mexican diversion. And in 1972, Minute 242 guaranteed that the United States would deliver water to Mexico at approximately the level of salinity of the water delivered to Mexico at the last U.S. diversion point, Imperial Dam (Dregne 1975; Holburt 1975; Friedkin 1988).

Implementation of the salinity minutes to the treaty required measures within the United States to reduce and reverse the amount of salt being added to the river. For many years now, the United States has bypassed large quantities of brackish water beyond the Mexican intake. This has resulted in delivery of 125,000 to 140,000 acre-feet of water per year to Mexico beyond the amounts promised in the treaty—a concept objectionable to the basin states. In addition, the United States had to construct elaborate facilities to deal with naturally saline waters destined for the river and to prevent salt loading from irrigation return flows. To accomplish these ends, the Colorado River Basin Salinity Control Act was passed in 1974, authorizing major expenditures of federal money (*Colorado River Basin Salinity Control Act* 1974).

Today, another international conflict over the river is brewing because of the destruction of some of the most important ecosystems of the Colorado River Delta region—ecosystems that are shared by the United States and Mexico. Scientists and environmental organizations in both countries have become alarmed over the problem (Luecke et al. 1999; Pitt et al. 2000). The Cucupá Indians in Mexico, who rely on the water flowing to the delta for their livelihoods, which historically have been based on agriculture and fishing, are threatened as well (Valdés-Casillas et al. 1998). Recent flows in excess of the required minimum delivery to Mexico have revived ecosystems in the delta, and this has dramatically illustrated the sensitivity of the area to climate variability. Without a change in the way water is managed, however, the delta inexorably will again confront losses to fish and wildlife and their natural habitats.

## 4. THE NEXT CRISIS IN U.S.-MEXICO RELATIONS CONCERNING THE COLORADO RIVER

As Aldo Leopold experienced in 1922, the Colorado River Delta was one of the world's great desert estuaries, supporting 780,000 hectares of wetlands at Rio Hardy, El Doctor, and Ciénega de Santa Clara and remarkable riparian habitat (Luecke et al. 1999; Glenn et al. 1996). The delta historically was sustained by periodic floods and tides. Before Glen Canyon Dam was closed, it was not unusual for 4 to 6 million acre-feet to flood across the area (see Fig. 2).

The delta and its associated wetlands have provided habitat for large numbers of shorebirds and migratory waterfowl, and this area is known for bird hunting. The tidal area periodically becomes estuarine habitat during high tides; it provides spawning areas for the totoaba and habitat for shrimp and corvina.

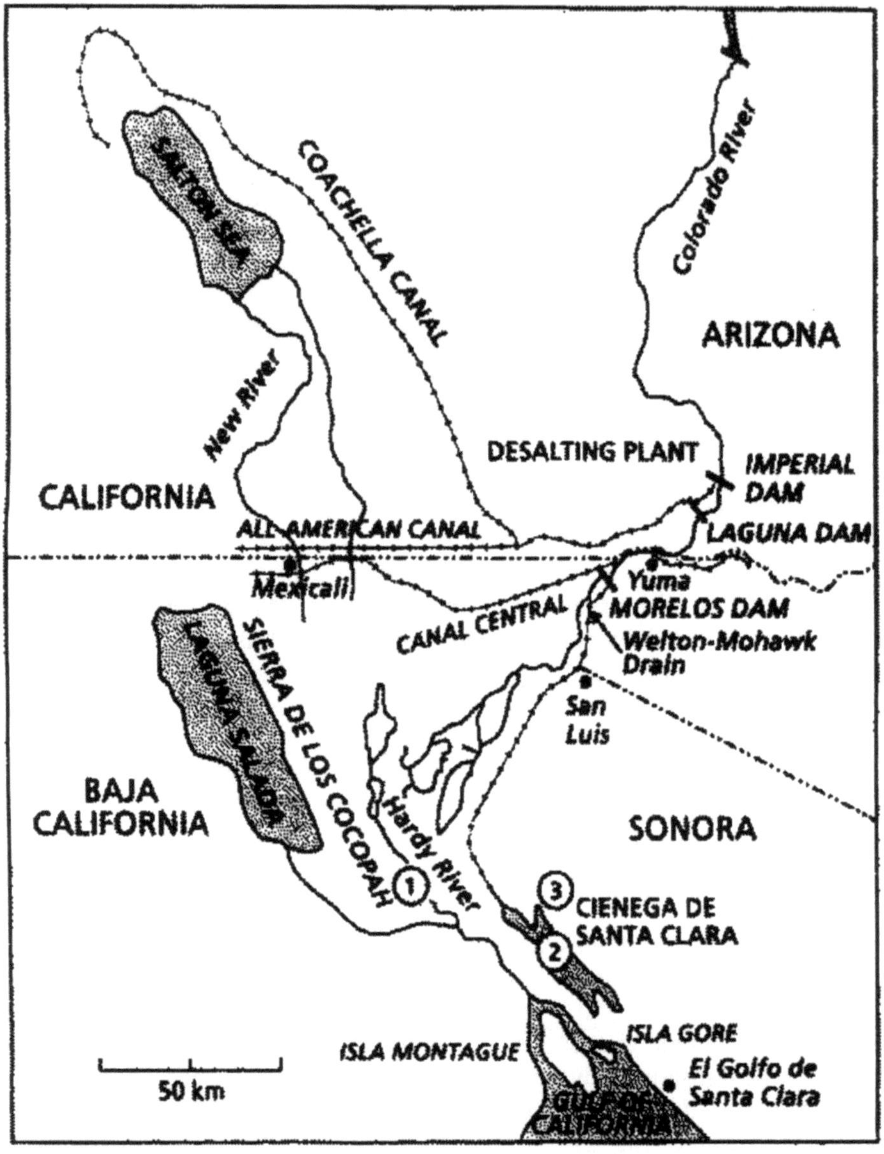


**1 = Rio Hardy wetlands, 2 = Cienega de Santa Clara, 3 = El Doctor wetlands**

*Figure 2.* Map of the Colorado River Delta Region

Because of the reduced flows of water in the Colorado River, dramatic changes occurred in habitat and the viability of flora and fauna throughout the area. Significant wetland areas nevertheless survived; others grew. About 2.5 million acres of wetlands historically were created by nutrient-rich water and silt from the river. Today, the Ciénega de Santa Clara remains the largest and most important wetland area. It has been sustained and enlarged by the Wellton-Mohawk Outlet Drain Extension (Glenn et al. 1992, 1996; Zengel et al. 1995). The drain was constructed as part of the salinity control effort; by putting the brackish irrigation return flows in a canal that bypassed the point on the river where Mexican diversions are made, the United States was able to comply with its commitment to reduce the concentrations of salts. This vitally important wetland area depends on these discharges.

By the mid-1980s, the Colorado River ended as a trickle in barren mud or salt flats (Pitt et al. 2000). The old river channels of the Colorado River and Rio Hardy became highly saline from tail waters and agricultural drainage. Desiccation of the area's ecosystems was caused by upstream management practices in the United States and by Mexico's complete utilization of its 1.5 million acre-feet on farms in the Mexicali area (Pitt et al. 2000). The consequences could be seen in the loss of wetlands, changes in the vegetation of the delta's riparian corridor, and declining fisheries in the estuary at the Sea of Cortez. Several species were driven to the brink of extinction. The reduction of habitat coupled with overfishing in the Upper Gulf has seriously depleted the once-abundant totoaba fish (Wilson 1994). The vaquita, a tiny porpoise inhabiting the gulf, is the world's most endangered sea mammal. Two species of birds that depend on delta ecosystems, the Yuma clapper rail and the willow flycatcher, are also endangered.

All of the elements of the ecosystem of the delta began to revive in 1983 thanks to several years of high runoff (Luecke et al. 1999). The wet years of 1983 to 1987 have been attributed to El Niño. There were also high flows in the 1990s. The recent flood years revitalized riparian areas that include gallery forests of cottonwood and willow in a corridor covering some 82,000 acres, running approximately 70 miles from Morelos Dam to Rio Hardy. Although the ecosystems of the delta have not been restored to their original condition, the surplus flows occurring in the past 17-year period resulted in a major revival of ecosystems.

The revitalization of the region's ecosystems also has given hope to small communities in Mexico that economically viable enterprises could survive there. Impoverished Cucupá Indians could again pursue traditional fishing and gathering. The rich ecosystems might also support tourism by visitors interested in bird-watching, hunting and fishing, and other recreational pursuits.

Official recognition for the ecological importance of the delta and the upper gulf area came in 1993 with the designation of the Alto Golfo de California y Delta del Rio Colorado as a biosphere reserve under a United

Nations program that recognizes such extraordinary areas as deserving protection (Getches 1997). The biosphere reserve encompasses 2.3 million acres, and a significant part of it is in the Colorado River Delta. Rules for the core area limit use to small-scale shellfish harvesting and low-impact tourism. The Mexican government provides substantial funding for the biosphere reserve; the United States participated in celebrating and supporting its establishment.[2]

## 5. CONTINUING U.S. NEGLECT OF MEXICAN INTERESTS IN THE COLORADO RIVER DELTA

The United States' continuing neglect of Mexican interests in its Colorado River management decisions increases the likelihood of a crisis over the situation in the delta. In 1999, a consortium of university scholars and environmental organizations from both nations prepared a major study of the delta. The study concluded that "The principal threats to delta habitats are insufficient and unreliable water supplies and their relatively low water quality" (Luecke et al. 1999). The environmental impacts on the delta appear to be primarily a function of water management in the United States. Although almost 90% of the consumptive use of Colorado River water occurs in the United States, activity in Mexico is also to blame for the delta's plight. Mexico annually uses all the water allocated to it, and Mexican farmers have built levees to limit the extent of flooding of cultivated lands, which has caused the elimination of wetlands and curtailment of riparian vegetation.

The 1999 study, based on limited fieldwork and the data then available, determined that annual flows of 32,000 acre-feet and periodic pulse flows of about 260,000 acre-feet could maintain the most important habitat zones in the delta. Given the magnitude of annual flows and the extensive

---

[2] It should be noted that there are multiple tensions between the United States and Mexico concerning the groundwater resources that are shared along the California-Mexico border (Mumme 2000). One significant dispute involves the lining of the All-American Canal, which will deprive the transboundary aquifer of the seepage that has been a major source of recharge (Hayes 1991). There is also a major issue over Mexico's obligations to ensure deliveries of water in the Rio Grande in Texas. A treatment of these important transboundary water issues is beyond the scope of this chapter, but solutions to the problem of the delta might appropriately be linked to finding solutions to them.

consumptive use of the river, these streamflow objectives appear relatively modest and attainable without major dislocations.

United States activities and programs, however, persist in ignoring the delta and its ecology. In particular, the delta is at the mercy of the U.S. Bureau of Reclamation's plans for operating the dams on the Colorado River. Other programs in the United States could offer opportunities for creative solutions to the delta problem but are likewise proceeding without considering the issue.

## 5.1. Operating Criteria for Federal Dams

The Bureau of Reclamation, within the U.S. Department of the Interior, operates the dams on the Colorado River according to technical criteria governing releases of water and regimes for storing water. These criteria have been designed primarily to optimize power production and assure deliveries according to interstate compacts. The goal for operating criteria is to assure that water is delivered consistent with the law of the river.

For many years, California has been using water far in excess of its entitlements under the law of the river in order to meet the demands of the growing southern California area. Its right to take 4.4 million acre-feet a year is based on the 1922 Colorado River Compact and the Boulder Canyon Project Act as interpreted in the case of *Arizona v. California* (1963). But California has taken as much as 5.2 million acre-feet per year in recent years (Pitt et al. 2000). The other states in the basin have been concerned that if and when they reach the point that they need to use their full entitlement of the river's water, California will have become so dependent that it will be impossible for the other lower basin states to reclaim their allocations, as specified by the law of the river. For many years, this was merely a concern for the lower basin states. That is, neither Arizona nor Nevada used its full entitlement and so, while California was using more than its share, the aggregate use in the lower basin did not exceed the 7.5 million acre-feet allocated by the 1922 compact. Now, Arizona and Nevada are nearing full utilization of their entitlements and they want to limit California's use. At the same time, the upper basin states resist allowing the lower basin states to encroach on their unused legal entitlements. Thus, all of the other states brought pressure on the Secretary of the Interior to take steps to ensure that California reduces its consumption to within its 4.4 million acre-feet entitlement (Lochhead 2001).

According to the law of the river, in order for California to continue using more than its share, the Secretary of the Interior must declare a "surplus" of water available for use (*Colorado River Basin Project Act* 1968, sec. 602). The secretary has discretion to set the quantity of water that ought to be stored in the reservoirs of the Colorado River for future deliveries. If the secretary's estimates of future rainfall and inflow to the reservoirs justify such a determination, the secretary can allow reservoirs to be drawn down in any particular year in anticipation of their being replenished sufficiently to meet future demands of the various states. California is then allowed to use the "surplus" water. In recent years the secretary has several times declared a "surplus" condition to allow California to continue exceeding its entitlement.

In 2001, the secretary adopted rules to guide future decisions on whether a surplus exists (Colorado River Interim Surplus Guidelines 2001). The purpose of the guidelines is to make it easier for California to get surplus water and at the same time to motivate the state to phase out its excessive use, giving it 15 years to come into compliance with the law of the river. In making these rules, the secretary had an opportunity to consider environmental issues and to rule specifically that the environmental needs of the Colorado River be met before a surplus is declared. All the states of the basin ultimately supported the idea that the secretary should adopt surplus guidelines, but they adamantly opposed including any environmental factors in the determination of surplus. Secretary Babbitt took the position that the United States should not mitigate the impacts its actions may have in a foreign country, and resisted the idea that he should use the rulemaking as anything other than a way to help California overcome its dependence on using more than its legal entitlement to Colorado River water (D'Amours 2001).

Immediately after the rules were adopted, Mexico protested. A "diplomatic note" was delivered by a Mexican diplomat to the United States Department of State. The United States did not respond to the note for nearly 1 year, then rejected Mexico's request that it reconsider the decision.

## 5.2 The California Plan

California completed a major plan dealing with the larger question of how it will meet its demands for water consumption without exceeding its share of the Colorado River water. The California Plan describes methods California can use to fulfill its water demands without exceeding its share of the water under the law of the river. The plan includes measures such as trades of water among Colorado River water users, conservation efforts, construction of new facilities, and reuse of municipal water (Marcum 1998).

The California Plan and the surplus guidelines are interdependent. Under the guidelines, California must successfully implement an effective plan for reducing Colorado River water use. If it fails, the guidelines allow the secretary to refuse to declare that "surplus" water is available for California under the liberal guidelines. The California Plan does not anticipate any new environmental demands on the river. In this sense, its success depends in part on continuing to keep water away from the delta of the Colorado River until all of California's demands are met.

## 5.3. Salton Sea Planning Process

The Salton Sea Reclamation Act was passed in 1998 to deal with an environmental crisis in California's largest lake (*Salton Sea Reclamation Act of 1998*). The Salton Sea is a body of water sustained largely by agricultural wastewaters discharged from the Colorado River and used in irrigating Imperial Valley crops. The fundamental problem is that salts and other contaminants have become concentrated in the sea. Although the Salton Trough was accidentally filled with Colorado River water almost 100 years ago, during a major flood episode on the Colorado, it is now essential habitat for migratory birds. Much of the habitat on which birds once relied has been destroyed by development (Reisner 1986; Pitt et al. 2000).

The state and federal governments are committed to maintaining the level of the sea and restoring its quality so that it can sustain fish populations and remain a viable part of the ecosystem, on which hundreds of thousands of migratory birds depend. They have committed hundreds of millions of dollars to the effort. The plan may require a continued supply of water from the Colorado River. Although this would seem to provide an opportunity for developing a coordinated plan for sustaining the nearby and ecologically related Colorado River Delta, the mission of the current program does not include measures to protect the delta. This leaves the delta theoretically in competition with the Salton Sea for environmental water.

## 5.4. Lower Colorado River Multi-Species Conservation Program

Endangered species problems have plagued most of the Colorado River Basin. Almost the entire river and its major tributaries have been designated critical habitat for several species of endangered fish. Species that are in danger of extinction are designated as "threatened" or "endangered" by the Secretary of the Interior under the Endangered Species Act of 1973 (ESA). The ESA requires all federal agencies to consider the impacts of their decisions on endangered species and their habitats. If a federal action will jeopardize the continued existence of an endangered species, agencies cannot proceed with a federal project or with permitting a private development but must consider "reasonable and prudent alternatives" to the proposed action. Attempts to avoid jeopardy without hampering development plans for further use of the Colorado River have produced four different processes to plan for the recovery of the fish in four different areas of the basin (Wigington and Pontius 1996).

The most recently initiated, and probably the most ambitious, of the plans to recover endangered species is the Multi-Species Conservation Program (MSCP) (*Federal Register* 1999). The MSCP was sparked by concerns that if new endangered species were listed, it could limit further development and use of Colorado River water and power, especially in southern California. Local and state agencies responded to these concerns by initiating a large planning effort in cooperation with the U.S. Fish and Wildlife Service. The participants initially included the lower basin states and other federal agencies. Representatives from environmental groups were not invited at first. The scope of the process is enormous. It includes the habitat of approximately 100 different species of fauna and flora. The program aims to protect not only species listed as endangered under federal and state laws but also candidate species and those that are sensitive and likely to be listed if conditions worsen. The idea is to avoid serious population declines by attending to the concerns of these species before they became endangered (D'Amours 2001).

At first, the MSCP proceeded without any environmental participants. Then two national environmental organizations were invited to the table. The environmentalists promptly objected to excluding habitat in Mexico from the MSCP, because the program covers the mainstream of the lower Colorado River from the Glen Canyon Dam but stops at the international boundary with Mexico. When the other participants refused to expand the process to include habitats that extended into Mexico, the environmental groups quit the process in 1998 (Getches 1997). After leaving the MSCP process, environmental interests continued to advocate expansion to include habitat in Mexico. The U.S. Fish and Wildlife Service held scoping hearings for the MSCP in 1999 but eventually concluded they would not expand the scope of the program.

The Defenders of Wildlife and other environmental organizations in the United States and Mexico sued the Department of the Interior in the summer of 2000 under the ESA. They claimed that the federal government was obligated to consult with the U.S. Fish and Wildlife Service concerning its activities that would affect endangered species habitat in Mexico. The consultation requirement exists under Section 7 of the ESA. In 1996–97 the Bureau of Reclamation consulted with the Fish and Wildlife Service regarding the operation of its dams and reservoirs during the estimated 5 years it would take to complete the MSCP, but it did not consider the impacts on endangered species in Mexico. The suit challenges the continuing position of the Department of the Interior that agency action that may affect endangered species within another country is not covered by the consultation requirement of the ESA, even when the government action that would jeopardize the continued existence of the species occurs in the United States. The environmental groups allege that this is contrary to the ESA (Pitt et al. 2000).

## 5.5. CALFED

The enormously complicated problem of protecting water quality in the San Francisco Bay Delta has grown into a statewide water planning process that indirectly affects the use of the Colorado River and therefore potentially affects the distant Colorado River Delta. The issue was triggered several years ago by impacts traceable to massive withdrawals of water for export to the Central Valley of California that were instead redirected to southern California municipal providers (Rieke 1996).

The Bay Delta is an area just above San Francisco Bay that serves as the conduit for deliveries of water released from huge reservoirs filled by northern California rivers. It is a sensitive and narrow bottleneck where water transfer and pumping operations can dramatically affect the intrusion of salt water from the bay into the estuary and river corridors, impact the rich farmlands that have been artificially created behind levees in the delta area, affect the quality of municipal and industrial water, and determine the fate of significant fish populations that depend on the related ecosystems. Addressing these problems requires adjusting the quantity and timing of deliveries of waters through the Bay Delta to meet the demands of the heavily water-dependent agricultural users in the Central Valley and the growing municipalities of southern California that are connected by aqueducts to water sources in the north. Therefore, any program for solving the Bay Delta water problem is a concern for southern California as well.

Because of the interdependence of uses and the diversity of the interest groups affected by the complicated Bay Delta problem, a multistakeholder process was created and has become essentially a major statewide water planning effort. It involves state and federal agencies, water users, and environmental interests. The intergovernmental nature of the process caused it to be dubbed "CALFED." The success of CALFED is important to users of the Colorado River because the potential demands of southern California on the Colorado River depend in part on the availability of northern California water that is the subject of CALFED's programs. CALFED has proceeded, however, without any consideration of possible delta needs (Michel 2000).

## 6. CLIMATE: THE ELEPHANT IN THE CLOSET

Climate variability has always been the "elephant in the closet" in efforts to allocate and manage the Colorado River. The law of the river itself is based on a major error in estimates of long-term average flows. When the negotiators of the 1922 compact allocated 15 million acre-feet among the seven states, they anticipated that natural flows would, on average, far exceed the total allocation. Indeed, they provided for another 1 million acre-feet of water to be allocated and indicated that any future allocations to Mexico—which were later quantified at 1.5 million acre-feet—would also be available. This total of 17.5 million acre-feet was within the reasonable expectations for river flows based on the historical record from 1896 through 1921 (Getches 1985).

Since the Colorado River Compact, however, dendroclimatological studies have reconstructed river flows over a 400-year period. These analyses of tree rings in the Southwest revealed that average annual flows, taking into account episodic drought periods, have been closer to 13.5 million acre-feet (Gregg and Getches n.d; Stockton and Jacoby 1976). Therefore, the river is grossly overallocated. The states of the Colorado River Basin nevertheless continue to hold fast to their interstate allocations and insist that they will be fully developed. They refuse to recognize that to do so is unrealistic and risky given the likely availability of water in the river over the long term.

There are several potential manifestations of climate variability that would exacerbate the Colorado River Delta problem. One such phenomenon is the enormous annual variation in flows. During the period of historical records for the Colorado River, virgin flows have been as low as 5.5 million acre-feet and as high as 24 million acre-feet. The response of water planners has been to construct storage facilities to bridge periods of low flows by taking advantage of above-average inflows in other years.

The presence of huge storage facilities with the exposure of large reservoir surfaces to evaporation has resulted in considerable evaporative loss. Today, evaporation from Colorado River reservoirs consumes approximately 2 million acre-feet. This is, of course, more water than is allocated annually to Mexico and more water than Utah, Wyoming, and Nevada consumptively use. Evaporation is a constant demand on water in storage and therefore it limits how long effective storage can be achieved. Experts have shown that the level of water development on the Colorado far exceeds the useful limits of reservoirs to provide drought protection (Hardison 1972).

The possibility of a severe, sustained drought in the basin brings these paradoxes into sharp relief. Prehistoric data based on tree-ring studies predict a future recurrence of droughts that are longer and more severe than those experienced during the period of record (Gregg and Getches n.d.). The reconstructed record of flows in the Colorado River Basin shows multiyear episodes of flows lower than any experienced in the twentieth century. These scenarios were not available when plans were made for development and use of the Colorado River. Yet, even today, although the data are no longer new, severe, sustained drought is still not being considered in Colorado River institutional arrangements, planning, or management.

Another climate variable is long-term climate change. Researchers agree that there will be long-term changes in the timing and quantities of flows in the Colorado River Basin, but they differ on the location and extent of impacts (Frederick and Gleick 1999). While reliable and precise predictions are not possible, the parameters suggested by different models and projections at least could be calculated and reported to those involved in the decision-making process. Along with the other, more demonstrable climate variability factors, the possibility of long-term climate change is ignored, like the elephant in the closet.

Climate variations of one or more of the kinds described will surely occur. One of the most palpable effects in the Colorado River Basin, unless a scheme is developed to sustain the area, will be a change in the ecosystems of the Colorado River Delta. Although the timing and severity of climate events are unpredictable, their occurrence is virtually certain, and the adverse consequences for ecosystems of the delta have been demonstrated in low-flow periods in the past. If the United States is forced to respond in the midst of crisis, the options may be more limited and more difficult to implement.

## 7. FINDING WATER IN AN OVERALLOCATED RIVER

If official resistance to addressing the delta issue is to be overcome, it will be necessary to find water and dedicate it to the delta. There are incipient efforts by nongovernmental organizations to restore physical habitat in Mexico (Pitt et al. 2000; Valdés-Casillas et al. 1998) Ultimately, however, an assured water flow will be needed to maintain ecosystems. The potential sources of the water are many. Among them are:

- Purchasing and transferring agricultural water rights in the United States from irrigation districts, Indian tribes, or others
- Purchasing and transferring water use rights from irrigators in Mexico
- Fallowing lands in either country to free up water
- Committing water to the delta that is now "lost" by the Bureau of Reclamation system in the United States because of operational imprecision
- Releasing water from dams in the United States
- Conserving water in Mexico's irrigation system and dedicating it to the delta
- Redirecting brackish water from agricultural return flows in both countries to delta ecological needs
- Diverting water from the New River, which now carries wastewater from Mexico into the United States
- Using treated water from new sewage plants in Mexico
- Pumping groundwater

In 2001 a group of researchers organized by Michael Clinton Engineering and funded by the Packard Foundation prepared a report assessing the feasibility of acquiring water immediately for ecological uses in the delta (Clark et al. 2001). After studying the physical, political, and legal impediments to many of the solutions listed above, they identified sources in both countries that could address the delta's needs with a minimum of transactional delays. The report showed that it was feasible to pursue temporary or long-term purchases of water in Mexico under existing law and use the water in the Rio Hardy riparian and wetland areas. In the United States, brackish water from drains in the Yuma area could be delivered by a new pipeline to delta wetlands, and releases of small amounts of water from main stem reservoirs could replace the brackish water that now is put into the river and delivered to Mexico. The total amounts of water that could be acquired under the proposal approximated the annual quantities that Luecke et al. (1999) estimated would sustain key ecosystems, although the researchers recognized that further study would be needed to determine whether the water from the particular sources was targeted optimally to achieve ecological goals. The study also advised that delivery of water acquired in both countries should be facilitated by a legal agreement between the two countries in the form of a minute to the 1944 treaty. This mechanism is discussed below in the section "Programs of the International Boundary and Water Commission" (IBWC).

The identification of immediate options for acquiring water suggests that, with sufficient capital, deliveries of water could begin soon. If financial support were not forthcoming from government agencies, the funds could come from private donors and foundations or from bilateral or multilateral institutions. If a program of ecologically targeted water deliveries were undertaken temporarily, it could aid scientific research on the effectiveness of measures to protect delta ecosystems. Given the lack of official engagement, a role for private efforts, facilitated by nongovernmental organizations, could be an important ingredient in addressing the delta issue. Nevertheless, the success of any program to secure water for the delta will depend on support within the institutional framework that governs the Colorado River.

## 8. LEGAL AND INSTITUTIONAL FRAMEWORKS FOR SOLUTIONS

No official government entity or official in either country has yet embraced the ideal of protecting the Colorado River Delta by assuring an appropriate delivery of water. There are current opportunities, however, for addressing the delta's ecological problems within established institutional contexts. Progress in each of these contexts is possible, but the most important existing institutional framework for addressing the delta issues comprehensively is the IBWC.

### 8.1. Transboundary Application of the Endangered Species Act

As was described earlier, environmentalists commenced a lawsuit in the United States District Court for the District of Columbia under the ESA. Section 7 of the Endangered Species Act requires federal officials to consult with the Fish and Wildlife Service to determine if their actions would jeopardize the continued existence of endangered species or their habitat and, if so, to find alternatives to such actions. The federal government is involved in several activities that could come within the scope of this claim and therefore, if the claim is sustained by the courts, it could implicate numerous U.S. actions.

Although the delta provides habitat for a number of endangered species in Mexico, the ESA has not yet been held to require consideration of the impacts on species in foreign countries. The courts, however, have never had to apply the ESA to situations where actions in the United States by a federal agency directly impacted habitat just below the border, even when it may be interconnected with habitat within the United States. The claim appears strong. If the environmental groups prevail in their claim, it would require the agencies to consult with the Fish and Wildlife Service, and if their proposed actions were found to jeopardize the continued existence of endangered species or their habitats, the agencies would have to find reasonable and prudent alternatives to their actions. This could open the door to development of a plan for getting water flows to the delta.

## 8.2. Revised Operating Criteria to Require Ecological Water for Mexico

The interim surplus guidelines that were adopted by the United States Secretary of the Interior disregard the delta, and the relaxed rules for declaring surpluses for the benefit of California could deprive the delta of water that would otherwise flow to it. (*Federal Register* 2000, 2001, 2002). A change in these rules could provide a source of water in higher than average flow years and prevent the harm to the delta. In practice, this would have little impact on the success of the California Plan to reduce water use. The small difference in the amount of surplus water available to California would provide significant protection to the delta but would not seriously inhibit California in its water use, because the amounts of water needed to benefit the delta are tiny compared to California's huge demands. A change in the guidelines could be provoked by continued Mexican objections, a new policy in the United States, or by a ruling in the ESA litigation that extends endangered species protections to extraterritorial impacts.

## 8.3. Expanded Multi-Species Conservation Program to Include Habitat in Mexico

Whether or not the ESA lawsuit forces the issue, the government or the other parties to the MSCP could decide to expand the planning process to include species and habitat south of the border. It was this omission that led to the refusal of environmental organizations to participate in the MSCP. The parties may reconsider their position in order to attain the credibility that the process would gain from the presence of those groups. It is unlikely, however, that the government will expand the MSCP study area to include any territory or species outside the United States without a court order requiring it to do so.

## 8.4. Binational Environmental Cooperation Efforts

Several binational institutional arrangements exist for improving environmental management of the Colorado River Delta. These arrangements are considered below.

### 8.4.1. Joint Declaration

On June 14, 2000, Secretary of the Interior Bruce Babbitt, and Mexican Secretary of the Environment, Natural Resources and Fisheries Julia Carabias signed a joint declaration between the United States and Mexico. The agreement pledges the countries to coordinate "policies leading to the conservation of natural and cultural resources," including pilot projects in Mexico. Projects to restore or protect the biological resources of the delta could be within the scope of this agreement. Such projects would also be consistent with, and perhaps gain financial or institutional support under, several programs or initiatives that deal with environmental concerns of the United States and Mexico in the border region.

### 8.4.2. Border Environment Cooperation Commission and the North American Development Bank

Efforts to deal with environmental concerns in the border region include the Border Environment Cooperation Commission (BECC) and the North American Development Bank (NAD Bank). The BECC is charged with the responsibility of certifying proposed border environmental infrastructure projects. To be eligible for BECC certification, a proposed project must comply with all applicable environmental laws and satisfy BECC criteria, including financial sustainability, technical feasibility, environmental sustainability, community participation, and promotion of public health (Milich and Varady 1999). Once a project is certified by the BECC, it becomes eligible for NAD Bank financing. The type of assistance that is necessary must be defined by governmental or nongovernmental organizations, or by additional binational accords. Presumably, BECC and the NAD Bank could participate in financing needed to support transactions or infrastructure that would make water available to the delta ecosystems.

### 8.4.3. Border XXI

The Border XXI Framework provides a vehicle for binational cooperation between various federal agencies on substantive environmental issues that affect the border region. The La Paz Agreement of 1983 established several workgroups under the Border XXI Framework. The Water Working Group is composed of the IBWC, the U.S. Environmental Protection Agency (EPA), the BECC, and Mexico's National Water Commission (CNA), and has focused on the development of BECC-certified projects (Mumme 1999).

### 8.4.4. North American Free Trade Agreement

The North American Free Trade Agreement (NAFTA) was ratified amidst objections that it would result in or perpetuate environmental problems. Thus, it was accompanied by a side agreement—the North American Agreement on Environmental Cooperation (NAAEC)—intended to ensure that liberalization of trade between the nations did not create an incentive to lower the threshold of environmental protection (Mumme and Duncan 1996). In particular, concerns were expressed and efforts were made to deal with economic activity in the border region. In 1994, the Commission on Environmental Cooperation (CEC) was created under the authority of the NAAEC to "promote environmental protection in the North American region, support NAFTA's environmental purposes, strengthen cooperation between the parties in the development of environmental policies and procedures, and enhance compliance with environmental regulations within the region" (Mumme and Duncan 1996). As is the case with the other border environmental initiatives, the Colorado River Delta has not yet been a subject of CEC attention.

### 8.4.5. Programs of the International Boundary and Water Commission

The IBWC is a binational entity that assists in monitoring and facilitating the fulfillment of the 1944 treaty between the United States and Mexico. The IBWC's duties include resolution of disputes related to observance of the treaty's water delivery requirements. Although IBWC is not thought of as performing environmental protection functions, its mission has gradually expanded to include not only water deliveries but also related environmental issues like Colorado River salinity, sediment removal, and wastewater treatment (Pitt et al. 2000). These water quality responsibilities have, at least indirectly, raised the environmental consciousness as well as the technical capacity of the commission.

The IBWC would be an appropriate vehicle for addressing the problems of the Colorado River Delta. On December 12, 2000, the two countries, acting through the IBWC, adopted Minute 306, recognizing a shared interest in "the preservation of the riparian and estuarine ecology" of the delta. It promised cooperation in using a binational task force to study the effects of flows on the delta's ecology, and promised an exchange of information that includes nongovernmental organizations. Most significantly, the countries committed to joint studies of "possible approaches to ensure use of water for ecological purposes" in the delta and formulation of recommendations for cooperative projects. Motivated by Minute 306, IBWC sponsored a Symposium on the Colorado River Delta in Mexicali on September 11–12, 2001, which was attended by hundreds of invited participants from both countries. Although representatives of state governments were cautious about accepting any United States obligation to solve the delta problem, all seven Colorado River Basin states participated in order to expand their knowledge of the issues and possible approaches to them.

Because solving the delta problem will require that water be delivered to ecosystems there, and because water deliveries are governed by treaty, the two nations must agree on the framework for providing and ensuring such deliveries as intended. A formal agreement through a minute to the treaty would pledge the cooperation of the two countries and define the actions necessary to ensure the proper use of water committed to ecological purposes in the delta. The minute could also specify the means for administering any water deliveries to the delta and deal with how any transboundary conveyances of water are to be treated under the 1944 treaty.

## 9. CONCLUSIONS

The tenuous situation of the Colorado River ecosystems is the latest in a series of transboundary issues between the United States and Mexico over the Colorado River. It has not yet ripened into conflict, in part because climate patterns have allowed delta ecosystems to flourish temporarily. In the past, Colorado River issues created serious stress on U.S.-Mexico relations before they were resolved. The approach to the delta question can be different, however. First, although it almost surely will require a commitment of some water, the quantities needed are relatively small. Second, nongovernmental organizations and university researchers have taken the lead on the issue, first bringing it to public notice, and, now, pressing an agenda that could result in designing rational and effective responses.

The two countries are not specifically engaged on the Colorado River Delta issue beyond a rather stern and nonpublic diplomatic exchange—Mexico objected to the United States' decisions to intensify consumptive use of the river without consideration of the delta, and the United States denied any responsibility for solving the problem.

It is reasonable to predict that as climate variability inexorably deprives the delta of flows and ecosystems and dependent species die out, Mexican protests will intensify. By then, the United States' water demands will be even firmer and the negotiation climate will likely be less favorable than it is now. Solutions to the delta problem should also be stimulated by knowledge of the reality of climate variability.

Although neither country has focused significant official attention on the delta matter, the United States and Mexico are engaged in several environmental quality programs that require joint efforts. Moreover, in IBWC Minute 306 they have committed themselves to binational cooperation at least in research on the issue. This is a promising start, but much more than research needs to be done. In particular, dedicating water to ecological uses will require financing, local understanding of the issue in Mexico, legal assurances of actual delivery, and creative application of the law of the river in the United States. Nongovernmental entities can play an important role in designing practical approaches, but government support or approval will be required to accomplish a final solution. States may resist using even a drop of water in a way that does not comport with their traditionally inflexible interpretation of the laws allocating water within the United States. That is now the position of some of them. Other basin states are showing interest in and understanding of the delta issue, and they may provide support for a reasonable solution. If, however, the states are intransigent and their position is allowed to drive U.S. policy, it could become the largest obstacle to the United States' acceptance of a solution. Then, it will again take a crisis in international relations to provoke a practical solution to a U.S.-Mexico problem concerning the Colorado River.

## 10. REFERENCES

Bates, S.F., D.H. Getches, L.J. MacDonnell, and C.F. Wilkinson. 1993. *Searching Out the Headwaters: Change and Rediscovery in Western Water Law*. Washington, D.C.: Island Press.

*Boulder Canyon Project Act*. 1928. *U.S. Code*. Vol. 43, sec. 617.

Bloom, P.L. 1986. Law of the River: A critique of an extraordinary legal system. In, Weatherford, G.D., and F.L. Brown (eds.), *New Courses for the Colorado River: Major Issues for the Next Century*. Albuquerque: University of New Mexico Press.

Clark, J., M. Clinton, P. Cunningham, D.H. Getches, J.L. Lopezgamez, M.H. McKeith, L.O.Martinez Morales, B. Bogado, J. Palafox, and C. Valdés-Casillas. 2001. *Immediate Options for Augmenting Water Flows to the Colorado River Delta in Mexico*. Report prepared for The David and Lucile Packard Foundation, Los Altos, California.

*Colorado River Basin Project Act*. 1968. U.S. Public Law 537. 90th Congress, 2nd sess., 30 September 1968.

*Colorado River Basin Salinity Control Act*. 1974. U.S. Public Law 320. 93rd Congress, 2nd sess., 24 June 1974.

*Colorado River Compact of 1922*. Colorado Revised Statutes, Sec. 37-61-101.

Culp, P.W. 2001. *Feasibility of Purchase and Transfer for Instream Water for Instream Flow in the Colorado River Delta, Mexico*. Tucson, Arizona: Udall Center Publications.

D'Amours, P. 2001. The Colorado River Delta. *Colorado Journal of International Environmental Law and Policy*. Yearbook 2000: 183–191.

Dregne, H.E. 1975. Salinity aspects of the Colorado River Agreement. *Natural Resources Journal* 15(1): 43–53.

*Federal Register*. 2002. Review of Existing Coordinate Long Range Operating Criteria for Colorado River Reservoirs. 67: 1986–1988.

*Federal Register*. 2001. Colorado River Interim Surplus Guidelines. 66: 7772–7782.

*Federal Register*. 2000. Colorado River Interim Surplus Criteria. 65: 48531–48537.

*Federal Register*. 1999. Multi-Species Conservation Program (MSCP) for the Lower Colorado River, Arizona, Nevada, and California. 64: 27000–27002.

Fradkin, P. 1995. *A River No More*, 2d ed. Berkeley: University of California Press.

Frederick, K.D, and P.H. Gleick. 1999. *Water and Global Climate Change: Potential Impacts on U.S. Water Resources*. Arlington, Virgina: Pew Center on Global Climate Change.

Friedkin, J.F. 1988. The international problem with Mexico over the salinity of the Lower Colorado River. In, Getches, D.H., *Water and the American West: Essays in Honor of Raphael J. Moses*, Boulder, Colorado: Natural Resources Law Center.

Getches, D.H. 1985. Competing demands for the Colorado River. *University of Colorado Law Review* 56(3): 413–479.

Getches, D.H. 1997. Colorado River governance: Sharing federal authority as an incentive to create a new institution. *University of Colorado Law Review* 68(3): 573–658.

Glenn, E.P., R.S. Felger, A. Búrquez, and D.S. Turner. 1992. Cienega de Santa Clara: Endangered wetland in the Colorado River Delta, Sonora, Mexico. *Natural Resources Journal* 32(4): 817–824.

Glenn, E.P., C. Lee, R. Felger, and S. Zengel. 1996. Effects of water management on the wetlands of the Colorado River Delta, Mexico. *Conservation Biology* 10(4): 1175–1186.

Glennon, R.J., and P.W. Culp. 2002. The Last Green Lagoon: How and why the Bush Administration should save the Colorado River Delta. *Ecology Law Quarterly* 28(4): 903–992.

Gregg, F., and D.H. Getches. n.d. *Severe, Sustained Drought in the Southwestern United States: Phase I Report*. A report on research supported by the U.S. Department of State, Man and Biosphere Program.

Hardison, C.H. 1972. Potential United States water-supply development. *Journal of the Irrigation and Drainage Division*. 1972 Proceedings, American Society of Civil Engineers (September): 479–491.

Hayes, D.L. 1991. The All-American Canal Lining Project: A catalyst for rational and comprehensive groundwater management on the United States–Mexico border. *Natural Resources Journal* 31(4): 803–828.

Holburt, M.B. 1975. International problems of the Colorado River. *Natural Resources Journal* 15(1): 11–26.

Hundley, N., Jr. 1975. *Water and the West: The Colorado River Compact and the Politics of Water in the American West*. Berkeley: University of California Press.

Hundley, N., Jr. 1992. *The Great Thirst: Californians and Water, 1770s–1990s*. Berkeley: University of California Press.

Leopold, A. [1949]. *A Sand County Almanac and Sketches Here and There*. Reprint, New York: Oxford University Press.

Lochhead, J.S. 2001. An Upper Basin perspective on California's claims to water from the Colorado River. Part I: The Law of the River. *University of Denver Water Law Review* 4(2): 290–330.

Luecke, D.F., J. Pitt, C. Congdon, E. Glenn, C. Valés–Casillas, and M. Briggs. 1999. A Delta Once More: Restoring Riparian and Wetland Habitat in the Colorado River Delta. Environmental Defense Fund Report, available from EDF Publications, 1875 Connecticut Avenue, N.W., Washington, D.C. 20009.

Marcum, C. 1998. "River Master" Takes Charge. *South Carolina Environmental Law Journal* 7(1): 123–128.

Martin, R. 1989. *A Story that Stands Like a Dam*. New York: Holt.

Meyers, C.J. 1966. The Colorado River. *Stanford Law Review* 19: 1–75.

Meyers, C.J., and R.L. Noble. 1967. The Colorado River: The Treaty with Mexico. *Stanford Law Review* 19: 367–419.

Michel, S.M. 2000. Defining hydrocommons governance along the border of the Californias: A case study of transbasin diversions and water quality in the Tijuana–San Diego metropolitan region. *Natural Resources Journal* 40(4): 931–972.

Milich, L., and R.G. Varady. 1999. Openness, sustainability, and public participation: New designs for transboundary river basin institutions. *Journal of Environment & Development* 8(3): 258–306.

Miller, T.O., G.D. Weatherford, and J.E. Thorson. 1986. *The Salty Colorado*. Washington, D.C: The Conservation Foundation and the John Muir Institute.

Mumme, S.P. 1999. Managing acute water scarcity on the U.S.-Mexico border: Institutional issues raised by the 1990's drought. *Natural Resources Journal* 39(1): 149–166.

Mumme, S.P. 2000. Minute 242 and beyond: Challenges and opportunities for managing transboundary groundwater on the Mexico-U.S. border. *Natural Resources Journal* 40(2): 341–378.

Mumme, S.P., and P. Duncan. 1996. The Commission on Environmental Cooperation and the U.S.-Mexico border environment. *Journal of Environment & Development* 5(2): 197–215.

Pitt, J., D.F. Luecke, M.J. Cohen, E.P. Glenn, and C. Valdés-Castillas. 2000. Two countries, one river: Managing for nature in the Colorado River Delta. *Natural Resources Journal* 40(4): 1–33.

Rieke, E.A. 1996. The Bay-Delta Accord: A stride toward sustainability. *University of Colorado Law Review* 67(2): 341–369.

Reisner, M. 1986. *Cadillac Desert: The American West and Its Disappearing Water*. New York: Viking Press.

*Salton Sea Reclamation Act of 1998*. U.S. Public Law 372. 105th Congress, 2nd sess., 12 November 1998.

Stockton, C.W., and G.C. Jacoby, Jr. 1976. Long-term surface water supply and streamflow trends in the Upper Colorado River Basin based on tree-ring analysis. *Lake Powell*

*Research Project Bulletin 18*, Institute of Geophysics and Planetary Physics, University of California, Los Angeles.

Treaty Respecting Utilization of Waters of the Colorado and Tijuana Rivers and of the Rio Grande. February 3,1944. *U.S. Statutes at Large* 59: 1219–1267.

*Upper Colorado River Basin Compact*. 1949. *U.S. Statutes at Large* 63: 31–43.

Valdés-Casillas, C., E.P. Glenn, O. Hinojosa-Huerta, Y. Carillo-Guerrerro, J. García-Hernandez, F. Zamora-Arroyo, M. Muñoz-Viveros, M. Briggs, C. Lee, E. Chavarría-Correa, J. Riley, D. Baumgartner, and C. Condon. 1998. *Wetland Management and Restoration in the Colorado River Delta: The First Steps*. Special publication of CECARENA-ITESM Campus Guaymas and NAWCC, Mexico.

Ward, E. 1999. Two rivers, two nations, one history: The transformation of the Colorado River Delta since 1940. *Frontera Norte* 11: 113–140.

Wigington, R., and D. Pontius. 1996. Toward range-wide integration of recovery implementation programs for the endangered fishes of the Colorado River. In, *The Colorado River Workshop: Issues, Ideas, and Directions*. Sponsored by the Grand Canyon Trust under a cooperative agreement with the Bureau of Reclamation.

Wilson, F. 1994. A fish out of water: A proposal for international instream flow rights in the lower Colorado River. *Colorado Journal of International Environmental Law and Policy* 5(1): 249–272.

Zengel, S.A., V.J. Meretsky, E.P. Glenn, R.S. Felger, and D. Ortiz. 1995. Ciénega de Santa Cara, a remnant wetland in the Río Colorado Delta (Mexico): Vegetation distribution and the effects of water flow reduction. *Ecological Engineering* 4: 19–36.

9

# RESERVOIR MANAGEMENT IN THE INTERIOR WEST

*The Influence of Climate Variability and Functional Linkages of Water*

Andrea J. Ray
*NOAA Climate Diagnostics Center, Boulder, Colorado, U.S.A.*

**Abstract** Two trends in the Gunnison Basin of western Colorado are increasing the sensitivity of reservoir systems to climate variability. These trends are increasing water utilization within the basin, especially for environmental purposes, and the increasing importance of functional linkages of the basin's water to other places in the U.S. West. There is a potential for seasonal climate forecasts to provide advance guidance of wet or dry years, which could allow managers to better plan for dry conditions, or to take advantage of wet conditions. Trends in this basin mirror similar trends across the interior West, and thus it is valuable to examine the Gunnison Basin to understand how climate variability interacts with critical water issues facing the larger region, and how this information might be incorporated into decision making for reservoir management.

## 1. INTRODUCTION

The Gunnison River Basin in western Colorado is a microcosm of water issues in the interior West, where water use is intensifying to meet uses that were unforeseen in original water resources planning. Intensification of water management refers to increasing (or changing) the purposes for which the same water is used rather than developing new water resources (White 1961; MacDonnell 1999). As intensity of use increases, sensitivity to climate variability may also increase, because there is less capacity to absorb shortages in the system (Riebsame 1990) compared to the water avail-



*Henry F. Diaz and Barbara J. Morehouse (eds.), Climate and Water*, 193-217.

able. As an externality, beyond the control of water resources managers, climate variability may create or exacerbate water issues within the basin.

Two societal trends are intensifying water management. The first is a move towards using water for environmental sustainability purposes, including recovery of endangered species, another is greater use of instream water for recreational and other purposes. Another factor is that the Gunnison Basin is being increasingly linked to human activities in other places far outside its hydrologic boundaries. Water originating in the Gunnison Basin is a source of salinity that contributes to water quality problems downstream. It is in demand for municipal use in the Front Range of Colorado, it is used for the production of hydropower marketed elsewhere, and it makes a significant contribution to some downstream ecosystems. Reservoir management in the Gunnison Basin is evolving in response to policies that promote uses of water both within the basin as well as areas outside its hydrologic boundaries.

Use of water beyond the physical boundaries of a basin is called a functional linkage by White (1961), who suggested that functional linkages between internal and external resource use are a key aspect of water planning. To the extent that functional linkages increase the demands on the Gunnison Basin's water, these linkages are another aspect of the intensification of water management. This intensification requires that water managers acknowledge that there are actions and consequences transcending many temporal and spatial scales, referred to as "cross-scale issues" by Pulwarty and Melis (2001).

Climate is an overarching physical process that can be thought of as a functional linkage: Climate affects the basin directly, as well as affecting places to which the Gunnison's water is linked. Thus, the basin's water management may be sensitive to climate variability due to climate anomalies that could be either internal to, or external to, the basin. Within the basin, climate variability increases competition for water, and may jeopardize implementation of environmental policies and lead to increased conflict over implementation. External to the basin, a hot summer in a distant place, for example, may create demands for hydropower created here.

The influence of climate variability on this basin, as well as on areas to which it is functionally linked, has been overlooked with respect to how that variability interacts with evolving reservoir management practices. This chapter focuses on reservoir management as a subset of water management, the climate impacts on the use of water stored in reservoirs, and the functional linkages of that water. Reservoir management is sensitive to climate variability both temporally (i.e., the wet and dry periods affecting the Gunnison Basin itself) and spatially (in any year climate may have different effects on the hydrology and reservoir capacity of the basin, as well as in other areas to which it is linked via water uses). Seasonal forecasts of climate variability may provide information on increased risks of wet or dry conditions up to a year in advance. Thus, there is a potential to use these forecasts to plan for water allocation during dry periods or to take advantage of wet periods to satisfy more uses. A critical water problems approach (Wescoat, 1991) was used to identify water management problems confronting the region, which are sensitive to climate variability. Key water and reservoir management issues that are sensitive to climate variability and that might benefit from improved climate information and products include:

- interstate and international water obligations (the Colorado Compacts, and treaties with Mexico) by allowing for better long-range planning for use of water in the Colorado Basin as a whole
- hydropower generation planning
- management of salinity and other water quality problems, by anticipating periods of low flow in which water quality problems are exacerbated
- environmental sustainability: ecosystem restoration and endangered species recovery programs. These policies require careful management of water so that both traditional and environmental uses of water can be met.

This study analyzes how climate variability affects these critical water issues and what climate information can be used in the reservoir management decision process to meet multiple and expanding water uses in the basin. Potential uses of climate information in the reservoir management decision process are presented as they relate to the evolution of reservoir management to meet environmental sustainability goals. Use of this climate information may lower the vulnerability of reservoir management to climate variability.

## 2. GEOGRAPHIC CONTEXT OF THE GUNNISON BASIN

The Gunnison River is formed by the confluence of the Taylor and East Rivers, with their headwaters in the Elk Mountains and Sawatch Range on the Continental Divide in central Colorado (Fig. 1). The two major tributaries, the North Fork of the Gunnison and the Uncompahgre River, drain the Elk and San Juan Mountains. Elevations range from over 14,000 ft (4,260 m) to 4,550 ft (1,387 m) at its confluence with the Colorado River near Grand Junction, Colorado. The Gunnison River contributes about 40% of the flow of the Colorado River at the Colorado-Utah border, an average annual discharge of 2.016 million acre-feet (maf) per year (1977–96), ranging from 0.061 to 3.460 maf (U.S. Fish and Wildlife Service [USFWS] 2001). Most of the runoff is due to snowmelt: Over 70% of the annual flow occurs between April and July (USBR Western Colorado Area Office, 1999). Three U.S. Bureau of Reclamation (USBR) reservoirs, collectively known as the Aspinall Unit, inundate about 40 miles of the river, and control about one-third of the total discharge of the river. The largest, Blue Mesa Reservoir, begins about 10 miles west of the town of Gunnison and has a capacity of approximately 940,000 acre-feet (af); Morrow Point and Crystal Reservoirs together have a live storage capacity of about 130,000 af (http://www.dataweb.usbr.gov).

About 71% of the 7,928-square-mile basin (20,530 $km^2$) is federal land (Knapp 1993). The largest areas are the Grand Mesa, Uncompahgre, and Gunnison National Forests, and Bureau of Land Management (BLM) areas. The National Park Service manages the land around the Aspinall Unit, known as the Curecanti National Recreation Area, and also the Black Canyon of the Gunnison National Park (BCNP). Downstream of the BCNP, the Gunnison Gorge National Conservation area is managed by the BLM. Ranching, agriculture, and recreation form the economic base of the basin. Recreation has increased in the basin, especially in the Upper Gunnison, where the annual value of water-based recreational uses is approximately $35 million (excluding skiing) and represents the primary source of revenue for Gunnison County (Curry and McClow 2001). Wildlife-related recreational activities include trout fishing on some rivers and reservoirs, as well as elk and deer hunting. Boating on the reservoirs and river rafting are also economically important recreational activities.

Large-scale water development in the Gunnison Basin was motivated by settlement of the land for ranching and agriculture. In 1903, The Uncompahgre Valley Project (UVP) was among the earliest projects of the newly established federal Reclamation Service (later the USBR). The UVP serves over 76,000 acres of project land in that valley; other USBR projects serve 52,000 acres in the Upper Gunnison Valley, and about 30,000 acres in the North Fork drainage (USBR, 2001). Federal funds were used to construct the Gunnison Tunnel, which diverts water from the Gunnison River just upstream from the Black Canyon to UVP canals near Montrose. In the 1930s, the USBR constructed Taylor Park Reservoir in the Upper Gunnison Basin to provide more water for the project. Water stored in the reservoir flows through the Taylor and Gunnison Rivers, and is diverted to the Uncompahgre Valley via the existing Gunnison Tunnel. Water is also diverted from the Upper Gunnison Basin to the east side of the Rocky Mountains: Three transmountain diversions transfer about 2,500 af/year to the Arkansas and Rio Grande Basins (USGS 1985).

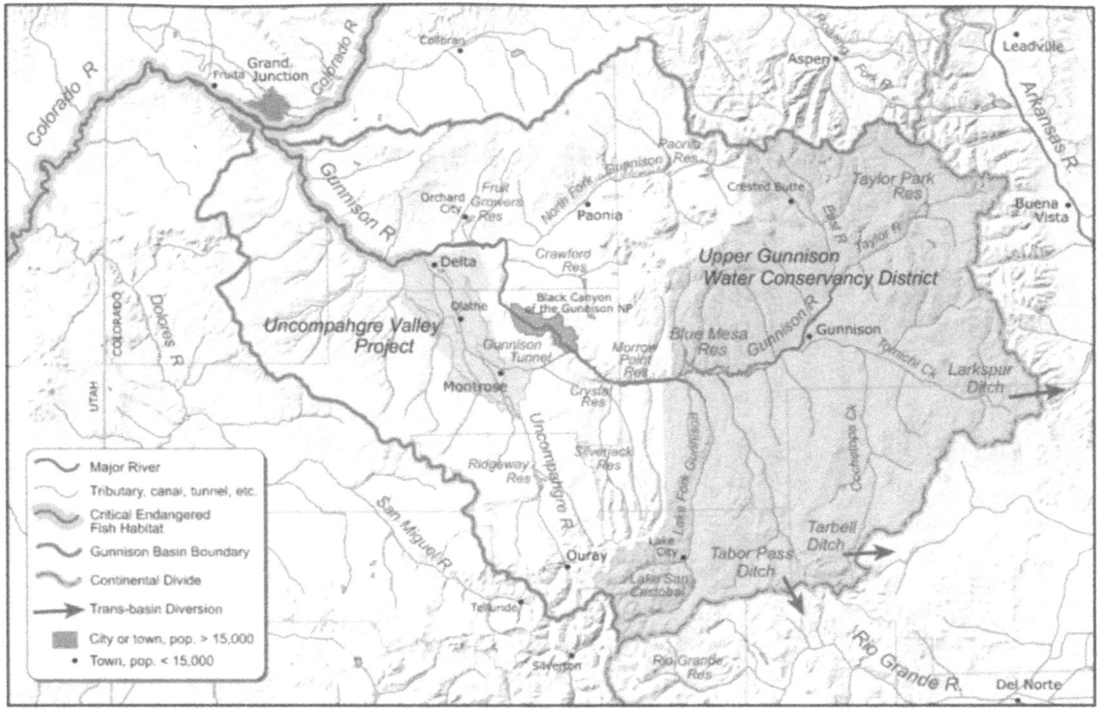


*Figure 1.* The Gunnison River Basin. The Uncompahgre Valley Project, the Upper Gunnison Water Conservancy District, and the Black Canyon of the Gunnison National Park are shaded in gray. Sources: Colorado Division of Water Resources, Office of the State Engineer; Utah Automated Geographic Reference Center; U.S. Fish and Wildlife Service; U.S. Bureau of Reclamation; U.S. Geologic Survey. Map by Thomas Dickinson, University of Colorado-Boulder.

The USBR Aspinall Unit was the next step in large-scale water development, authorized in 1956 as part of the Colorado River Storage Project (CRSP). The CRSP was intended to furnish long-term regulatory storage needed to permit states in the Upper Basin to meet their flow obligation to the Lower Basin, as defined in the Colorado River Compact, while still utilizing their allotment of water under the compact. The dams were constructed between 1962 and 1976. Benefits of the Aspinall Unit within the basin are control of flooding, development of recreation opportunities, production of electricity, and conservation of fish and wildlife. Revenues from hydropower generation fund repayment to the government for the costs of irrigation project development.

Virtually all of the water in the Gunnison River is now allocated. In 1991, in a case related to a proposed transmountain diversion from the Upper Gunnison Basin, a Colorado water court found that less than 20,000 af of water was available from the Upper Gunnison Basin for future appropriation (Board of Commisioners of Arapahoe County vs. Crystal Creek Homeowners Assn, 2000).

## 2.1 Critical Water Problems for the Gunnison Basin

Water issues in the Gunnison Basin are a microcosm of those in the state of Colorado and the Colorado Basin. Major themes in the state of Colorado include trans-basin diversions, environmental protection, water quality, and interstate obligations, including the Colorado Compact (Nichols et al. 2001). Major issues for the larger Colorado Basin include the potential for the Upper Basin states to fully develop their compact allotments; salinity control; Indian water rights; and ecosystem sustainability, including endangered fish recovery programs, Colorado River Delta ecosystem restoration, and the Grand Canyon Adaptive Management Program (Pontius 1997). Many of these problems are manifested in the Gunnison Basin: Its naturally saline soils are a source for salinity; it is a source of small trans-basin diversions and a potential source for more diversions; and it is important for management of interstate obligations due to its position close to the Colorado-Utah border. Ongoing environmental sustainability efforts include the Upper Colorado Recovery Implementation Plan for endangered fish, and an effort to restore the natural resources of the Black Canyon of the Gunnison. The Gunnison Basin also plays a role in these and other critical water problems at the larger scale of the Colorado Basin and interior West as a whole, via functional linkages of water.

# 3. CLIMATE AND CRITICAL WATER PROBLEMS

## 3.1 Climate of the Region

Stream flows supplying the Aspinall Unit are highly variable (Fig. 2). Since Blue Mesa was completed in 1966, the average annual April–July inflow has been about 700,000 af, but inflows vary from about 167,000 af (1977) to over 1,400,000 af with a standard deviation of about 280Kaf (USBR Western Colorado Area Office 1999). April through July inflows, fed by melting snowpack, represent over 70% of the inflows for the year. There are significant relationships between snowpack for the Gunnison region and large-scale circulation patterns such as the El Niño/Southern Oscillation (ENSO) (McCabe 1994). The ENSO phenomenon describes anomalous sea surface temperature (SST) conditions in the tropical Pacific, which in turn impact temperature and precipitation across the United States. This phenomenon has significant but subtle effects on the Gunnison region: The risk of warm anomalies is suppressed during La Niña winters (December–February), and the risk of warm anomalies is increased during El Niño winters (December–February) (Wolter et al., 1999). Clark et al. (2001) evaluated how extremes of the ENSO phenomenon influence the snowpack for sites in these river basins. For the Lower Colorado Basin, they found a strong correlation between El Niño and anomalously high snow accumulation (as indicated by snow water equivalent, SWE) and also annual streamflow at selected gages. The opposite was found for La Niña, which is associated with anomalously low SWE and low annual runoff. The presence or absence of ENSO conditions is also associated with shifts in the probabilities of extreme precipitation and streamflow across a large region including the Gunnison Basin (Serreze et al. 2001).

The impacts of ENSO on the Gunnison Basin represent only part of the spectrum of climate impacts on the larger scale Colorado Basin and the West. Spatially, the influence of ENSO varies, and the northern parts of the Colorado Basin are affected differently than the southern parts of the basin.

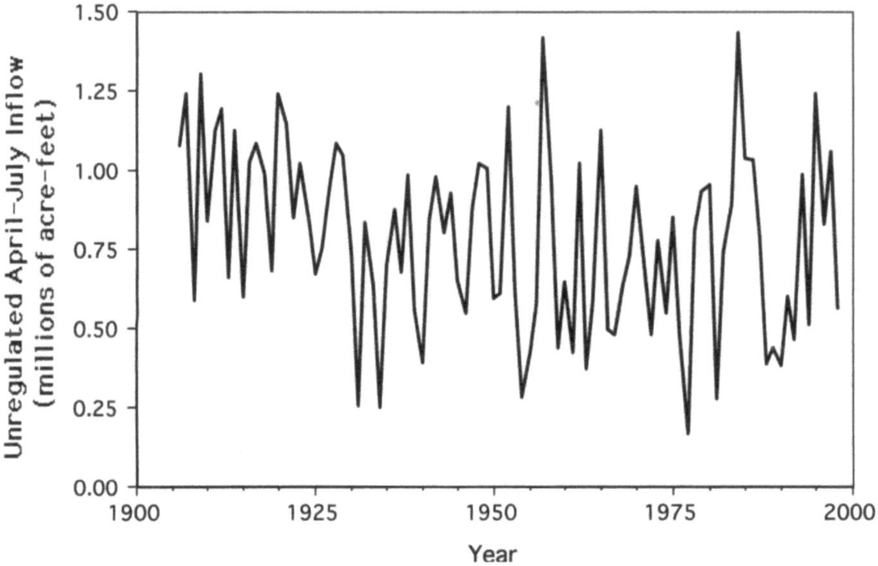


*Figure 2.* Variability of unregulated flows into Blue Mesa Reservoir, the main reservoir of the Aspinall Unit. Unregulated inflows are calculated by adjusting for upstream storage and diversions. Prior to the construction of reservoirs, the flows are calculated as those that would have entered the reservoir. Data from U.S. Bureau of Reclamation Western Colorado Area Office.

Within the northern part of the Colorado Basin, the Upper Green River Basin typically experiences anomalously dry conditions during an El Niño and wetter conditions during a La Niña (Clark et al. 2001). Downstream, the Lower Colorado Basin typically has stronger and opposite relationships with ENSO (Wolter et al. 1999), and higher risks of low/high snowpack and streamflows during cold/warm phases of ENSO (La Niña/El Niño) (Cayan et al. 1999). The Gunnison Basin has a similar response to the Lower Basin, with higher snowpack and reservoir inflows during strong warm phases of ENSO, and lower snowpack during cool phases (Ray 2002).

Climate variability, both temporal and spatial, is the physical process that links the basin to other places. Water shortages related to climate variability may intensify critical water issues, both within and outside the Colorado and Gunnison Basins. Four types of interactions related to climate are possible: (1) in-basin shortages limit the ability of water managers to meet all demands, thus intensifying the competition among in-basin water uses; (2) dry climate conditions and related water shortages within the Gunnison Basin stress external linkages, limiting flows for ecosystems in and downstream of the basin, dilution of salinity, and generation and export of hydropower; (3) dry conditions and shortages in one or more of the functionally linked, external regions create higher export demands on the Gunnison Basin, such as demands for hydropower; (4) shortages in both the area receiving water or benefits of the water intensify demands on the Gunnison Basin as well as within it. On the other hand, wet conditions in the basin can create opportunities; for example, in-basin surpluses may allow managers to supply flows for environmental purposes both within and downstream of the basin, or for hydropower export. In both wet and dry cases, as water management becomes more intensive in order to meet additional uses, the sensitivity of water management to climate variability within the basin also is likely to increase. These cross-scale interactions are likely to increase in intensity as demands for each use increase within the Gunnison Basin and in the other places linked to the basin's water.

## 3.2 Climate and Water Development

Temporal and spatial climate variability has influenced development of water in the Colorado Basin. Climate variability was ignored when the Upper Basin and Lower Basin divided the river's water in the 1922 Colorado River Compact. The compact assumed 7.5 maf would be available for the Upper Basin in addition to 7.5 maf allocated to the Lower Basin (Hundley 1975). However, this amount was not available in most years—the compact's authors had miscalculated the long-term average flows of the Colorado River, based on an anomalously wet period in the early twentieth century. The Colorado River Storage Project (CSRP) was conceived to remedy this, and authorized in 1956 to allow the Upper Colorado Basin states to develop their water while still meeting their Colorado Compact obligations during dry years (Hundley 1975). The CSRP reservoirs, including Lake Powell, Flaming Gorge, Navajo Reservoir, and the Aspinall Unit, store water which can be released during periods of low streamflow to meet the Lower Basin obligation. Dry years in series result in lowering of water levels in these reservoirs to meet the downstream obligation. Wet years allow more water to be stored in the reservoirs.

The 1948 Upper Colorado Compact, an agreement among the Upper Basin states, does address variable streamflows associated with climate variability by allocating water among the states by percentages of the flows available. Colorado was granted 51.75% of the Upper Basin's share, recognizing the large contribution of the state of Colorado to the Colorado River.

## 3.3 Transmountain Diversions and Climate

Transmountain diversions are based both on climate differences and compact allocations. Rain shadow effects of the Rocky Mountains are a regional feature of climate, which creates differences in river runoff. This spatial variation in climate results in much higher river runoff from the combined Colorado River tributaries in the state of Colorado, more than 10 maf/year, compared to the east slope river basins, where the population and demand are higher. The annual runoff of the South Platte River is about 1.44 maf, and that of the Arkansas River is about 875,000 af. Transmountain diversions are also driven by the Colorado Compact because the state has not been able to use most of its compact allocation (3.079 to 3.855 maf in any year; Colorado Water Conservation Board 2000) on the west slope where it arises. Transmountain diversions, a functional linkage of water, were developed in order to use the water within the state on its more populous east slope. These included a diversion of 110,000 af from the Colorado Basin to the Arkansas, and nearly a half million acre feet from the Colorado to the Platte (USGS 1985).

The Gunnison River's water has long been coveted for transmountain diversions, although only three small diversions exist, transferring a total of about 2,500 af/year from the Upper Gunnison Basin to the Arkansas and Rio Grande Basins, a small amount compared to diversions from other tributaries. In recent years, there has been increasing interest in moving water from the Upper Gunnison to the Front Range, including a proposal for a 900,000 af reservoir in Union Park, and 110,000 af average annual diversions out of the upper basin (Curry and McClow 2001). The proposers planned to fill the reservoir during wet years only, taking advantage of climate variability. The implication of taking water only in wet years is that it reduces the "buffer" in the system, and thus limits Aspinall Unit reservoir inflows in those years. Higher inflows in wet years often replenish the reservoir after dry periods. This buffer is now important for ensuring ecosystem sustainability, but it is also important for the generation of hydropower and the dilution of salinity, which will be discussed below. Although the Union Park Project proposes to take advantage of climate variability, that advantage is likely to have negative effects on other parts of the system by minimizing the buffer provided by wet years.

## 3.4 Hydropower

Hydropower links the Gunnison Basin to other areas through the power exported, not the water itself. Power is generated in response to out-of-basin demands and is sold to power companies that are often far away from the generating area. This demand is related to the weather and climate in the regions where the power is used, but the supply—reservoir storage—is related to climate in the basin where it is generated. The Aspinall Unit dams have a combined hydropower generation capacity of 248,000 kilowatts (kW), marketed by the Western Area Power Administration (WAPA) to wholesale power companies in the desert Southwest, intermountain West, and upper Great Plains service areas (WAPA 2001). Hydropower often supplies "peaking power," or the extra power above base power from other generation methods that is needed when demands peak in the late afternoon or during hot or cold periods. Other kinds of power generation cannot ramp up and down quickly and in a cost-effective manner to meet these needs, so hydropower is particularly valuable for this purpose. WAPA is particularly concerned about maintaining the flexibility in the Aspinall Unit for peak power production because its flexibility at Glen Canyon and Flaming Gorge has been limited by changes in operations to sustain ecosystems below those dams (USBR Western Colorado Area Office, August 1994 and January 1996).

Hydropower is related to climate variability in two important ways: Temperature extremes in regions using the power drive the demand for hydropower, but seasonal precipitation variability affects the supply. If reservoirs are low during dry years, power production capacity decreases linearly with the reservoir head, but during wet years, there may be extra capacity.

## 3.5 Salinity

Salinity is an important water issue between the United States and Mexico, where it causes damage to crops south of the U.S.-Mexico border, more than a thousand miles downstream. Salinity is related to climate variability in that salinity is inversely related to streamflows: When flows are high, salinity is diluted, but during dry years not only is there less dilution, but increased irrigation may leach more minerals from soils. During low-flow periods, salinity negatively impacts crops in the Lower Gunnison Valley and Grand Valley orchards. The Colorado River is naturally somewhat saline, but the problem is exacerbated by irrigation and by evaporation from reservoirs. Irrigation return flows contribute up to 37% of the salinity in the river. In the 1960s, Mexico recognized a growing problem of salinity in the water flowing from the United States as specified in agreements in the 1944 Mexican Water Treaty. Mexico protested to the United States that high salt concentrations were a major problem for crops in the Mexicali Valley. In 1974, the United States and Mexico signed the Minute No. 242 amendment to the treaty, which established salinity standards for water upstream of Mexico's Morelos Dam (Pontius 1997). The U.S. response to this treaty is salinity control efforts. Early salinity control efforts focused on the Grand and Uncompahgre Valleys; these efforts are ongoing today (MacDonnell 1999).

Although the Gunnison River supplies a relatively small proportion of the water in the Colorado River as a whole (average about 2 maf vs 14 maf for the Colorado River), it supplies a disproportionate amount of the salinity related to USBR projects. Due to their naturally saline soils, the Uncompahgre Valley and the Grand Valley immediately downstream contribute almost one-third of the salinity contributed by USBR projects, and about 10% of the salt load at Hoover Dam (Pontius 1997).

## 3.6 Ecosystem Sustainability

As noted, two critical water problems in the basin relate to the changes in the natural hydrograph that have occurred in part because of dams: the decline of native fishes and environmental changes to the Black Canyon of the Gunnison National Park. Ironically, the dams were built to minimize effects of climate variability, and in doing so they affect ecosystems dependent on that variability. The dams minimize flows and dampen the natural hydrograph in the BCNP, negatively impacting the downstream critical habitat for endangered fish. Historically, high spring flows, especially those associated with wet years, have been important to maintaining the habitat of these fish. The National Park Service (NPS) and the U.S. Fish and Wildlife Service now are both seeking to take advantage of wet years to achieve spring peak flows to help recover the fish, and to improve habitat in the Black Canyon.

Native fish have declined in the Gunnison Basin, as well as the Colorado Basin as a whole. The Gunnison River from the city of Delta to its confluence with the Colorado (Fig. 1) is designated as critical habitat under the Recovery Implementation Plan (RIP) for Endangered Fish of the Upper Colorado (known as the RIP; USFWS 1999), which was formalized in 1987 (MacDonnell 1996). This reach of the river receives water from other tributaries, but is still subject to the dampening of the natural hydrograph by the Aspinall Unit. The RIP recommends modifying the water management practices in the basin in order to improve these critical reaches as habitat. The reservoirs are viewed as an asset to assist in the effort (USFWS 1999). The Endangered Species Act (ESA) requires that federal agencies such as the USBR operate projects "consistent with its responsibilities under Section 7 of ESA" and requires agencies to consult with the Fish and Wildlife Service and "utilize their authorities in furtherance of the purposes of the ESA, by carrying out programs for the conservation of threatened or endangered species."

Before the Blue Mesa Reservoir was completed in 1967, spring peak flows of the Gunnison River at Grand Junction averaged over 8,000 cubic feet per second (cfs) and flows of 15,000 cfs were common; the dams have reduced the magnitude of spring runoff, and only 4 years between 1965 and 1996 reached this magnitude (USFWS 2000). The higher flows move sediment, scour spawning bars, and reshape the channel to improve and maintain desirable habitat for endangered fish (Smith and Wilson 1999; Pitlick et al. 1999). Flows of this magnitude, known as "habitat restoring and building flows," are therefore focus of the recovery effort. Reservoirs and diversions have also changed instream flows at other times of the year; in particular, minimizing late summer flows. Changes in both the spring flows and those during other times of the year affect the survival of native fish as juveniles and adults, which had adapted to the unregulated hydrograph.

The second ecosystem sustainability issue is maintaining the ecosystem of the Black Canyon of the Gunnison National Park. The National Park Service claims a federal reserve right for water (Getches 1997) to attain and preserve the "recreational, scenic, and aesthetic conditions" existing when the monument was founded, as well as ensure the continued existence of fish inhabiting the waters or introduced thereafter (U.S. Department of Justice 2001). The federal reserve right was decreed in 1983, after a long court battle (Sheildt 1999). This decree has a priority date of 1933, when the national park was originally created as a national monument. The federal reserve right is junior to (later than) UVP rights in the Uncompahgre Basin, but senior to many other water rights in the basin, so many water users are concerned about how it will be implemented.

The implementation of the RIP and flows for the Black Canyon both depend on water in the reservoirs and on changes in reservoir operations. Both programs have flow recommendations that are based on releases from the Aspinall Unit. The RIP's flow recommendations, released in draft form in 2000, are based on mimicking the natural hydrograph, including spring peak flows for habitat restoration, late summer minimum in streamflows to provide enough water for the fish to live, and avoidance of sudden increases in flows in the winter that trigger migration and spawning responses

(USFWS 2001). Because the natural hydrograph historically has varied with wet and dry years, the flow recommendations are for six hydrologic categories. During a year categorized as dry, recommended flows would provide a small peak to be used as a spawning cue by fish, although it would not contribute to habitat building. Larger peaks and base flows during the rest of the year are recommended for average to wet hydrologic conditions (USFWS 2001). The hydrologic category of a year is defined by the late spring runoff volume forecast. Water releases from the Aspinall Unit will be timed to augment the natural peak in the critical reach.

The BCNP water right application requests minimum base flows year-round of 300 cfs or more to ensure survival of aquatic life in the canyon; an annual peak between May 15 and June 15 of 3–14 days in duration, flows of 3,500–12,000+ cfs, and ramping rates of 250–500 cfs per day or 10% per day; and shoulder flows on each side of peak. The flows are also expected to vary, based on the availability of water in a given year (U.S. Department of Justice 2001). Peak flows are calculated as a function of the forecasted inflows on May 1.

Both activities acknowledge that there is natural variability of inflows, although the BCNP proposal accounts for continuous variability based on forecasted inflow rather than the six hydrologic categories used in the flow recommendations for endangered fish. Both activities determine annual flow requests based on river forecasts made by the NOAA Colorado Basin River Forecast Center, which are based on accumulated snowpack. However, these forecasts do not currently use forecasts of climate variability to anticipate wet or dry conditions.

The Aspinall Unit dams influence ecosystems outside the basin as well. Habitats for endangered species and other fish and wildlife extend far downstream in the Upper Colorado Basin, the Grand Canyon, and finally, the Colorado Delta ecosystem in Mexico (Fig. 3). All of these systems are influenced by the volume of water flowing out of the Gunnison Basin. The Gunnison River supplies a large proportion of the available flow to critical habitat between the Gunnison-Colorado confluence and the Green River. The Colorado Delta ecosystem has been starved for water by upstream development, but supporters of restoration efforts calculate that supplying as little as a total of 40,000 af/year could improve the ecosystem (Getches 2003). Getches suggests that this amount is within the operational "imprecision" of the USBR system, and that small amounts of water might be found in other parts of the system. Thus, increasing the instream flows leaving the Gunnison Basin may benefit the Colorado River Delta ecosystem far downstream.

## 4. CLIMATE VARIABILITY AND WATER MANAGEMENT

Climate variability both directly and indirectly impacts multi-purpose water management and reservoir operations. There is significant variability of the inflows to the reservoirs (Fig. 2), and both the endangered fish and Black Canyon water right policy issues described depend on water from the Aspinall Unit reservoirs. Therefore, the potential for these policies to meet their goals is sensitive to climate variability. However, forecasts of reservoir inflows are not considered sufficiently reliable for operational use until late spring. As was described above, the USBR must begin planning reservoir releases in the winter (or even the previous fall) before late-spring inflows are well known. Given the uncertainty, reservoir managers plan operations conservatively. In the past, there has been just enough "slack" or buffer in the system to meet operational requirements. However, the addition of new uses and expanding operational objectives requires more intensive planning and management. The additional pressure of new uses is leading the USBR to consider how climate variability intersects with the combined reservoir management–environmental sustainability issue.

Therefore it is valuable to consider what climate products would be useful and useable in the reservoir operations decision process. Below, this planning and decision process is analyzed from the standpoint of incorporating the new ecosystem sustainability uses, in order to identify needed climate products.

### 4.1 Reservoir Operations Planning

Current operations of the USBR Aspinall Unit reservoirs follow a hierarchical planning and decision framework (Zagona et al. 2001). Planning over 1–2 years involves a reservoir operations model based on known reservoir contents and expected inflows. Scenarios are generated from this model for operations based on operational requirements, including contracts, maintenance plans, and operating requirements, some of which are set by the Law of the River. This plan, known as the "24-month study," is updated at least monthly to determine daily operations. At each time step, plans may be revised iteratively based on new information, such as updated streamflow forecasts or other changes in conditions. For example, if outflows must be reduced to accommodate repairs, the reservoir operations model can be used to calculate how that period of low flows affects reservoir levels and releases to downstream users. Similarly, the reservoir operations model is used to calculate operations based on inflow forecasts. These midterm models provide operational targets for short-term scheduling models run on a daily time step over 4–6 weeks, which provide information for systems control. Longer-term planning models are run on a monthly time step over a time span of decades.

To illustrate the multipurpose reservoir management problems that managers face, the annual cycle of key reservoir operations decisions made by the USBR can be integrated with environmental restoration goals. This cycle begins in October, the beginning of the water year. Reservoir operations models are run using information on current reservoir levels and operations requirements. Operations plans at this point in the year are considered preliminary because the only information available on potential April–July reservoir inflows is in the historical records of annual snowpack and resulting inflows. River levels are low in the late summer and early fall, and irrigation withdrawals may still be occurring. Reservoir releases are limited in order to retain water in the event of a dry winter. However, some reaches of the river can be too low for the survival of aquatic species, so maintaining minimum instream flows is a management concern.

In January, early observations of snowpack and the first water supply outlook are used to update the operating planThe key issue in January is estimating the "start of fill" elevation, or reservoir level, for Blue Mesa. This elevation target is set to balance the goal of filling the reservoir with that of the goal of flood control; i.e., capturing a large peak. Reservoir managers plan to lower the reservoirs to the target elevation by April 1. If this level is set too low, reservoir managers risk not filling the reservoir in the spring. If it is set too high, they risk spilling water (spilling refers to water that is released without passing through power turbines). The start of fill target must be determined a month or more in advance: Evacuation of a large amount of water is limited by the size of the turbine outlets and can take a number of weeks. Ecosystems concerns in the January–March period relate to avoiding sudden increases in flows, which may give a false cue to fish to migrate or spawn. In late February–March the "start of fill" target elevation is refined based on new water supply outlooks, and the reservoir elevation is adjusted to the revised target.

In early April, most of the snowpack has accumulated, but the inflows are still low. Based on streamflow forecasts by the NOAA Colorado Basin River Forecast Center (CBRFC), detailed plans are made for managing the spring runoff, which usually peaks in the Gunnison River between mid-May and mid-June. The recommended peak for endangered fish, based on the anticipated hydrologic category for that year, is incorporated into the operating plans. Streamflow forecasts are updated through the runoff period based on observed changes in snowpack, but not on seasonal climate forecasts. During the April–July runoff period, snowpack and runoff observations are tracked daily. Operational plans can be adjusted daily based on these outlooks and conditions and on short-term weather forecasts. For example, spring storms may increase the anticipated total runoff, or a warm (cool) spring may correspond with an early (late) peak runoff. Releases are made from the Aspinall Unit for a peak in the Gunnison River critical reach based on the hydrologic category in May.

By the end of July, the spring runoff has tapered off, and the operational issue is releasing water for high summer irrigation demands. Hydropower releases are also highest in the mid- to late summer. Maintaining minimum flows in the BCNP and the critical reach downstream is the ecosystem concern, because much of the water released may be diverted through the Gunnison Tunnel. Instead, flows occur in the Black Canyon only if power releases are larger than the diversion through the Gunnison Tunnel. The USBR has a contract with USFWS to provide water to augment late summer flows to maintain minimum instream flows in the Gunnison River critical habitat. In the fall, the operations and decision cycle begins again: After each year's runoff, managers plan conservatively to manage the water available until the next year, with only climatological information to guide the range of possible conditions for the next year. In fact, reservoir managers say that they must assume in their planning that a multiyear drought is beginning.

## 4.2 Potential Uses of Forecasts

This reservoir management decision process is stepwise and iterative: Plans are updated monthly or more often from January until the end of high runoff, based on snowpack observations and updated runoff volume forecasts. The iterative nature of this decision process provides an opportunity for incorporating other new information, such as information on climate forecasts. The decision process for the Aspinall Unit reservoirs is similar to that for other large Upper Colorado Basin reservoirs, although there are somewhat different operating criteria and policy issues for other reservoirs. Reservoirs across the Colorado Basin are facing similar challenges of new demands and expanding operational objectives. Climate variability is a factor that influences not only the Gunnison Basin, but other basins and places to which the Gunnison Basin is linked. However, climate forecasts are not currently utilized in reservoir operations planning. Climate and weather forecasts may be useful at several points in the decision process to improve the efficiency of reservoir operations. Seasonal and subseasonal forecasts are potentially useful for runoff season planning.

Seasonal climate forecasts could be used to improve runoff and inflow forecasts, based on better estimates of seasonal snowpack. If available in mid-March or earlier, the "start of fill" target could be improved. A seasonal forecast of spring (April–June) temperatures is important: Whether the spring is warm or cool influences whether the runoff is more likely to be early or late. If this forecast were available in March–early April, managers could improve the detailed planning to manage the runoff, including whether there is enough water to augment peak flow, but still have enough water to meet other contracts later in the year. A seasonal forecast in March or April of summer conditions could also indicate a hot, dry (cool, wet) summer in which irrigation demands are likely to be high (low). The 2000 USFWS flow recommendations and the Black Canyon water right application both request that varying peak flows and base flows be calculated according to spring forecasts of April–July reservoir inflow volumes (USFWS 2000; Department of Justice 2001). The flows requested for the Black Canyon of the Gunnison also are based on the forecasted inflows on May 1.

Subseasonal forecasts could improve the ability to forecast the timing of peak flow periods, which would be useful in augmenting the natural peak. Temperature and precipitation forecasts with 10 days' through 2 weeks' lead time could assist in predicting the timing and magnitude of the natural spring peak. These forecasts could improve the CBRFC short-term forecast product of river conditions, which is intended to provide a general outlook for significant flooding events, including peak flows in the spring. This product incorporates NOAA National Weather Service forecasts of up to 8 days. A longer lead time forecast of river conditions could allow better planning to meet USFWS peak flow recommendations in three ways. First, several days' lead time is needed to "ramp up" the flows (a gradual increase is desired by both USFWS and NPS); second, the travel time is 2–3 days from the reservoir to the critical reach; and third, the USBR also prefers to give several days' notice to anglers and boaters of significant changes in river volume.

A third potential use of climate information relates to how climate forecasts might influence perceptions of water availability in the basin, and thus the potential to manage reservoirs for multiple uses. The USBR must operate its reservoirs in a complex arena of stakeholders. Although they have the final authority to manage the reservoirs, they often need the cooperation of other organizations, as they seek to balance multiple objectives. Water users in the Uncompahgre and Gunnison Valleys are thus concerned about how the RIP 2000 flow recommendations and Black Canyon water right will be implemented, because their water rights and water use practices may be affected. In particular, they fear that releases for peak augmentation in May or June may limit the water available later in the summer and fall.

However, they are more likely to agree to spring peak flows if they feel assured that their late-summer water needs will be met as well. Seasonal or subseasonal forecasts of summer precipitation and temperatures might help convince USBR stakeholders that all water needs are likely to be met. Thus, the climate perceptions of these stakeholders in the basin are critical to the ability of reservoir managers to meet environmental sustainability goals.

## 4.3 Climate and Managing Functional Linkages

The Gunnison Basin is linked to other areas of the West via its water, and climate variability may intensify and exacerbate these functional linkages. Linkages (Fig. 3) include transbasin diversions to the Arkansas and Rio Grande Basins; hydropower that is primarily exported elsewhere in the intermountain West and to the desert Southwest and upper Great Plains; and salinity in Mexico. Ecosystem sustainability linkages include flows for critical habitats in the Lower Gunnison Basin and the Colorado River above Lake Powell, and flows for the Colorado Delta in Mexico. There is a potential to use climate forecasts in the management of these functional linkages.

Anomalously dry conditions in the Gunnison Basin, for example, limit flows for ecosystems both in the basin and downstream, and for dilution of salinity. Critical habitat for endangered fish in the Colorado River downstream of the Gunnison River may be affected, because the Gunnison Basin supplies a large percentage of the water available in the critical reaches of the Colorado River upstream of the confluence of the Colorado and the Green Rivers. The Aspinall Unit also accounts for most of the storage above this confluence, and the only storage, that can provide flows to the Black Canyon. With respect to salinity, there is less dilution of dissolved minerals during low-flow periods, and the salinity problem is exacerbated by increased irrigation in response to drier conditions, which flushes more minerals from the soils. The potential to generate hydropower also is limited in dry periods. Advance warning, one to three seasons in advance, of a dry winter or a hot, high water demand summer, could enable reservoir management decisions to maximize the potential flows for ecosystems, in particular to save water for late summer flows. WAPA might also optimize its plans for hydropower generation, including the use of other power generation methods. On the other hand, anomalously wet conditions in the Gunnison Basin create an opportunity to meet wet year flow recommendations for downstream ecosystems, as well as the Black Canyon and critical

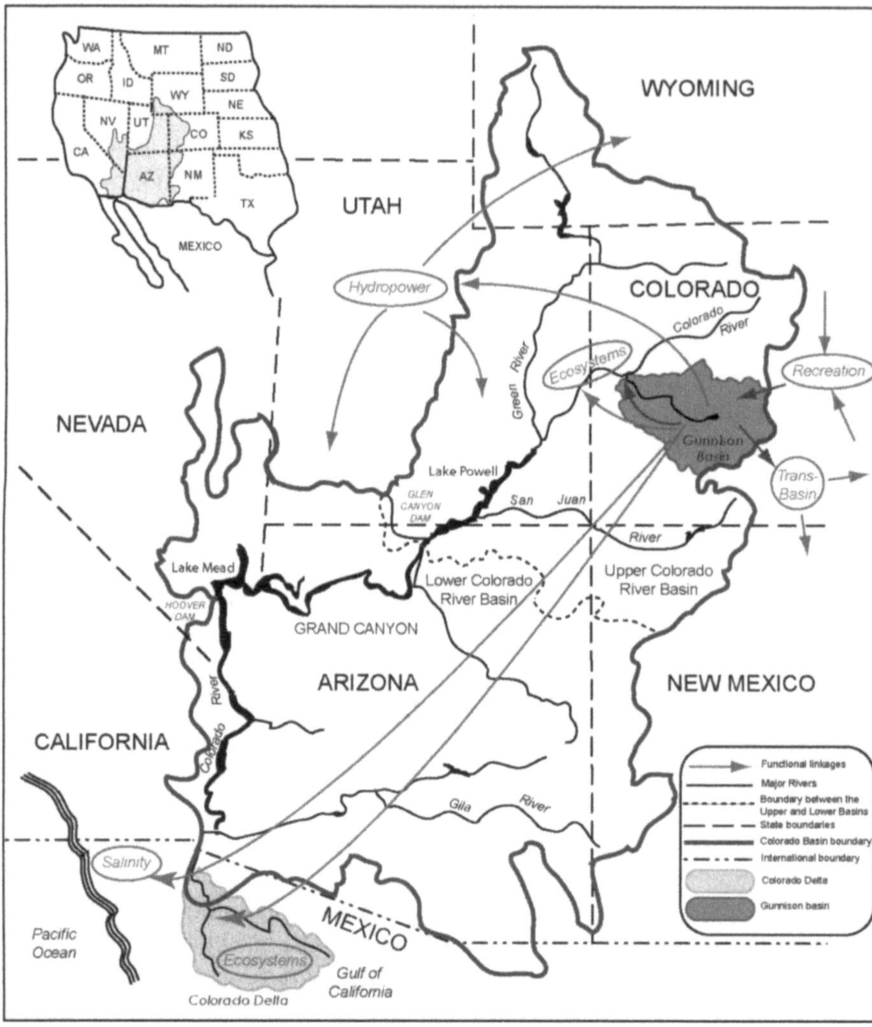


*Figure 3.* Out-of-basin functional linkages of Gunnison River water to other parts of the western region, indicated by arrows. The Gunnison Basin is indicated in dark gray. Linkages include transbasin diversions, hydropower, salinity, and flows for ecosystems.

habitat reaches. Efficiencies in water management here also could provide water for the Colorado Delta ecosystem in Mexico, if it is protected as it flows downstream (Getches 2003). Forecasts of anomalously wet conditions in the basin could facilitate planning and agreements to take advantage of this opportunity.

Dry conditions and shortages in one or more of the places receiving benefits of the water create higher demands on the Gunnison Basin. Dry conditions in the Southwest are associated with La Niña, and may increase the Lower Basin demand for water and hydropower, especially when reservoirs are low in other parts of the West. This situation occurred in the summer of 2000 (USBR Western Colorado Area Office, April 2001). Again, advance warning of anomalous conditions could allow for coordinated water resource planning across the Gunnison and Colorado Basins and the West, in order to optimize use of water for ecosystems and hydropower while still meeting contracts and obligations.

Shortages in both external areas receiving benefits of Gunnison Basin water and in the Gunnison Basin itself will intensify demands for the basin's water. This situation is less likely than situations in which the Gunnison Basin has an anomaly of opposite sign from other areas, such as the Lower Colorado Basin, because the ENSO signal varies across the basin. Transbasin diversions may also interact with climate variability: A specific diversion volume represents a larger proportion of natural flows during a dry year, exacerbating downstream effects of shortages. The Union Park proposal planned to store water for diversion only in wet years. This strategy would lower flows in the Gunnison Basin in wet years, creating higher salinity, less water for hydropower, and less water for ecosystems in wet years.

## 5. CONCLUSIONS

Two trends in water management in the Gunnison Basin are increasing the sensitivity of water resources management in the basin, and reservoir management in particular, to climate variability. These trends are intensification of water use within the basin, in particular for environmental purposes, and the increasing importance of functional linkages of the basin's water to other places. Both trends affect sensitivity to climate variability because there is less capacity to absorb shortages.

Two critical water problems within the Gunnison Basin require changes in operation of the reservoirs: the RIP 2000 flow recommendations and the federal reserved right for water for the Black Canyon of the Gunnison. Reservoir operations rules can be a constraint to the use of climate information (Pulwarty and Redmond 1997), but a change in operations can become an opportunity to try new tools and procedures. Analysis of the reservoir operations decision calendar reveals points at which climate information, if available, might improve the efficiency of reservoir management as well as mitigate sensitivity to variability. The NPS and the USFWS might ultimately benefit from climate information if it assists them in meeting their environmental restoration goals; the USBR and its stakeholders may benefit if reservoir management is improved with less conflict among stakeholders.

Several water issues functionally link the Gunnison Basin to other areas in the West. These include production of hydropower marketed around the West; export of salinity that is a water quality problem downstream; demand for transmountain diversion of water by users in the Front Range of Colorado; and water for downstream ecosystems. These issues all interact with climate variability. In addition to the effects of in-basin climate anomalies on water management within the basin, three types of external climate interactions are possible: (1) Dry climate conditions and related water shortages within the Gunnison Basin limit export of hydropower, water dilution of salinity, and downstream flows for ecosystems; (2) dry conditions and shortages in one or more of the places *receiving* the benefits of the Gunnison Basin's water create higher demands on the basin, such as demands for reservoir releases to produce hydropower; and (3) shortages in both the Gunnison Basin and the areas receiving benefits of its water intensifies demands on the Gunnison Basin.

Climate information and forecasts have the potential to minimize the negative impacts of these anomalies. Because the influence of ENSO varies across the Colorado Basin, and other regions that receive the flows that arise in this basin, there are likely to be many years in which there are opposite anomalies in the Upper and Lower Colorado Basins, and between these areas and other parts of the West. The differences in the anomalies provide opportunities to plan to use water for its maximum benefit across the basin. Forecasts of climate conditions one or more seasons in advance have the potential to improve reservoir management within the basin and also to reduce conflicts as uses of water increase across the West. Seasonal forecasts, and the spatial variability of climate impacts, might be taken into account in order to better manage water across the Colorado Basin as a whole, taking advantage of advance warning of wet or dry conditions in different parts of the basin. By recognizing how water resource management in this basin is linked to resource use in other areas, it may be possible to develop strategies that maximize the benefits across a larger scale.

Climate variability, in its temporal and spatial dimensions, affects reservoir management, and thus affects the environmental sustainability goals related to that management. This analysis illustrates that there is potential for reservoir management to take advantage of climate information to intensively manage water to meet expanding operational goals.

*Acknowledgements:* Support for this project was provided by NOAA-OAR funding to the Climate Diagnostics Center. Thanks to Coll Stanton of the U.S. Bureau of Reclamation Western Colorado Area Office for data on the Aspinall Unit. This paper benefited greatly from discussions with James L. Wescoat, Jr. Robert Webb, Douglas Kenney; Barbara Morehouse, and Henry Diaz provided helpful comments on the manuscript. Barbara Deluisi of CDC created Figure 3 and provided other technical assistance with the manuscript.

## 6. REFERENCES

Board of County Commissioners of County of Arapahoe v. Crystal Creek Homeowners' Association, 14 P.3d 325-352 (Colo. 2000).

Cayan, D.R., K.T. Redmond, and L.G. Riddle. 1999. ENSO and hydrologic extremes in the western United States. *Journal of Climate*. 12: 2881–2893.

Clark, M.P., M.C. Serreze, and G.J. McCabe. 2001. Historical effects of El Niño and La Niña events on the seasonal evolution of the montane snowpack in the Columbia and Colorado River basins. *Water Resources Research* 37: 741–757. Colorado Water Conservation Board 2000. Interstate water allocation. Available at: www.cwcb.state.co.us

Curry, K.E., and J.H. McClow. 2001. Transbasin diversions: A view from the basin of origin: *Proceedings of the Conference on Transbasin Water Transfers*, U.S. Committee on Drainage and Irrigation, Denver, Colorado, June 27–30, 2001.

Getches, D.H. 1997. *Water Law in a Nutshell*. St. Paul, Minnesota: West Publishing Co., 456 pp.

Getches, D.H. 2003. Impacts in Mexico of Colorado River management in the United States: A history of neglect, a future of uncertainty. In, Diaz, H.F., and B.J. Morehouse (eds.), *Climate and Water: Transboundary Challenges in the Americas*. (This volume.)

Hundley, N., 1975. *Water and the West: the Colorado River Compact and the politics of water in the American West*. University of California Press, Berkeley. 375 pp.

Knapp, L.K., 1994. A task analysis approach to the visualization of geographic data. Dissertation, University of Colorado, Boulder, CO, 376 pp.

MacDonnell, L.J. 1999. *From Reclamation to Sustainability: Water, Agriculture, and the Environment in the American West*. Niwot, Colorado: University Press of Colorado, 385 pp.

MacDonnell, L., 1996. The Upper Colorado Basin, Colorado. In: *Natural Resources Law Center (Editor), Restoring the West's waters: Opportunities for the Bureau of Reclamation*. Univ. of Colorado School of Law, Boulder, CO, pp. 5-1 to 5-55.

McCabe, G.J.Jr. 1994. Relationships between atmospheric circulation and snowpack in the Gunnison River basin, Colorado. *Journal of Hydrology* 157: 157–175.

Nichols, P.D., M.K. Murphy, and D.S. Kenney. 2001. Water and growth in Colorado: A review of legal and policy issues. Boulder, Colorado: University of Colorado School of Law, Natural Resources Law Center, 191 pp.

Pitlick, J. and VanSteeter, M.M., 1998. Geomorphology and endangered fish habitats of the upper Colorado River, 2, Linking sediment transport to habitat maintenance. *Water Resources Research*, 34(2): 303-316.

Pontius, D. 1997. *Colorado River Basin Study: Report to the Western Water Policy Review Advisory Commission*, Tucson, Arizona, SWCA, Inc, Environmental Consultants, p. 127.

Pulwarty, R.S., and T.S. Melis. 2001. Climate extremes and adaptive management on the Colorado River: Lessons from the 1997–98 ENSO event. *Journal of Environmental Management* 63: 304–324.

Pulwarty, R.S. and Redmond, K.T., 1997. Climate and Salmon Restoration in the Columbia River Basin: the role and useability of seasonal forecasts. *Bulletin of the American Meterological Society*, 78(3): 381-397.

Ray, A.J., 2002. Climate, institutions, and resources: The new story of multi-purpose reservoir management in the Gunnison Basin, Colorado. Dissertation, University of Colorado, Boulder, CO, in preparation.

Ray, A.J. 2000. Climate and water management in the Interior West: A critical water problems approach. *Proceedings of the 16th Annual Pacific Climate Workshop on Climate Variability of the Eastern North Pacific and Western North America* (PACLIM), pp. 171–182.

Riebsame, W.E., 1990. Anthropogenic climate change and a new paradigm of natural resource planning. *Professional Geographer* 42: 1-12.

Schieldt, W. 1999. U.S. Park Service Proposal for Quantification of Federal Reserved Rights in Black Canyon National Monument, Colorado Stream Lines, accessed online, August 21, 2001, water.state.co.us./streamlines/parkservice991.htm.

Serreze, M.C., M.P. Clark, and A. Frei. 2001. Characteristics of large snowfall events in the montane western United States as examined using snowpack telemetry (SNOTEL) data. *Water Resources Research* 37: 675–688.

Smith, G.R. and M. Wilson, 1999. Upper Colorado River Recovery Implementation Program Coordinated Reservoir Operations Program, *Proceedings of the American Water Resources Association Annual Meeting*, Seattle WA.

U.S. Bureau of Reclamation (USBR). 2001. Project Dataweb. Dataweb.usbr.gov.

U.S. Bureau of Reclamation (USBR), Western Colorado Area Office. 1993–2002. Minutes of the Aspinall Unit Coordination Meeting, 1993–2002. Available from the USBR Western Colorado Area Office, Grand Junction, Colorado

U.S. Bureau of Reclamation (USBR), Western Colorado Area Office. 1999. Data on Aspinall Unit inflows, personal communication.

U.S. Department of Justice. 2001. Concerning the Application of the United States of America to make Absolute a Conditional Water Right, Case No. 01CW005

U.S. Fish and Wildlife Service (USFWS). 1999. Recovery implementation program for endangered species in the Upper Colorado River basin. Denver, Colorado: U.S. Fish and Wildlife Service, 52 pp.

U.S. Fish and Wildlife Service (USFWS). 2000. Flow recommendations to benefit endangered fish species in the Colorado and Gunnison Rivers. Draft Final Report. Grand Junction, Colorado: U.S. Fish and Wildlife Service.

U.S. Geological Survey (USGS). 1985. *Estimated Use of Water in Colorado, 1985*, Water-Resources Investigation Report 88-4101.

Upper Gunnison River Water Conservancy District. 2001. The Upper Gunnison River Water Conservancy District: Its historical perspective, accomplishments, the future., 18 pp.

Wescoat, J.L.J. 1991. Managing the Indus River basin in light of climate change: Four conceptual approaches. *Global Environmental Change*, December 1991: 381–395.

Western Area Power Administration (WAPA). 2001. *Annual Operations Report, 2000.* (Available on www.wapa.gov)

White, G.F. 1961. The choice of use in resource management. *Natural Resources Journal* 1: 23–40. Reprinted in Kates, R., and I. Burton (eds.). 1986. *Geography, Resources and Environment*. Chicago, Illinois: University of Chicago Press, pp. 143–165.

Wolter, K., R.M. Dole, and C.A. Smith. 1999. Short-term climate extremes over the continental United States and ENSO: Part I: Seasonal temperatures. *Journal of Climate* 12: 3255–3272.

Zagona, E.A., T.J. Fulp, R. Shane, T. Magee, and H.M. Goranflo. 2001. RiverWare: A generalized tool for complex reservoir system modeling. *Journal of the American Water Resources Association*, 37: 913-929.

SECTION B

# Climate, Hydrology, and Ecosystem Processes

10

# BIOMES, RIVER BASINS, AND CLIMATE REGIONS

*Rational Tools for Water Resources Management*

Henry F. Diaz
*Climate Diagnostics Center, NOAA/OAR, 325 Broadway, Boulder, Colorado 80305*

**Abstract** This chapter makes a case for the use of climate-based boundary definitions for regions where the management of water and biotic resources is of great concern, using as examples the Colorado River system and the southwestern U.S. desert biomes. Dam construction in the U.S. portion of the Colorado River beginning in about 1930 reduced natural flows to the Colorado River Delta to nearly zero by the late 1960s. A delta ecosystem encompassing nearly 2 million acres was nearly extinguished; today it exists within fewer than 200,000 acres. It is suggested that impacts of such magnitude to a major continental river system occurred because of the presence of an international political boundary, and also due to the absence of a decision-making framework for water resources management that takes into account ecological values. Incorporation of data and information regarding possible climatic variations in the future is also emphasized.

## 1. INTRODUCTION

Managing transboundary water resources can be complex and difficult, given the diverse spectrum of human interests and the often differing goals of the institutions responsible for the management of those resources. The issue has taken on increased urgency because population growth and water demand have increased across the globe, and issues such as the potential impact of climate change on precipitation and streamflow now constitute important national and international challenges.



*Henry F. Diaz and Barbara J. Morehouse (eds.), Climate and Water*, 221-235.

This book is concerned with issues related to transboundary water resources management in the Americas. Recent notable contributions to the general literature on this subject include a special volume of the *Natural Resources Journal* (*Water Issues in the U.S.–Mexico Borderlands*), published in the fall of 2000, and a book edited by J. Blatter and H. Ingram (2001) (*Reflections on Water: New Approaches to Transboundary Conflicts and Cooperation*). This volume addresses these issues from different perspectives. In particular, the chapter by B. Morehouse argues for consideration of new integrated approaches to water resources management in which sociopolitical boundaries are acknowledged, but where the physical characteristics, such as climate and landscape, are taken more specifically into account as fundamental factors in the decision-making process.

This chapter, which serves as an introduction to Section 2, "Climate, Hydrologic, and Ecosystem Processes," explores some ideas for managing transboundary water resources in the context of the fundamental biogeophysical characteristics of such regions, and in particular in the Southwest United States and northern Mexico. Three methodological approaches are used to define homogeneous regional characteristics irrespective of sociopolitical boundaries. The first uses the concept of a biome—defined as an entire community of living organisms in a single major ecological region; a second one follows a statistical approach, focusing on key aspects of the seasonal characteristics of the prevailing climate in an area; and the third one defines a region based on the physical drainage boundary of a river system (in this case the Colorado River Basin).

## 2. DEFINING HOMOGENOUS REGIONS

### 2.1. An Ecosystem Approach

Two major desert regions transcend the international boundary between Mexico and the United States—the Sonoran and Chihuahuan Desert provinces. The Sonoran Desert covers an area of approximately 110,000 square miles (280,000 square kilometers) in North America. It is situated in parts of the Mexican states of Baja California and Sonora, and covers portions of the states of Arizona, California, and Nevada (Fig. 1). Most of the Chihuahuan Desert—the largest desert in North America, covering more than 200,000 square miles (518,000 square kilometers)—lies south of the international border. In the United States, the Chihuahuan Desert extends into parts of New Mexico, Texas, and sections of southeastern Arizona. Its minimum elevation is above 1,000 feet (300 meters), but most of the desert lies at elevations between 3,500 and 5,000 feet (1,100 and 1,520 meters). More information on the Sonoran and Chihuahuan Desert environments can be found at http://www.desertmuseum.org/sonora.html and http://www.desertusa.com/du_sonoran.html.

The physical environment—the topography, the climate, and the hydrology—determines the distribution of plant and animal species on the landscape. Throughout much of postglacial history, physical factors have also exerted considerable influence on human ecology in these desert regions. In the last few hundred years, and in particular the last 150 years, the impact of human beings on the physical environment and the living world has strongly modified the natural ecological balances of these regions. Human intervention, particularly since about the mid-1800s, has changed the distributions, abundances, and ecological functions of plant and animal life in these areas. Artificial boundaries associated with political divisions often tend to exacerbate the impacts of humans on their environment. It is now recognized that a more enlightened approach to the stewardship of natural resources is necessary to ensure sustainable improvements in human welfare (National Research Council [NRC] 1999).

One could argue that a decision-making framework for natural resources management should include, as a matter of priority, preserving ecosystem integrity at the scale of, say, a desert biome or a major river basin. From a climatological perspective, the geographical boundaries of such a region also make eminent sense, because the large-scale atmospheric circulation patterns that govern the seasonal distribution of precipitation and temperature tend to respond to similar large-scale climate features, such as, for instance, the El Niño/Southern Oscillation (ENSO) phenomenon. Figure 2 illustrates the seasonal cycle of precipitation and temperature for the Arizona portion of the Sonoran Desert region (upper panels). Also shown in the lower panels are the annual changes in total precipitation and in mean temperature for this region of Arizona since 1905. One can note that monthly mean precipitation exhibits a bimodal distribution, with one peak occurring in midwinter and another occurring during the summer. It is the presence of the additional winter moisture that makes the Sonoran Desert flora much richer than that of the Chihuahuan Desert to the east, which receives largely summer monsoon moisture.

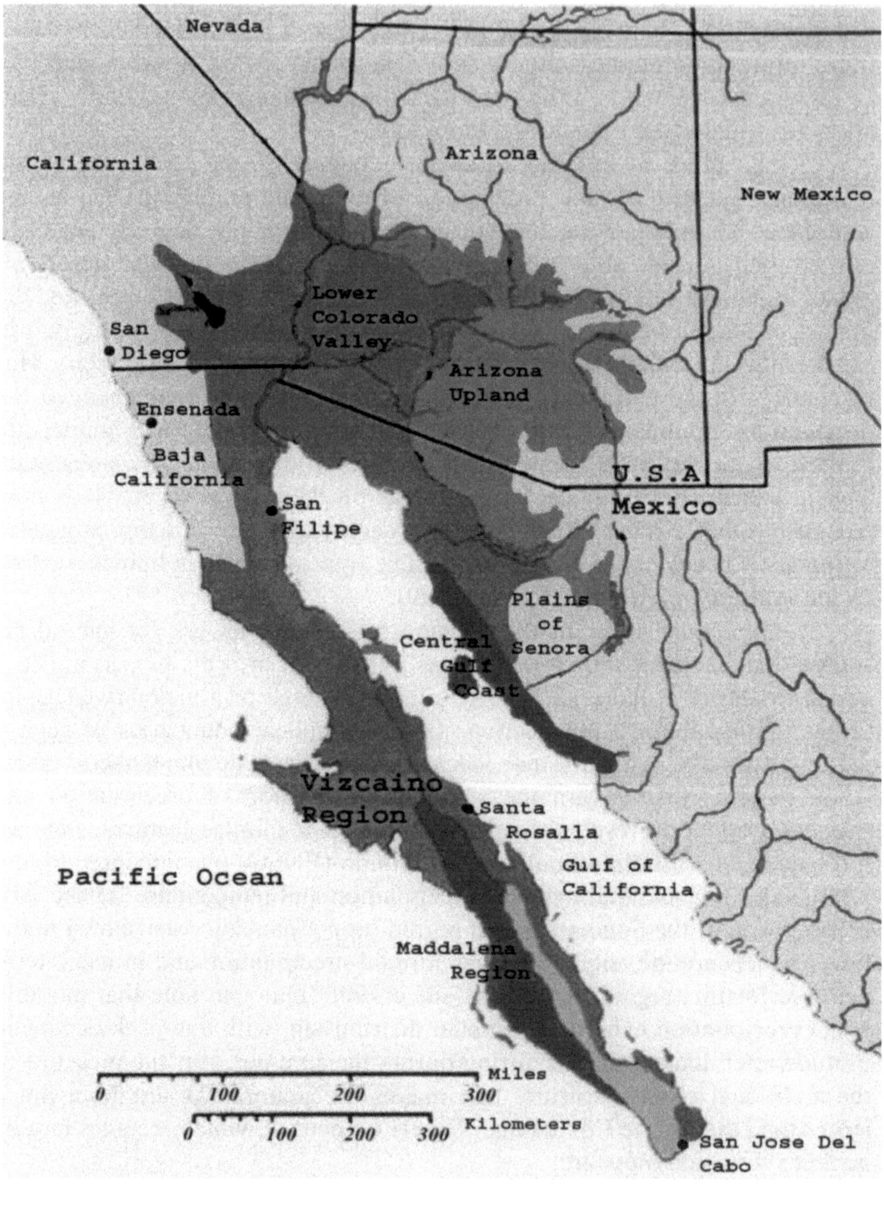


*Figure 1*. Map showing the nominal boundary of the Sonoran Desert. Similar animals and plants are found on both sides of the international border between Mexico and the United States.

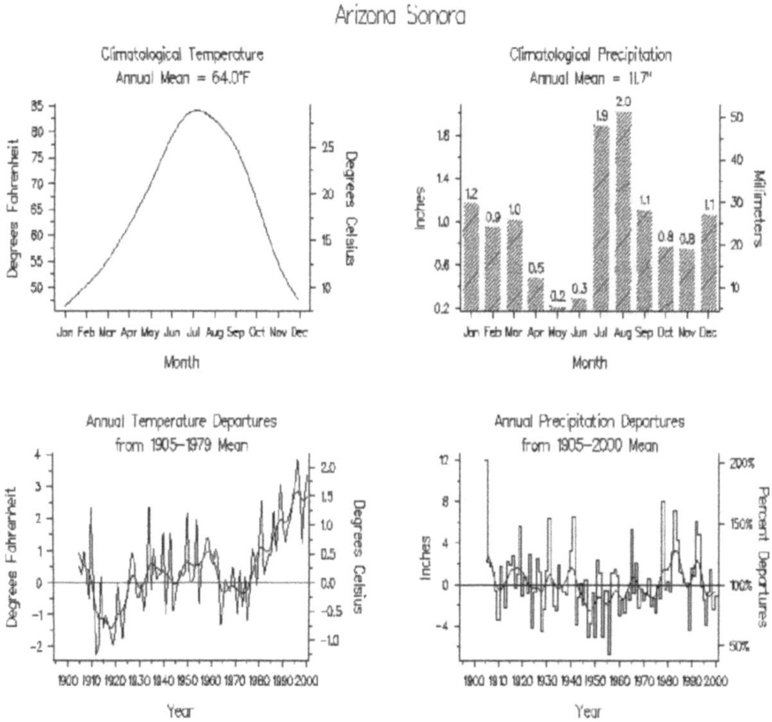


*Figure 2.* Top panels: Climatological annual cycle of the regions encompassed by the Arizona side of the Sonoran Desert. Bottom panels: Time series of annual departures from the long-term mean for the same region.

The characteristic pattern of interannual and decadal-scale variability in annual precipitation indicates that, on these timescales, temperature and precipitation have varied oppositely to one another, so that drought episodes are also characterized by higher than normal temperatures—a fact that enhances the aridity of the region. Among the more interesting features of the temporal characteristics of mean annual temperature and total precipitation are the large year-to-year variability, and a pronounced warming trend of about 2°–3°F (~1.5°C) over the past 20 years. Another interesting feature of the last 20 or so years is that for the most part, precipitation during this warm interval has also tended to be above average. This increase in precipitation appears to have had an impact on vegetation in the Southwest United States, including significantly higher growth rates among conifers as well as an increase in the numbers of trees on the landscape (Swetnam and Betancourt 1998). The impact of warmer temperatures has also been documented for other parts of the American West (Cayan et al. 2001), where it has been manifested, for example, by the earlier spring flowering of different shrubs. Further studies of the impacts of recent climatic changes in the Southwest have documented changes in bird breeding patterns (Brown et al. 1999) and ecosystem reorganizations, whereby changes have been seen in both the fauna and flora at a Chihuahuan Desert site in southeastern Arizona (Brown et al. 1997).

The modern climate record exhibits substantial decadal-scale variability, as is illustrated in Figure 2. In the post-World War II period, increased temperatures, a severe drought in the 1950s, and unprecedented population growth have left their mark on the western landscape (Regensberg 1996; Swetnam and Betancourt 1998). Considering a longer temporal context, ecological studies have established that significant changes also have occurred in vegetation structure over the past 200 years, due in part to human-induced factors (Betancourt et al. 1993); and paleoclimate records indicate there may have been even more pronounced episodes of drought in the past 1,000 years, as compared with the last hundred years (Grissino-Mayer 1996).

Extreme climatic anomalies, whether periods of severe and sustained drought or periods of heavy precipitation that lead to severe flooding, produce large and measurable responses on the landscape through changes in erosion processes. Human actions often exacerbate these natural shocks; they are often taken because an analysis of the potential impacts of such actions either is lacking or is ignored. Furthermore, in cases where natural biomes cut across national boundaries, managing the impacts of climate and humans on nature tends to be complicated by political divisions. A change in outlook and of perspective by decision makers to more fundamentally take into account the interconnectedness of natural systems (NRC 1999) is a crucial step toward more effective stewardship of our natural resources.

It is recognized that change and variability are natural attributes of the earth's climate as they are of its biosphere. Therefore, in a longer-term perspective, natural biome boundaries are subject to modification, linked to variations in climate and other natural processes. What should be clearly understood is that the natural rates of change of climate and ecosystems are much slower than the corresponding rates of change forced by human activity. Hence, in a rapidly changing climate, such as is expected in the next century in response to human-induced alterations in atmospheric greenhouse gases (carbon dioxide, methane, nitrous oxides, and other gases), ecosystems will be unable to adapt fast enough to the projected changes in climate, with potentially disastrous consequences. Likewise, when humans modify their environment directly through actions such as deforestation, introduction of nonnative species, and dam construction, the abruptness with which those changes occur precludes nature from developing balanced responses to those changes. As was noted earlier, one way to minimize the impacts of humans on ecological systems is to incorporate information about basic ecological processes and the factors that control them into the decision-making framework. Some possible approaches to accomplishing this task are discussed next.

## 2.2. A River Basin Approach

An integrated river basin approach to transboundary water resources management is a long-established practice (this view was advocated for the Colorado River by Maj. John Wesley Powell in the nineteenth century, see Stegner 1954). In the case of the Colorado River Basin (CRB), while a number of legal instruments and institutional arrangements exist to manage the river (see a review by Getches, this volume), political boundaries have artificially divided the basin, contributing greatly to harming the riparian and estuarine ecosystems in the region. While progress has been made towards a more sustainable administration of the Colorado River system, damage to a range of ecosystems will remain a threat because the annual renewable yield of the river is below the amount that has been allocated annually for use among the parties involved (Pitt et al. 2000). This means that during severe and sustained drought there will not be enough water to satisfy all needs. The presence of sustained drought conditions is not unusual for the CRB, having being extensively documented in observational studies (Dracup and Kendall 1991), and from numerous tree-ring records throughout the basin (Meko et al. 1991). Under the push and pull of competing interests for the water, riparian habitats, particularly those on the Colorado River Delta (CRD) in Mexico, could be greatly damaged.

The CRD originally covered over 1.9 million acres (768,574 hectares), and supported an astonishing richness of plant, bird, and marine life in its broad tidal wetlands. After decades of dam construction upriver, this fecund delta ecosystem was nearly extinguished. Increased flows over the past couple decades have restored the CRD to a meager, though viable, 150,000-acre delta ecosystem (Pitt et al. 2000). The annual flow of the Colorado River below all major dams and diversions has steadily decreased since the early 1900s (Fig. 3). One could hardly conceive of a management history of a natural river system that is less in accordance with the precepts for sustainable development as described in the NRC (1999) report *Our Common Journey*.

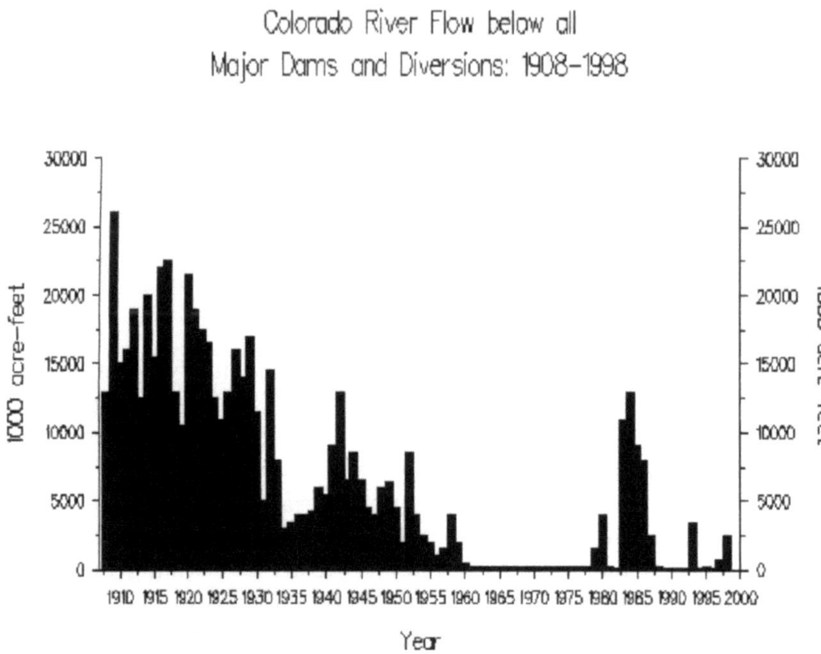


*Figure 3.* Graph depicting the annual flows of the Colorado River below all major dams and diversions since the start of record keeping. The reduction in flows beginning in the early 1930s is due to massive dam construction on the river in the United States.

While the storage capacity in the CRB is substantial and capable of ameliorating year-to-year fluctuations in precipitation and streamflow, the impact that a persistent period of aridity would have on the basin is generally not appreciated. Prolonged fluctuations are evident since 1900 in the decadally smoothed anomaly values (from the period-of-record mean) of the estimated virgin flow of the Colorado at Lees Ferry, Arizona (Fig. 4). The annual values were taken from the Fifty-Second Annual Report of the Upper Colorado River Commission, Salt Lake City, Utah (September 30, 2000). From 1934 through 1984, the 10-year progressive (cumulative) totals failed to exceed the estimated long-term mean flow at Lees Ferry of 150 million acre-feet (macf[1]). In two periods of up to 10 years, the flow was 120 macf or less (e.g., in 1931–40 and in 1954–63). Assuming that deliveries to the Lower Colorado River Basin and Mexico were to remain fixed, a recurrence of low flows in the Colorado similar to those that occurred in the historical record, as

[1] 1 million acre-feet = 1.2327 x $10^9$ $m^3$.

illustrated in Figure 4, would leave the Upper Colorado River Basin states with less than 20 macf for periods of a decade or longer to share among them. This figure represents less than 30% of the allocation that presumably is guaranteed to the Upper Basin states under the Colorado River Compact (see chapters by Getches, and Howe, this volume). The impact on riparian systems, hydropower generation, and instream uses, such as fishing and rafting, would be severe, and would likely lead to litigation among the nine basin states and possibly with Mexico.

A recognition on the part of decision makers is needed that (1) a river system is more than merely an aqueduct designed to pump an inexhaustible water supply to an ever increasing number of users; (2) that living things in a river basin depend on natural cycles of streamflow to maintain a proper ecological balance, and that we have an obligation to respect those natural balances; and (3) that humans benefit from the physical and biophysical functions that a river provides (not to mention aesthetic considerations). Thus, a critical goal of managing such systems must be its stewardship under long-term sustainable principles.

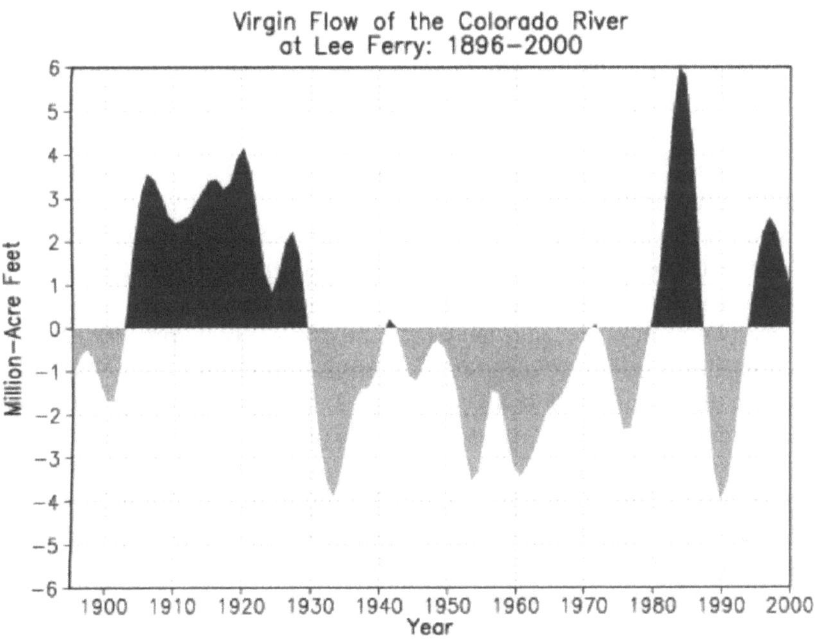


*Figure 4.* Graph showing a decadally smoothed series of annual unimpaired flow for the Colorado River as measured at Lees Ferry, Arizona. The graph illustrates the occurrence of a prolonged period of low flows, associated with western U.S. drought in the 1930s and 1950s.

## 2.3. A Climatological Approach

It is useful to define a set of internally consistent climatic regions that facilitate monitoring and assessment, and which are also useful for purposes of climate prediction and evaluation. A typical approach is to focus on homogeneous aspects of the hydrology and climatology within a broad area. This approach involves identifying those aspects of the climate that tie particular portions of the region together. Going through this exercise, one often finds that climate regions are quite similar to ecotones or biomes. The contiguous United States can be partitioned into a set of regions based on shared aspects of the annual cycle of monthly mean temperature and precipitation. The procedure can be done independently by using, first, the temperature field as the connecting variable, and then repeating the operation using the precipitation field.

A regionalization using the mean temperature field is shown in the top panel of Figure 5, while the result using the mean precipitation field is displayed in the bottom panel. The southwestern U.S. desert area emerges from this analysis as a unique region based solely on fundamental aspects of the long-term climate of the region. The annual cycles of monthly temperature and precipitation are illustrated in Figure 6 (top panels), and the bottom panels of Figure 6 give the decadal temperature and precipitation anomalies during the twentieth century. Note the close similarity of the graphs in Figure 6 with those given in Figure 2, for which the regional boundary for the Arizona portion of the Sonoran Desert biome was used to define the area and calculate the climatological values.

The graphs in Figures 2 and 6 illustrate an important point: The arid regions of the United States (and Mexico, data not shown) exhibit significant climatic variability on multiple timescales. Both these figures also show how the last two to three decades have been exceptionally warm and wet. As was noted earlier, studies have shown (Swetnam and Betancourt 1998) that vegetation in the Southwest United States has responded to this unusual warm/wet episode by expanding to higher elevations and becoming more densely distributed. Studies of changes in the hydrology of many western U.S. basins show that the seasonal streamflow peak associated with spring snowmelt has been occurring earlier during the last 20 years, compared with previous decades (Dettinger and Cayan 1995).

This analysis demonstrates that large areas of the southwestern United States, whether defined by robust statistical methodology of characteristic aspects of the local climate, or by natural processes on the landscape (such as an ecological biome), or by the boundaries of a large river basin, are generally subject to self-consistent physical constraints. Over the long course of time, this pattern has helped to shape the natural balances in the region. Below, some additional considerations on the subject of natural versus political boundaries are discussed within the context of climatic limits.

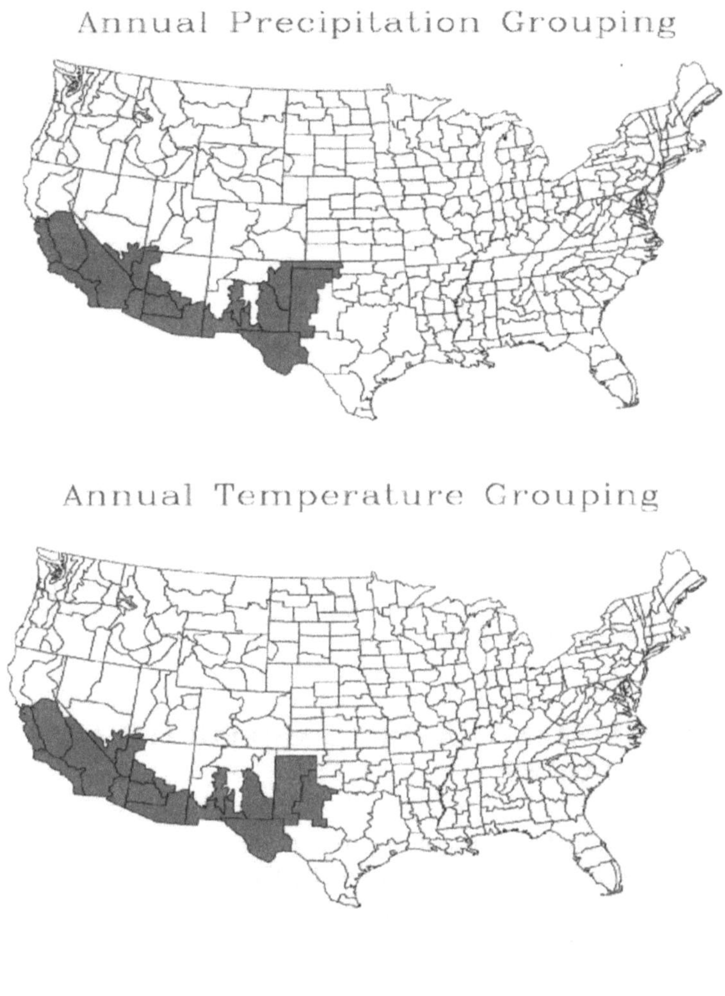


*Figure 5.* Regional representations of the Southwest desert region in the United States based on a regionalization technique that independently uses annual precipitation and annual temperature to develop a similarity metric. The results show very similar boundaries for the Southwest desert region of the United States.

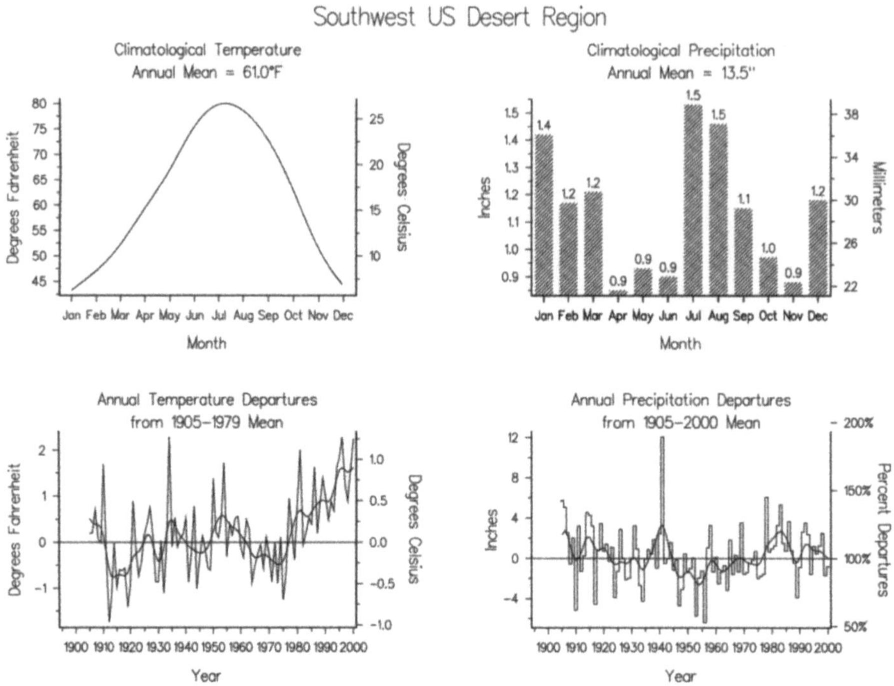


*Figure 6.* Same as Figure 2, except for the desert region shown in Figure 5 (using the precipitation-derived area). Note the strong warming trend and the generally wet conditions of the last 20–30 years.

## 3. RECONCILING NATURAL AND POLITICAL BOUNDARIES

The foregoing discussion has reiterated the case, made by others (e.g., Gleick 2000; Pitt et al. 2000; Michel 2000; Goodwin 2000) that natural resources, in general, and water resources in particular, need to be managed holistically. That means that a rational regional river basin unit would include the physical drainage area from the headwaters to the delta, including all tributary streams. To accomplish this goal, we must find a way to overcome the arbitrary impositions that political boundaries place on these natural systems. The Colorado River Basin is one of many basins where political boundaries impose irrational constraints on the natural system. The Colorado River Delta is a dynamic physical and ecological system that, before human influences became paramount, was subject to highly variable periods of low and high streamflow. The high variability in flow amounts was an intrinsic part of the functioning of the deltaic system, modifying the riparian habitat, but always assuring a sustainable ecosystem. At present, the sustainability of the CRD system has been seriously undermined by the retention in the United States of nearly all of the water that once flowed to the delta. If we are to develop sustainable goals for managing the entire CRB as a fundamental physical unit, then the process of natural streamflow variability must be restored, to a meaningful degree, to the CRD.

Improvements in climate prediction capabilities in the last few years offer one potentially useful tool for ensuring increased water flows to the CRD during times, such as during an El Niño event, when higher than normal precipitation may be expected in the Southwest United States. Many ideas have been proposed to provide sufficient water to the CRD in order to ensure its continuing viability as a rich, properly functioning ecosystem (see Pitt et al. 2000). One idea is to designate a high priority for surplus water flows from Lake Mead in the Lower Colorado during so-called surplus (wetter) years. Market mechanisms and water transfer agreements have also been suggested (see Howe, this volume). One suggestion is to charge the costs of ecosystem damages to Colorado River water users in the form of mitigation and restoration surcharges, implicitly recognizing the high-order function that these ecosystems play in sustaining a healthy environment (see also, Postel and Starke 1997).

In the case of the Colorado River, an arbitrary political boundary (from a biophysical viewpoint) between the United States and Mexico has severed the delta from its watershed. In the United States, water allocation allotments between the Upper and Lower Basins are based on state boundary and other political considerations. We also have a value system of water appropriation in which instream flows are not legally recognized as being of beneficial use. The Colorado is a nearly fully subscribed river; i.e., close to 90% of its renewable annual flow is being utilized. Consider then that paleoclimate data indicate that climatic variations over the past 500 years have resulted in prolonged periods of drought, during which reductions in annual streamflow of about 15% have occurred lasting for up to several decades (see Fig. 4).

It is clear that choices with long-term implications for the health of the Colorado River and its delta must be made soon. Obviously, the nature of the decisions will reflect the value our societies place on the multiple socioeconomic and natural uses that the river now supports—how much water for ecosystem health, how much for irrigation, how much for urban uses. The presence of the international border between the United States and Mexico presents unique difficulties for the future of the Colorado, but also unique opportunities to set an example of cooperation in order to achieve results for the greater good and long-term sustainable development of the region. The incorporation of climate information and analysis tools should be an integral part of the decision-making apparatus used in managing the Colorado River for the benefit, defined in the broadest of terms, of this and future generations.

## 4. REFERENCES

Betancourt, J.L., E.A. Pierson, K. Aasen-Rylander, J.A. Fairchild-Parks, and J.S. Dean. 1993. Influence of history and climate on New Mexico pinyon-juniper woodlands. *Proceedings: Managing Pinyon-Juniper Ecosystems for Sustainability and Social Needs*, Santa Fe, New Mexico, U.S. Department of Agriculture Forest Service General Technical Report RM-236, pp. 42–62.

Blatter, J., and H. Ingram (eds.). 2001. *Reflections on Water: New Approaches to Transboundary Conflicts and Cooperation*. Boston: Massachusetts Institute of Technology Press, 356 pp.

Brown, J.H., T.J. Valone, and C.G. Curtin. 1997. Reorganization of an arid ecosystem in response to recent climate change. *Proceedings of the National Academy of Sciences, USA* 94: 9729–9733.

Brown, J.L., S.-H. Li, and N. Bhagabati. 1999. Long-term trend toward earlier breeding in an American bird: A response to global warming? *Proceedings of the National Academy of Sciences, USA* 96: 5565–5569.

Cayan, D.R., S.A. Kammerdierner, M.D. Dettinger, J.M. Caprio, and D.H. Peterson. 2001. Changes in the onset of spring in the western United States. *Bulletin of the American Meteorological Society* 82: 399–415.

Dettinger, M.D., and D.R. Cayan. 1995. Large-scale atmospheric forcing of recent trends toward early snowmelt runoff in California. *Journal of Climate* 8: 606–623.

Dracup, J.A., and D.R. Kendall. 1991. Climate uncertainty: Implications for operations of water control systems. In, National Research Council, *Managing Water Resources in the West Under Conditions of Climate Uncertainty*. Washington, D.C.: National Academy Press, pp. 177–216.

Gleick, P.H. 2000. The changing water paradigm: A look at twenty-first century water resources development. *Water International* 25: 127–138.

Goodwin, S.L. 2000. Conservation connections in a fragmented desert environment: The U.S.–Mexico border. *Natural Resources Journal* 40: 989–1016.

Grissino-Mayer, H.D. 1996. A 2129-year reconstruction of precipitation for northwestern New Mexico, USA. *Tree Rings, Environment and Humanity: Proceedings of the International Conference*, Tucson, Arizona, 17–21 May 1994, *Radiocarbon*, 191–204.

Meko, D., M. Hughes, and C. Stockton. 1991. Climate change and variability: The paleo record. In, National Research Council, *Managing Water Resources in the West Under Conditions of Climate Uncertainty*. Washington, D.C.: National Academy Press, pp. 71–100.

Michel, S.M. 2000. Defining hydrocommons governance along the border of the Californias: A case study of transbasin diversions and water quality in the Tijuana–San Diego metropolitan region. *Natural Resources Journal* 40: 931–972.

National Research Council (NRC). 1999. *Our Common Journey: A Transition Toward Sustainability*. Washington, D.C.: National Academy Press, 363 pp.

Pitt, J., D.F. Luecke, M.J. Cohen, E.P. Glenn, and C. Valdés-Casillas. 2000. Two nations, one river: Managing ecosystem conservation in the Colorado River Delta. *Natural Resources Journal* 40: 819–871.

Postel, S., and L. Starke (eds.). 1997. *Last Oasis: Facing Water Scarcity*, 2nd edition. New York: W.W. Norton and Co., 239 pp.

Regensberg, A. 1996. General dynamic of drought, ranching, and politics in New Mexico, 1953–1961. *New Mexico Historical Review* 71: 25–49.

Stegner, W. 1954. *Beyond the Hundredth Meridian: John Wesley Powell and the Second Opening of the West*. Boston: Houghton Mifflin Co.

Swetnam, T.W., and J.L. Betancourt. 1998. Mesoscale disturbance and ecological response to decadal climatic variability in the American Southwest. *Journal of Climate* 11: 3128–3147.

11

# THE TRANSBOUNDARY SETTING OF CALIFORNIA'S WATER AND HYDROPOWER SYSTEMS

*Linkages between the Sierra Nevada, Columbia, and Colorado Hydroclimates*

Daniel R. Cayan,[1,2] Michael D. Dettinger,[2,1] Kelly T. Redmond,[3] Gregory J. McCabe,[4] Noah Knowles,[1] and David H. Peterson[5]

[1]*Climate Research Division, Scripps Institute of Oceanography, La Jolla, California 92093*
[2]*U.S. Geological Survey, La Jolla, California 92093*
[3]*Desert Research Institute/Western Regional Climate Center, Reno, Nevada 89512*
[4]*U.S. Geological Survey, Denver, Colorado 80225*
[5]*U.S. Geological Survey, Menlo Park, California 94025*

**Abstract** Climate fluctuations are an environmental stress that must be factored into our designs for water resources, power, and other societal and environmental concerns. Under California's Mediterranean setting, winter and summer climate fluctuations both have important consequences. Winter climatic conditions determine the rates of water delivery to the state, and summer conditions determine most demands for water and energy. Both are dictated by spatially and temporally structured climate patterns over the Pacific and North America. Winter climatic conditions have particularly strong impacts on hydropower production and on San Francisco Bay/Delta water quality.

It is thus noteworthy that precipitation from winter storms in California is more variable than in neighboring regions. For example, annual discharge from the Sacramento–San Joaquin system has a coefficient of variation (standard deviation/mean) of 44% compared to 19% in the Columbia Basin and 33% in the Colorado Basin. Also, in California, multiyear droughts occur more often than would be expected by chance, but wet years do not exhibit such persistence. A crucial aspect of California's climate stresses is that they influence conditions over broad spatial scales. Climate patterns that cause the state's climatic fluctuations typically reach well beyond its boundaries. This breadth affects California because much of the energy and water used here is supplied by distant parts of the state as well as from the Northwest and Southwest. When dry winters occur in the Sierra Nevada, they also tend to occur in the Columbia and Colorado Basins.



*Henry F. Diaz and Barbara J. Morehouse (eds.), Climate and Water*, 237-262.

These regional scales, coupled with California's reliance on resources from an especially broad region, including power from the Columbia and Colorado Basins and water from the Colorado, make the state especially vulnerable to climate fluctuations. These vulnerabilities are likely to grow as the population and demands for resources in the region continue to grow.

## 1. INTRODUCTION

Environmental stresses such as climate fluctuations have the potential to cause ever-greater impacts on the western United States as the population grows. From 1990 to 2000, the population of the 11 western conterminous United States increased by 19.7%, from 51.2 to 61.3 million residents. Over the same period, the number of people in California (the most populous state in the nation) rose by nearly 14%—from about 29.8 million to about 33.9 million residents (United States Census data 2000). Among the multitude of stresses threatening, and associated with, this rapidly growing population, climate variations have large impacts on societal and ecological structures because they determine the amount of resources, such as water, supplied to the region. Furthermore, the stresses placed upon one region can affect conditions in others, partly because water and energy are traded or transferred across state and watershed boundaries.

As can be argued for a global scale, in the western United States there are compelling reasons to consider environmental and societal stresses at the scale of large watershed systems. In many cases, a region's populace depends upon processes and human activities within a watershed for substantial portions of its water supply, electrical power, ecological habitat, transportation, and recreation. Consequently, recent applied science programs to study and organize multidisciplinary climate and environmental information for the western United States have structured their efforts around watersheds; e.g., the U.S. Geological Survey's Place-Based Studies Program (http://access.usgs.gov/) and the National Oceanic and Atmospheric Administration Office of Global Programs (NOAA-OGP) Regional Integrated Science Assessments (http://www.ogp.noaa.gov/mpe/csi/risa/index.htm). California's largest watershed is the collection of river drainages from the west slope of the Sierra Nevada that combine to form the Sacramento and San Joaquin Rivers. These large rivers converge at the San Francisco Bay Delta and supply much of the state's water. This overall watershed and rivers system is termed the Sierra watershed in this chapter.

Hydroclimatic linkages of the Sierra to two other large watersheds, the Columbia River Basin (Pulwarty and Redmond 1997; Hamlet 2003, this volume) and the Colorado River Basin (Diaz and Anderson 1995; Harding et al. 1995; Lord et al. 1995), are analyzed in this study. These three watersheds are shown on the map in Figure 1, and the annual discharge (natural flow estimates) is plotted in Figure 2. Each of these systems is highly managed (Hamlet 2003, this volume; Lord et al. 1995), and in each, climate variability is recognized as a major stress (Roos 1991, 1994; Pulwarty and Redmond 1997; Hamlet 2003, this volume; Diaz and Anderson 1995; Harding et al. 1995; Lord et al. 1995; California Department of Water Resources 1998).

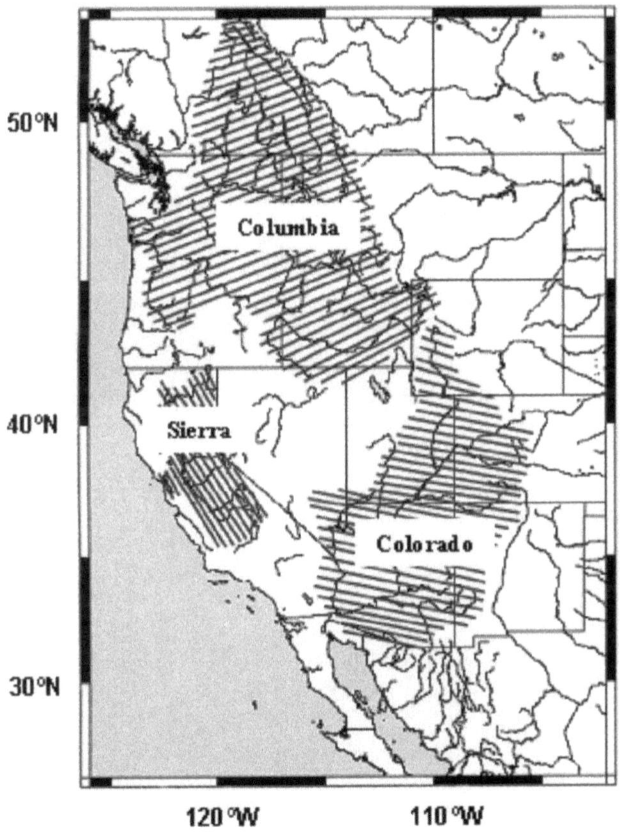


*Figure 1.* Columbia, Sierra, and Colorado watersheds.

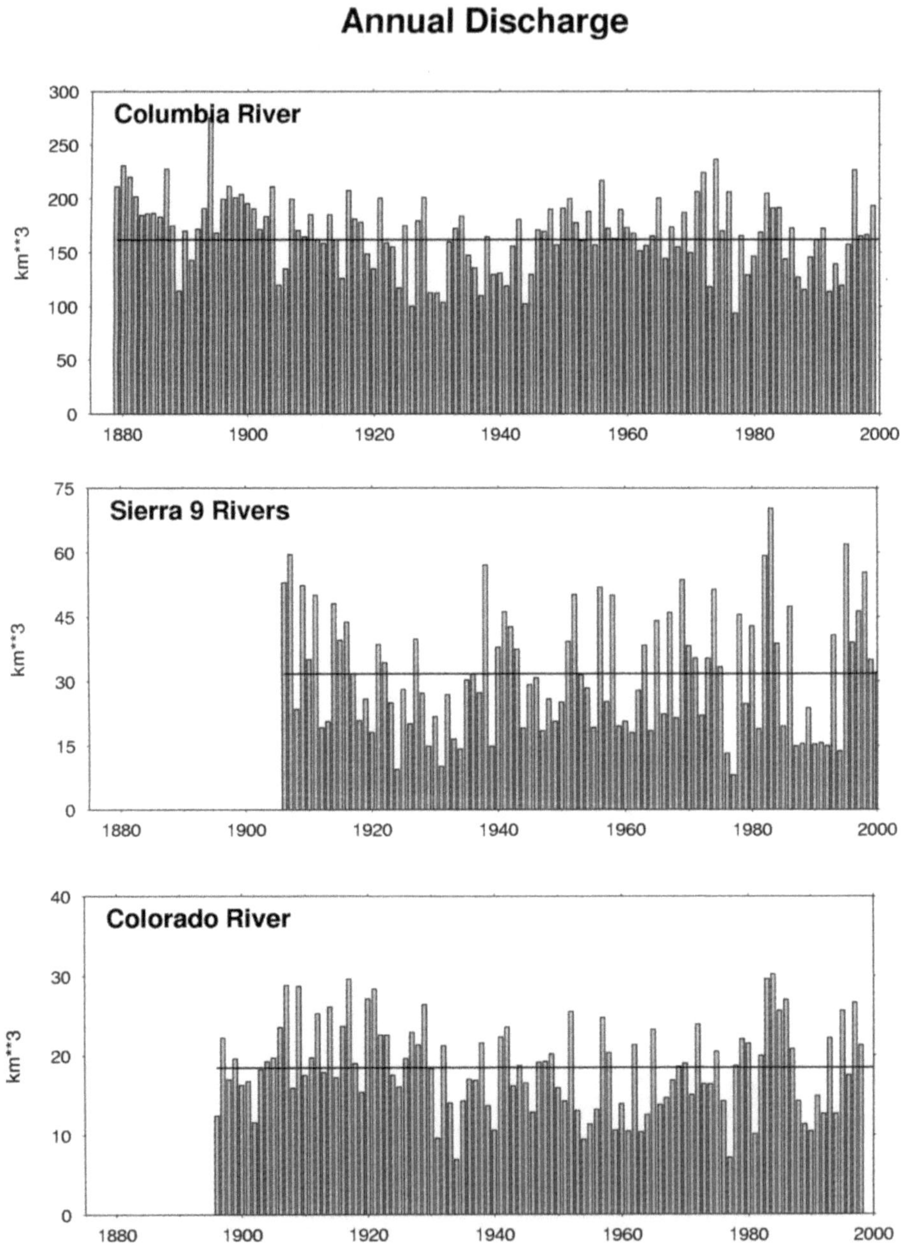


*Figure 2.* Water-year natural river-discharge totals (cubic kilometers) for the Columbia River at The Dalles, the Sierra watershed (nine largest rivers draining the west slope of the Sierra Nevada), and the Colorado River at Lees Ferry. Horizontal lines indicate 1906–99 mean (1906–99) discharge.

California's water and power supplies are intimately linked to the hydroclimates of all three watersheds. In an average year, California receives about 200 million acre feet (hereafter MAF; 1 MAF = 1.234 $km^3$ of water) of precipitation, of which only 71 MAF is left after evaporation and transpiration to form runoff. About 42 MAF of this runoff is used for nonenvironmental (agricultural or urban) consumption. Even within the state, water supplies link different regions. Approximately 75% of the state's runoff occurs north of San Francisco Bay, while 72% of nonenvironmental consumption occurs south of San Francisco Bay, supplied by massive federal and state water storage and conveyance systems (California Department of Water Resources 1998). These wholesale within-state water transfers affect water quality and ecosystems as well as water users. Water quality in San Francisco Bay, indicated by May monthly salinity anomalies at Suisun Bay, is very strongly correlated ($r = 0.96$) with freshwater flows from the Sierra watershed as shown in Figure 3. However, also illustrated in Figure 3 are freshwater withdrawals from the San Francisco Bay Delta southward to the San Joaquin Valley or Southern California. These exports have increased over the last few decades and have contributed to water quality and estuarine degradation in the bay and delta (Peterson et al. 1995; Knowles 2002; Knowles et al. 2002). Withdrawals from the delta are usually greatest in July and August and least in January and February (Knowles et al. 2002). Interestingly, and perhaps important from water and energy resources perspectives, is that withdrawals on the Columbia system are greatest in winter and least in summer (Hamlet 2003, this volume). These exports are determined by demand, supply, and perhaps the timing of the winter storm season, and thus exhibit a complex relationship to freshwater flows. From outside the state, during recent years California has imported about 5.4 MAF of water, most of it from the Colorado River and some from Oregon (California Department of Water Resources 1998). Also, about 1.2 MAF flows from California to Nevada from the east side of the Sierra Nevada. In summary, California depends upon the Colorado River to supply approximately 12% of its 42.6 MAF annual developed, nonenvironmental water supply. California is currently scrambling to assemble a workable plan to live within its legal yearly entitlement from the Colorado River of 4.4 MAF (Newcom 2002). Thus, California's water supplies include disparate sources, both within and beyond its boundaries, with particularly important linkages to the Colorado Basin.

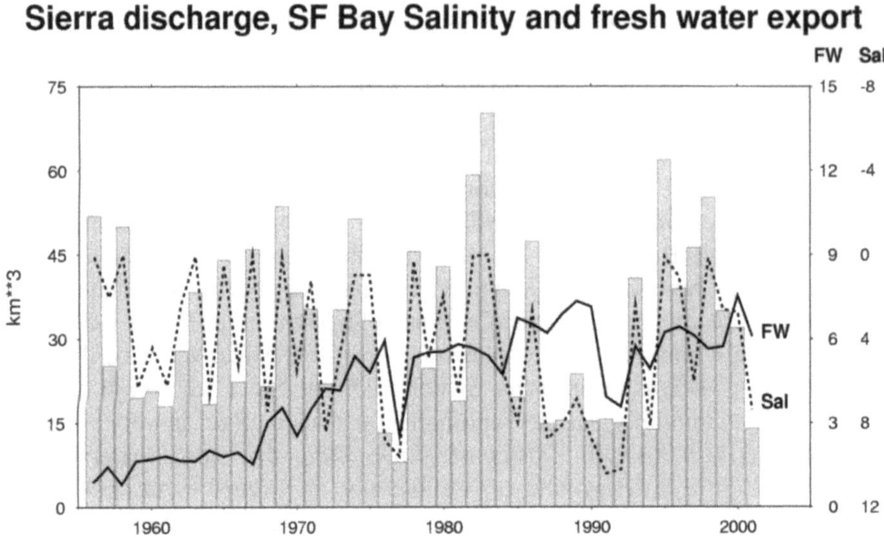


*Figure 3.* Sierra discharge (cubic kilometers) (bars), San Francisco Bay May salinity anomalies [per mille] in Suisun Bay (dotted line) and freshwater export (cubic kilometers) southward from the San Francisco Bay Delta (solid line). Salinity is estimated from the advection/diffusion model of Knowles (2002).

Meanwhile, California's electrical consumption, approximately 230,000 gigawatt-hours (GWh) per year, is approximately 40% of the total consumption of the 11 western states (Fisher and Duane 2001). This situation developed while annual consumption over the 11 western states increased from 350,000 GWh in 1977 to nearly 570,000 GWh in 1998, an increase of 63%. California's electrical system is closely connected to that of the 11 western states, and nearly 20% of California's electricity is imported, about equally from the Northwest and the Southwest regions of the United States (Fig. 4; California Energy Commission data). These imports are a necessary part of California's energy system today, because California/Mexico peak electrical demand is approximately 55,000 megawatts (MW) and California's electrical generation capacity is only about 44,000 MW. California's consumption has increased considerably, along with its electrical generation (Fig. 4). Consumption in the state rose from 160,000 GWh in 1977 to 230,000 GWh in 1998, an increase of 43%. The western region has a total capacity of 133,000 MW and a peak demand of approximately 130,000 MW (Fisher and Duane 2001). Important from a seasonal climate perspective is that in California, peak demand in summer is usually nearly 50% higher than it is in winter, while in the Pacific Northwest, peak demand in winter is about 20% higher than it is in summer. For the combined western states, peak demand is approximately 10% higher in summer than in winter, evidently because of the increased load caused by air conditioning, pumping water, and other seasonal activities. Peak demand in the western region has increased from 84,000 MW to over 130,000 MW from 1982 to 1998; this constitutes an increase of more than a 50% in two decades. Hydroelectric power generation within California averages about 35,000 GWh, about 15% of total electrical generation (Fig. 4). Notice, in Figure 4 and in Table 1, that year-to-year variations in hydropower generation in the Sierra Nevada have closely followed the year-to-year availabilities of Sierra discharge with a correlation of approximately 0.9. The importation of power to California, by contrast, has generally followed year-to-year fluctuations in the flow of the Columbia River, especially prior to 1995 ($r = 0.58$). The amount of imported electrical energy was particularly low in 2000, when the Columbia River Basin, along with most of the northwestern United States, was very dry. Thus California's electrical power situation typically responds to the hydroclimates of both river basins, as well as to the hydropower sources of the Colorado River system. Indeed, California's recent power crisis, during which electricity costs were driven up catastrophically in the summer of 2001, developed in part because of lower than expected hydropower production in the Columbia Basin due to prolonged dry conditions in the Northwest (e.g., http://www.sfgate.com/energy/).

Because of these interdependencies, a year during which any one of the three watersheds is drier than normal poses potential problems for California. Years in which two or more of the watersheds are dry are particularly threatening. These threats become especially problematic if neither of the remaining watersheds is wet enough to permit compensatory adaptations in California's water or, especially, power systems. Conversely, years in which one basin is wet and one of the others is dry may have compensating benefits. Droughts, as well as floods, are clearly normal facets of the modern climate, although they have posed particularly severe resource-management problems during recent decades (Roos 1994; Betancourt 2002; McCabe et al. 2002; Namias 1978, National Research Council [NRC] 1999). High-resolution paleoclimate measures indicate that the western region has enjoyed wet spells but has also suffered severe sustained drought during the last several centuries (Meko et al. 1995; Stahle et al. 2001; Meko et al. 2001; McCabe et al. 2002). Thus, in this chapter we attempt to clarify these linkages by investigating how high and low river discharges in the Sierra watershed have historically related to (coincided with) those in the Columbia and Colorado Rivers and how these relationships are determined by climate variability.

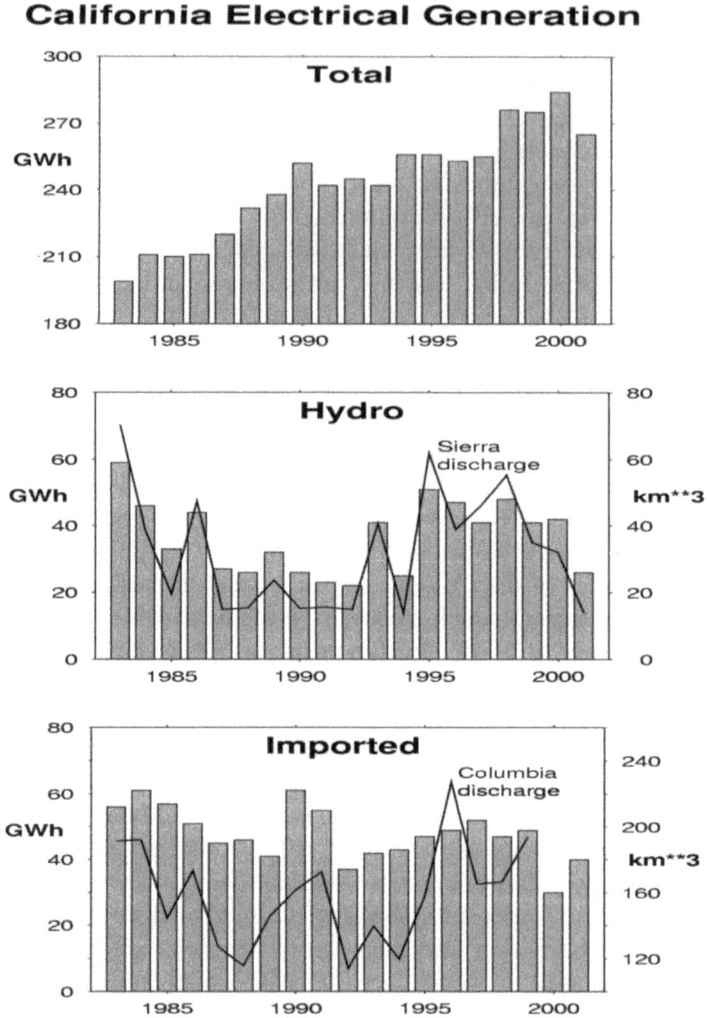


*Figure 4.* (*a*) Total generated electricity used in thousands of gigawatt-hours (GWh) in California (upper), (*b*) hydroelectric energy generated (thousand-GWh) in California (middle), and (*c*) net imported electrical energy used (thousand-GWh) in California (lower). For comparison, Sierra watershed water-year discharge (cubic kilometers) and Columbia River water-year discharge (cubic kilometers) are also plotted (solid lines) on the middle and lower panels. (Data from California Energy Commission).

*Table 1.* Correlations among annual hydroelectric generation totals and correlations between annual hydroelectric generation and annual discharge in the western United States, 1976-1994.

*(Hydroelectric data from Western Area Power Administration)*

| | *hydro-electric generation regions* | | | | *watersheds* | | |
|---|---|---|---|---|---|---|---|
| | Pacific NW | California-Nevada | Ari-zona-New Mexico | Rocky Mtns | Columbia River | Sierra Watershed | Colorado River |
| Pacific NW | | 0.27 | -0.08 | 0.19 | 0.92 | | |
| Califor-nia-Nevada | | | 0.33 | 0.76 | | 0.88 | |
| Arizona New Mexico | | | | 0.73 | | | 0.54 |
| Rocky Mtns. | | | | | | | 0.80 |

## 2. DATA

Natural discharge estimates are analyzed for the Columbia River at The Dalles, by using data from the U.S. Army Corps of Engineers, for the Colorado River at Lees Ferry, from annual reports of the Upper Colorado River Commission, and for the nine largest rivers (Upper Sacramento, Feather, Yuba, American, Stanislaus, Tuolumne, Merced, Upper San Joaquin, and Kings Rivers) draining the west slopes of the Sierra Nevada, based on data from the California Department of Water Resources. Discharge measurements from hundreds of additional gages around the conterminous United States, selected for their relative lack of human influences (Slack and Landwehr 1992), also are analyzed to provide regional hydrologic contexts for the behaviors of the three large watersheds considered here. Monthly precipitation totals for United States climate divisions (Karl and Knight 1985) from the National Climatic Data Center are used to characterize precipitation inputs to each of the three watersheds. For the Columbia watershed, the Canadian sector that comprises the upper part of the basin is not included. Daily precipitation from hundreds of cooperative and first-order stations over the conterminous western United States (Eischeid et al. 2000) were employed to characterize the variability within and between watersheds across the region. Monthly gridded 700 millibar (mb) height fields over the Northern Hemisphere obtained from the NOAA National Center for Environmental Prediction are analyzed to identify large-scale atmospheric circulations associated with the hydroclimatic variations. Finally, annual electrical energy generation and usage data were obtained from the Western Area Power Administration and the California Energy Commission.

## 3. RUNS OF HIGH AND LOW DISCHARGE

Water and power users in the three western river systems must contend with the year-to-year variations of flow in the three rivers, as illustrated by their time histories shown in Figure 2. For example, flows in the California Sierra have fallen as low as 25% (water year 1977) of the historical mean and have been as high as 221% (water year 1983) of the historical mean. The variability of the Sierra watershed is particularly high among the three watersheds analyzed in this chapter. Its coefficient of variation of the annual discharge is 0.44, compared to 0.19 for the Columbia and 0.31 for the Colorado (Table 2). The high Sierran variability is a reflection of a regional pattern of high coefficients of variation across the Southwest (Fig. 5). The Columbia watershed, like most rivers in the Northwest, experiences much less variability. The Colorado watershed drains parts of both the Southwest (with high variability) and the interior Northwest (with low variability), and thus as a whole is moderately variable. Also, the area of the Sierra watershed, at approximately 140,000 $km^2$, is less than one-fourth the size of the Columbia watershed (approximately 617,000 $km^2$) or the Colorado watershed (approximately 626,800 $km^2$). Consequently, there is not great opportunity for one portion of the Sierra drainage to compensate for the extremes that occur in another portion of the watershed (Cayan 1996).

*Table 2. Annual discharge statistics, 1906–1999.*

| | $\mu$ *[km³]* | $\sigma$ *[km³]* | *C.V* | max *[km³]* | min *[km³]* |
|---|---|---|---|---|---|
| Columbia | 162.4 | 31.0 | 0.19 | 236.9 *(1974)* | 93.9 *(1977)* |
| Sierra | 31.8 | 14.1 | 0.44 | 70.2 *(1983)* | 8.1 *(1977)* |
| Colorado | 18.6 | 5.7 | 0.31 | 30.2 *(1984)* | 6.9 *(1934)* |

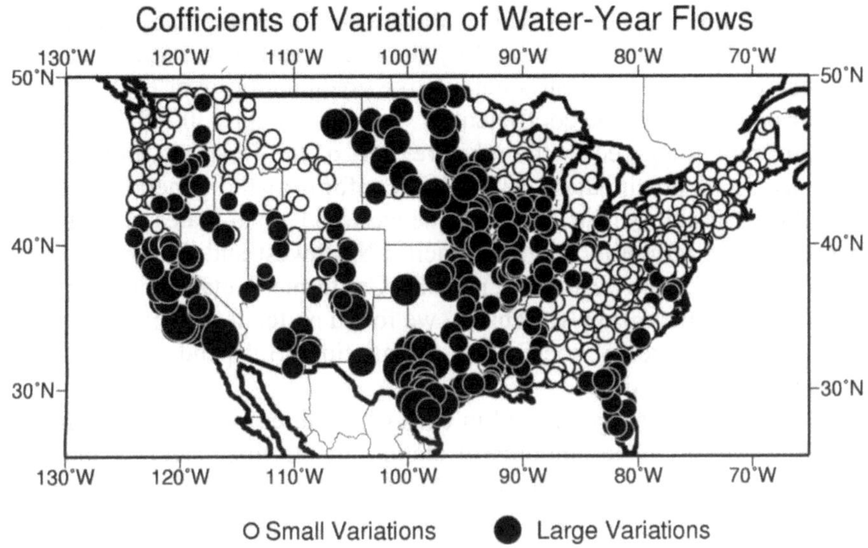


*Figure 5.* Coefficients of variation of annual discharges in streams in the conterminous United States for the periods of record at each gage. Circle radius is proportional to the magnitude of the coefficient. Values less than 0.37 and greater than 0.37 are plotted as open/filled circles, where 0.37 is the median value from all the gages shown.

Precipitation, which fosters this discharge variability in the three watersheds, falls over differing seasons and, especially, over differing fractions of the water year. The duration of the season over which the major fraction of annual total precipitation accumulates is particularly short in California, in accord with its Mediterranean precipitation regime. Figure 6a shows that L67, defined here as the number of days required to accumulate 67% of the mean annual precipitation, is a relatively brief period in the California region. L67 is calculated by using daily long-term mean precipitation data (Eischeid et al. 2000) to identify the period of the year, regardless of starting day, that accumulates 67% of the annual mean precipitation. L67 ranges from about 90 to about 120 days in California. In contrast, the wet seasons measured by L67 are longer (120 to 220 days) in the other two basins. This means that the Sierra watershed accumulates its yearly water supply in a relatively short time, on average. The narrow seasonal window that provides precipitation for California may also contribute to the high variability of the state's precipitation and runoff. The shorter the period within which a region typically accumulates its annual precipitation supply, the more vulnerable it will be to climate fluctuations. When the wet season is short, there are fewer chances to offset dry spells, should they occur during the core precipitation season. This is illustrated, for example, by the relative magnitude of variability, indicated by the coefficient of variation (standard deviation/mean) of each year's cumulative precipitation over each station's L67 period (Fig. 6b). The coefficient of variation is high (30–60%) throughout California compared to the other regions of the western United States. These levels of precipitation variability conform to the general pattern of streamflow variability (Fig. 5) across the western United States, in which the lowest relative variability is in the Pacific Northwest and the greatest variability is in the Southwest and especially California. Notably, in a separate analysis of the streamflow variability, we found no tendency for this pattern, or the absolute values of coefficients of variation, to depend on the sizes of the river basins considered.

Reservoir storage is used in each of the three watersheds to moderate this variability. The amounts of storage in the three river systems are quite similar, with 50, 32, and 60 MAF (60, 49, and 74 $km^3$) of storage in the Columbia, California, and Colorado systems, respectively. In terms of annual flow volume, however, these storages are remarkably dissimilar, at 30%, 154%, and 397% of average annual discharge, respectively. Thus, while the Colorado is distinguished by low variability, it has relatively little storage. The Colorado has moderately high variability but it has a large storage capacity. The Sierra system has high variability and a modest amount of reservoir storage.

Overall, there is little persistence of a given year's anomalous discharge to that of the next year. The 1-year autocorrelations of the Columbia, Sierra, and Colorado annual discharges are 0.06, 0.08, and 0.24, respectively. However, the annual discharge series (Fig. 2) for the three watersheds contain decadal to multidecadal variations. Spectral analysis, using the multitaper method (Mann and Lees 1996), identifies these low-frequency variations as marginally significant in the Colorado discharges, more convincingly significant in the Sierra, and strongly significant in the Columbia River series. For many concerns, the associated multiyear runs of high or low flows have greater societal and ecological impacts than isolated extreme years.

To investigate further, an analysis of "runs" of high or low extremes was conducted. In this chapter, we adopt relatively weak criteria for identifying wet spells and droughts: In the present analysis, extreme events are defined as years in which the annual discharge is in the upper/lower third or upper/lower sixth of its observed distribution. This analysis was performed by tallying the number (m) of above-high-threshold and below-low-threshold flows within each successive N-year interval of the flow series. In the experiment discussed here, the two thresholds considered were the upper and lower sextiles of the annual discharge series, although the upper/lower terciles and above/below median also were examined with qualitatively similar results. The interval length N = 10 yr was considered. The resulting distribution of m's was compared with those from a Monte Carlo experiment in which the annual discharge series were shuffled randomly over 1,000 independent trials (Fig. 7). As was suspected from visual inspection and the spectral analysis, this analysis demonstrates that anomalously low and sometimes high-discharge years cluster in time, more so than can be explained by chance. For example (Fig. 7), while the random series would on average produce only six 10-year intervals containing four or more lowest sixth flow volumes, the observed Sierra discharge series produced 17 such 10-year intervals in the 1906–99 historical record (Table 3). The low Sierra discharge sequences fell into two main episodes: beginning in the late 1920s and beginning in the mid-1980s through the early 1990s. The number of 10-year intervals with "runs" of 4 or more years of high flows is not as unusual as the number of low-flow runs, except for the Colorado River (Fig. 7).

*Table 3.* 10 *yr* intervals with 4 or more extremely high or extremely low annual discharge totals. Extremes are highest and lowest 16 years between 1906 and 1999. Years listed are beginning year of each 10-year interval.

| ***Columbia*** | | ***Sierra*** | | ***Colorado*** | |
|---|---|---|---|---|---|
| *high 10* | *low 10* | *high 10* | *low 10* | *high 10* | *low 10* |
| 1965 | 1921 | 1906 | 1924 | 1906 | 1952 |
| 1967 | 1922 | 1907 | 1925 | 1907 | 1953 |
| 1968 | 1923 | | 1926 | 1908 | 1954 |
| 1969 | 1924 | | 1927 | 1909 | 1955 |
| 1970 | 1925 | | 1928 | 1911 | 1956 |
| 1971 | 1926 | | 1929 | 1912 | 1957 |
| | 1928 | | 1930 | 1913 | 1958 |
| | 1929 | | 1931 | 1914 | 1959 |
| | 1985 | | 1982 | 1977 | 1985 |
| | 1986 | | 1983 | 1978 | 1986 |
| | 1987 | | 1984 | 1979 | 1987 |
| | | | 1985 | 1980 | 1988 |
| | | | 1986 | 1981 | 1989 |
| | | | 1987 | 1982 | |
| | | | 1988 | 1983 | |
| | | | 1989 | | |
| | | | 1990 | | |

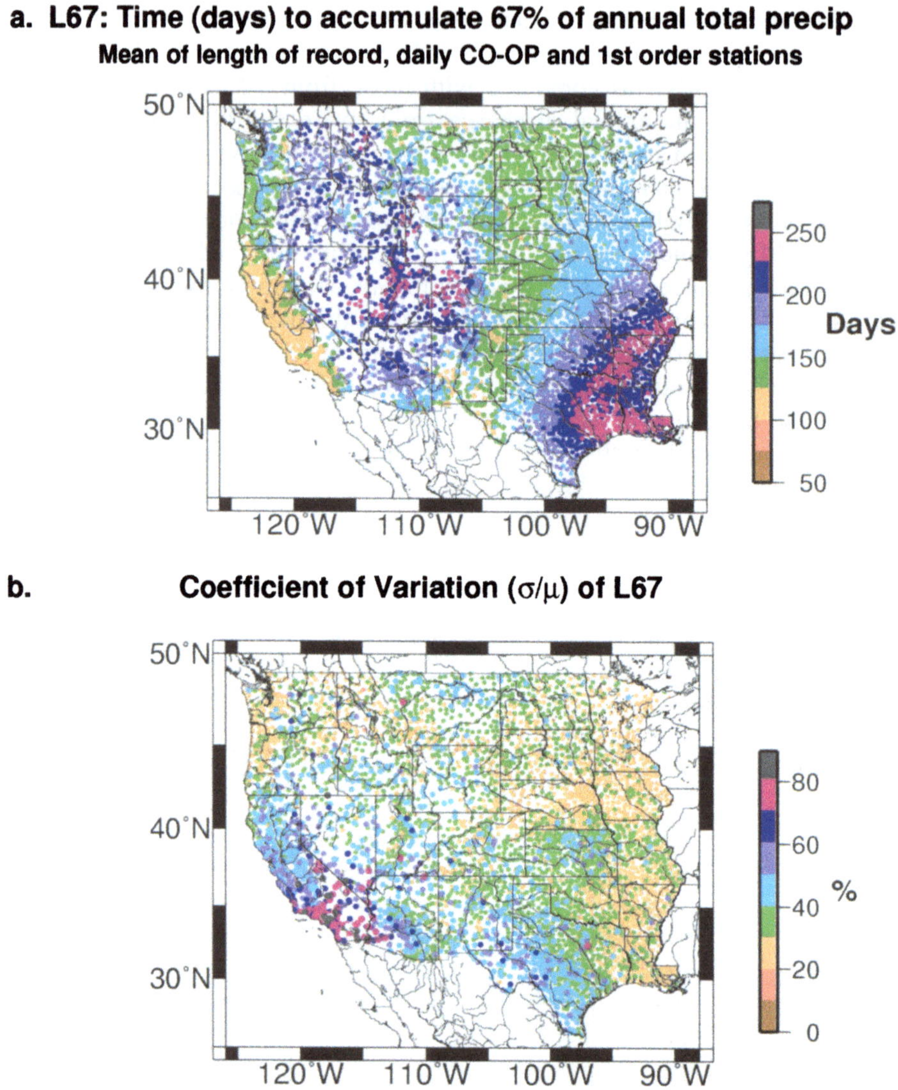


*Figure 6.*( *a*) Number of days (L67) required to accumulate 67% of the annual climatological total precipitation, calculated from long-term daily mean precipitation over the entire record available at each station. The beginning of the L67 "season" is the day for which L67 precipitation accumulation is at its minimum. The dot size becomes darker (see shading scale) and larger as L67 decreases. (*b*) Coefficient of variation (standard deviation/mean) of accumulated precipitation over the climatological L67 period of each year. The dot size becomes darker (see shading scale) and larger as the coefficient of variation increases.

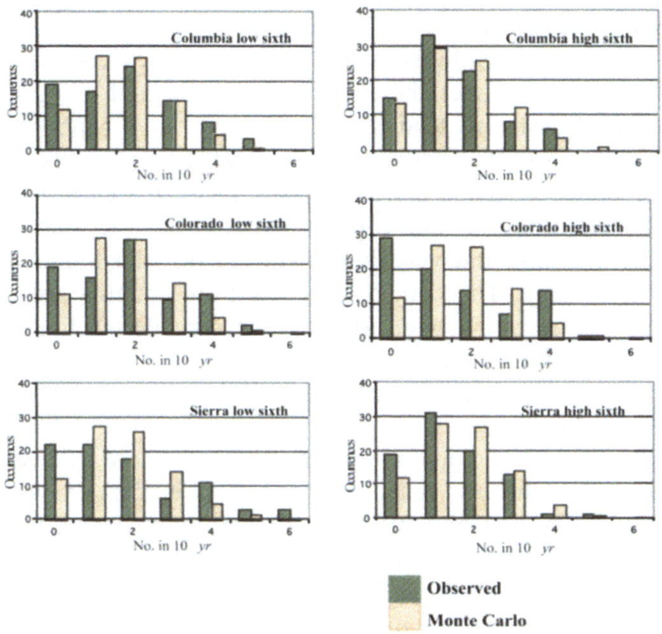


*Figure 7.* Observed (black) vs. Monte Carlo (gray) simulated occurrences of m years (x-axis) out of 10 years when annual discharges were in the lowest (upper) and highest (lower) sixth of the long-term flow distributions for the Columbia, Sierra, and Colorado watersheds.

## 4. REGIONAL CONNECTIONS OF HIGH AND LOW DISCHARGE

Although the climates of the Columbia, Sierra, and Colorado Basins differ from each other, their water-year hydrologic variations are often correlated (Cayan 1996; Dettinger et al. 1998). Figure 8 shows average discharge anomalies (as *t* scores) at U.S. Geological Survey stream gages during years with high (left side panels) and low (right) discharge in the three watersheds. High/low years are defined as those whose annual discharge was in the upper/lower third of its observed 1906–99 distribution. Each composite has the strongest anomalies in and near the targeted watershed, but as importantly, each has significant anomalies that spread on a regional scale, well beyond the watershed. Interestingly, some strong anomalies extend well to the east; e.g., when heavy flows occur in the Sierras and in the

Colorado River system, heavy flows also occur in rivers in the northern Great Plains and western Midwest. When the Colorado River experiences high or low flows, rivers over the Pacific Northwest and along a broad swath of the East Coast also tend to have low or heavy (opposite) flow statuses. Examination of the discharge series reveals several cases when the Colorado and Columbia Rivers are in opposite phase extremes (Table 4). There is a lesser tendency for out-of-phase structure for the Sierra vs. the Columbia River systems, and very little evidence for the Sierras vs. the Colorado River systems. The composite streamflow patterns are roughly symmetric over the western United States when considering high- versus low-flow years in the three watersheds.

However, an important distinction, which bears on the region's water and hydropower resources, is that the regional streamflow patterns associated with Columbia and Colorado low-flow years are more extensive than the patterns in the high-flow years. Notably for California's water and power resources, when the Columbia River is drier than normal, low flows occur in rivers extending southward into Northern California. Low-flow anomalies associated with dry years on the Colorado extend westward into most of California. The high-flow patterns associated with both of these basins are more constricted and not as strong over California.

The association between high and, especially, low flows in the three watersheds can also be seen by the numbers of co-occurring years with high and low annual discharges in the three river systems (Table 4). In each pair, especially for the Sierra-Colorado pair, high-high and low-low flow combinations predominate, and high-low or low-high combinations are relatively uncommon. These contingencies stand out as highly significant when compared to Monte Carlo experiments matching extreme-year pairings in 1,000 randomly shuffled versions of the three discharge series. The Monte Carlo exercise indicates that $7.6 \pm 2.0$ low Sierra/low Columbia flow years and $9.2 \pm 2.2$ low Sierra/low Colorado flow years would occur if the discharge series were randomly and independently arranged, while the observed series produced 13 and 16 such occurrences, respectively. High/high flow years are similarly accentuated in the observed series, while high/low and low/high coincidences across the pairs of basins occurred less often than chance would have them. Finally, if we consider cases where all three basins have low flows or all three have high flows, the Monte Carlo exercise produces $2.3 \pm 1.3$ and $1.1 \pm 1.0$ such years by chance, while the observed record has six low/low/low years (1931, 1939, 1977, 1988, 1992, 1994)[1] and four high/high/high years (1907, 1916, 1965, 1983). Thus, the three western

---

[1] 2001 was not included in the present analysis, but would qualify as another low/low/low year.

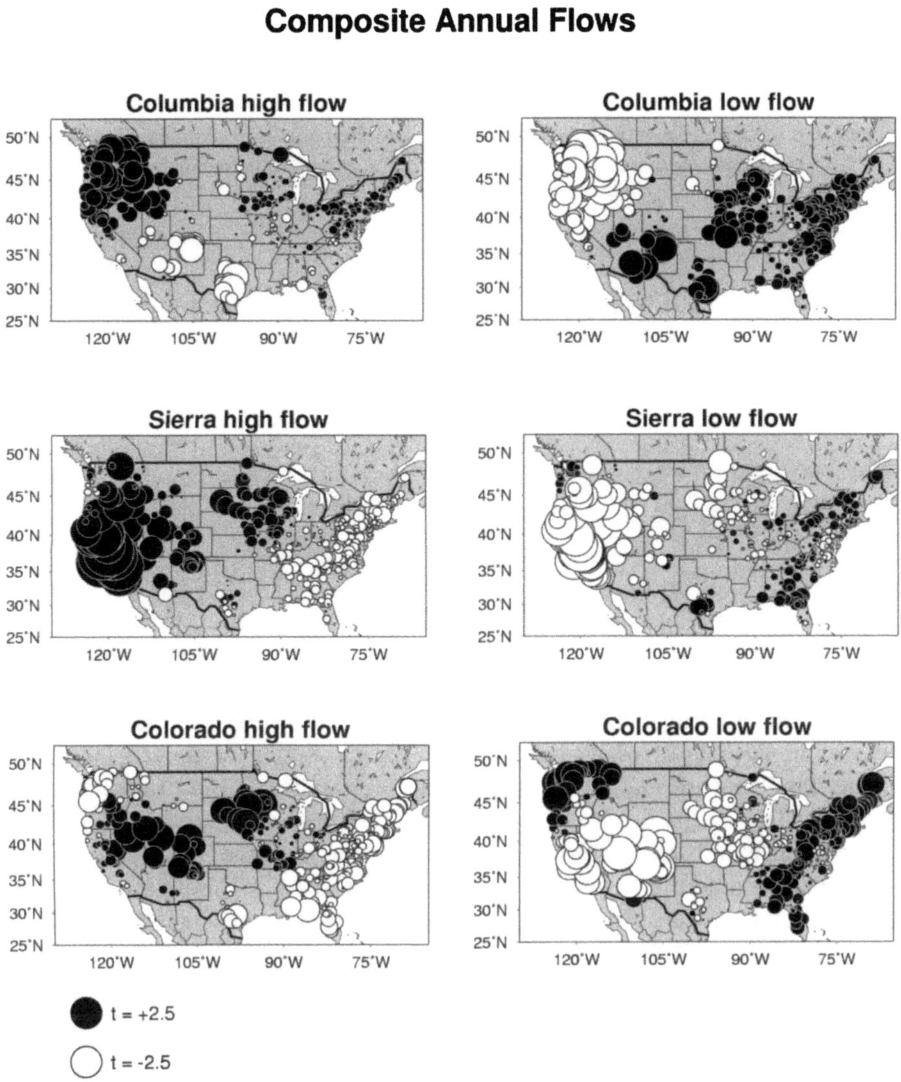


*Figure 8.* Average annual-discharge anomalies, as *t* statistics, for years when the Columbia, Sierra, and Colorado discharges exceeded their average discharges by 0.7 standard deviations or more (left-hand maps) or were less than their average by 0.7 standard deviations or more (right-hand maps). The filled circles indicate anomalously high flows, and open circles indicate anomalously low flows associated, on average, with flow anomalies in the three study watersheds.

watersheds share hydrologic extremes much more than would be expected by chance.

*Table 4*. Co-occurrences of high (*h*) and low (*l*) annual discharge for Columbia, Sierra and Colorado watershed pairs. Anomalies ≥ 0.7 σ, ≤ -0.7 σ define high and low discharge, respectively. From 1906–1999 data.

| | | Sierra | | | | Sierra | | | | Columbia | |
|---|---|---|---|---|---|---|---|---|---|---|---|
| | | *h* | *l* | | | *h* | *l* | | | *h* | *l* |
| Columbia | *h* | 8 | 3 | Colorado | *h* | 13 | 4 | Colorado | *h* | 7 | 4 |
| | *l* | 2 | 13 | | *l* | 2 | 16 | | *l* | 5 | 7 |

## 5. CLIMATE PATTERNS

Mechanisms that produce these watershed-to-watershed wet and dry coincidences are orchestrated by large-scale atmospheric circulations and climatic regimes (Namias 1978; Dettinger et al. 1998). In each of the three basins, the factor having the greatest impact on annual discharge is the winter pattern, as indicated by the winter precipitation anomaly (Fig. 9). Excesses or deficits of precipitation that build up during high- and low-flow years (and that ultimately generate those high and low flow rates) tend to begin in fall. In the Columbia watershed, precipitation excesses and deficits associated with high- and low-flow years are largely (and most reliably) established by about January. In the Sierra watershed, precipitation excesses and deficits typically (on average) are established by excesses and deficits that begin in late fall and continue to accumulate well into March. Deficits and, especially, excesses in the Colorado watershed are associated with precipitation anomalies from almost any month, either in the preceding or concurrent water year.

These differences in the seasons that ultimately contribute most to wet or dry years in the three watersheds result in differences in how well, and when, water managers in the watersheds can anticipate the eventual water-year discharge. Figure 10 compares the amount of information available about the eventual January–September and April–September discharges (on average) from knowing the accumulated precipitation to date in each of the three watersheds, as a function of the month of the water year. In this case, the amount of information is measured in variance of the January–

September and of the April–September discharge explained by the accumulated precipitation from the beginning of the water year to a given month. Precipitation is the climate division monthly precipitation averaged over each watershed. (For the Columbia River, the watershed region included was only the portion of the basin that lies within the United States.) Clearly, a water or hydropower manager in the Columbia watershed—by this measure—has an advantage, knowing relatively more about how wet or dry a given year will be, until February. In February, the manager in the Sierra is finally able to operate on equal footing. Thereafter, because of the more severely Mediterranean, wet-winter-only climate of the Sierra, compared to the Columbia and Colorado, the Sierran manager has a clearer idea of what the water-year total resource will be. A manager on the Colorado appears to be at an information disadvantage throughout the year, although there may be superior measures of precipitation than the divisional averages employed here. In each basin, a little more variance is explained for the January–September discharge than for the April–September discharge, but the month-by-month increases in variance explained for both of these discharge seasons are nearly the same.

The feature that provides the spatial coherence in the anomalously wet or dry precipitation patterns of the three watersheds is the atmospheric circulation. In winter, it is not unusual to find broad anomalous low- or high-pressure centers over the Pacific–North America sector that reflect the presence or absence of storm activity in one or more of the watersheds (Fig. 11). While prominent climate modes (El Niño/Southern Oscillation [ENSO] and the Pacific Decadal Oscillation [PDO]; Mantua et al. 1997; Gershunov et al. 1999) play strong roles in delivering wet or dry winters to the Columbia watershed, they are not very reliable determinants of wet or dry conditions in either the Sierra or the Colorado watershed (Tables 5 and 6). In the Columbia Basin, the La Niña phase of ENSO and the cool phase of the PDO favor high flows, and the El Niño phase of ENSO and the warm phase of the PDO favor low flows. Interestingly, though, all of the cases having the combination of higher than normal annual discharge on the Columbia River and lower than normal annual discharge on the Colorado River occurred during the PDO cool phase episode between 1947 and 1976.

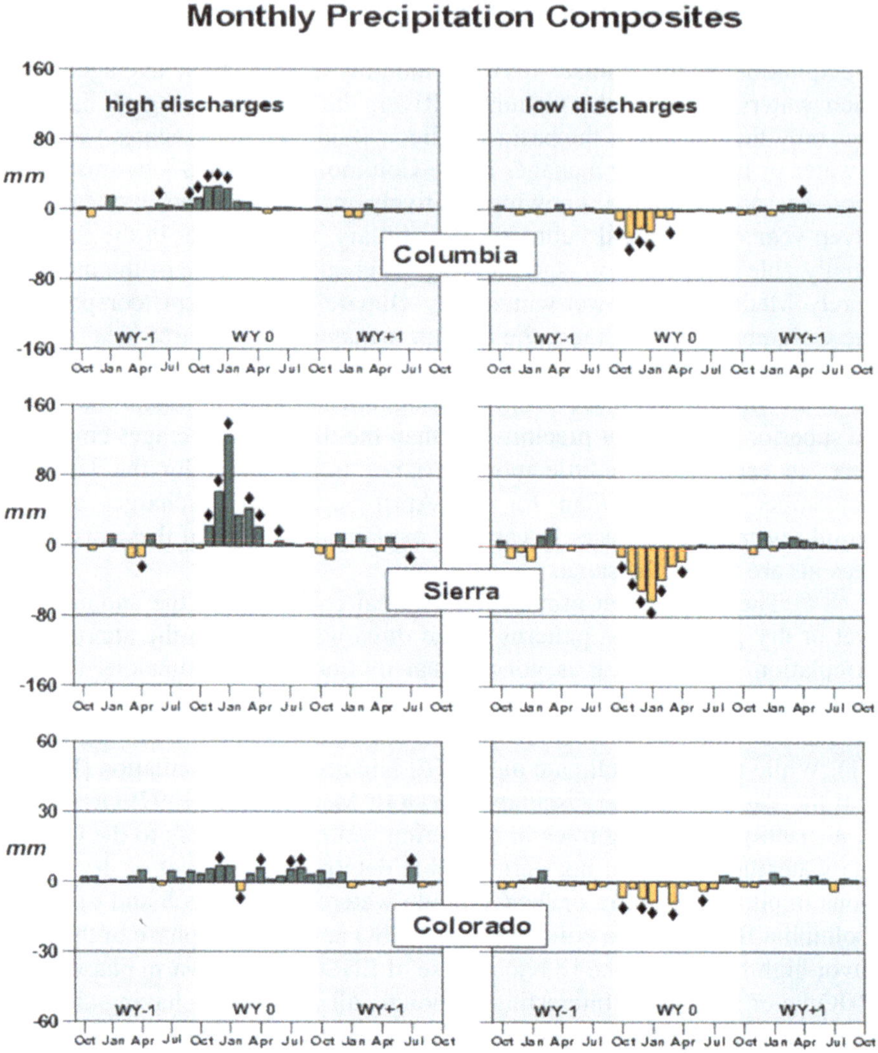


*Figure 9.* Average monthly precipitation anomalies from long-term monthly averages during years when Columbia, Sierra, and Colorado discharges were high (left) or low (right). Precipitation is from monthly climate division data. Criteria for high and low discharge are as in Figure 7. Water years (WY) –1, 0, and +1 designate the water years prior to, during, and following the year of high or low discharge (water year is October through September). Positive/negative anomalies are plotted above/below the zero line shaded dark/light gray. Months in which the composite anomaly was significantly different from zero at the 95% confidence level using a *t*-test are designated by diamonds. Note that the vertical scale (precipitation anomaly) is less for the Colorado than it is for the Columbia and Sierra.

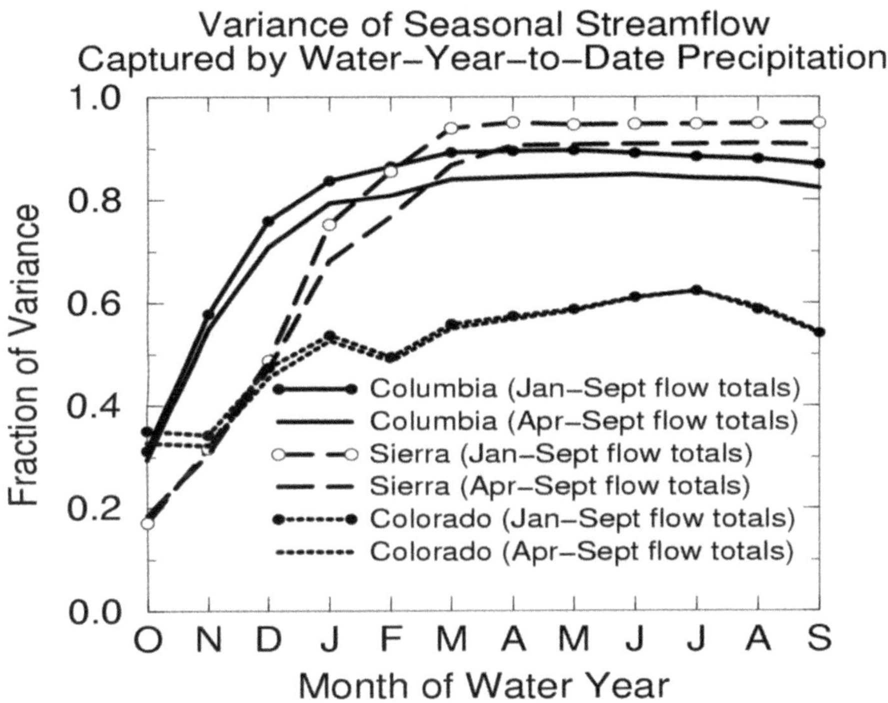


*Figure 10.* Variance of January–September and April–September discharges explained by accumulated precipitation anomalies beginning in October of the water year (see Figure 9) for the Columbia, Sierra, and Colorado watersheds. Precipitation is from United States monthly climate division averages, aggregated over each of the three watersheds.

*Table 5.* El Niño (*E*), Neutral (*N*) and La Niña (*L*) years with high/moderate/low (h/m/l) annual discharge totals, 1906-1999.

| | Columbia | | | | Sierra | | | | Colorado | | |
|---|---|---|---|---|---|---|---|---|---|---|---|
| | *h* | *m* | *l* | | *h* | *m* | *l* | | *h* | *m* | *l* |
| *E* | 4 | 10 | 15 | *E* | 12 | 7 | 10 | *E* | 11 | 8 | 10 |
| *N* | 14 | 17 | 14 | *N* | 13 | 14 | 18 | *N* | 14 | 15 | 16 |
| *L* | 13 | 5 | 2 | *L* | 6 | 11 | 3 | *L* | 6 | 9 | 5 |

*Table 6.* PDO warm (*w*) and PDO cool (*c*) years with high/moderate/low *(h/m/l)* annual discharge totals, 1906-1999.

| | Columbia | | | | Sierra | | | | Colorado | | |
|---|---|---|---|---|---|---|---|---|---|---|---|
| | *h* | *m* | *l* | | *h* | *m* | *l* | | *h* | *m* | *l* |
| *w* | 9 | 13 | 22 | *w* | 15 | 12 | 17 | *w* | 15 | 14 | 15 |
| *c* | 22 | 19 | 9 | *c* | 16 | 20 | 14 | *c* | 16 | 18 | 16 |

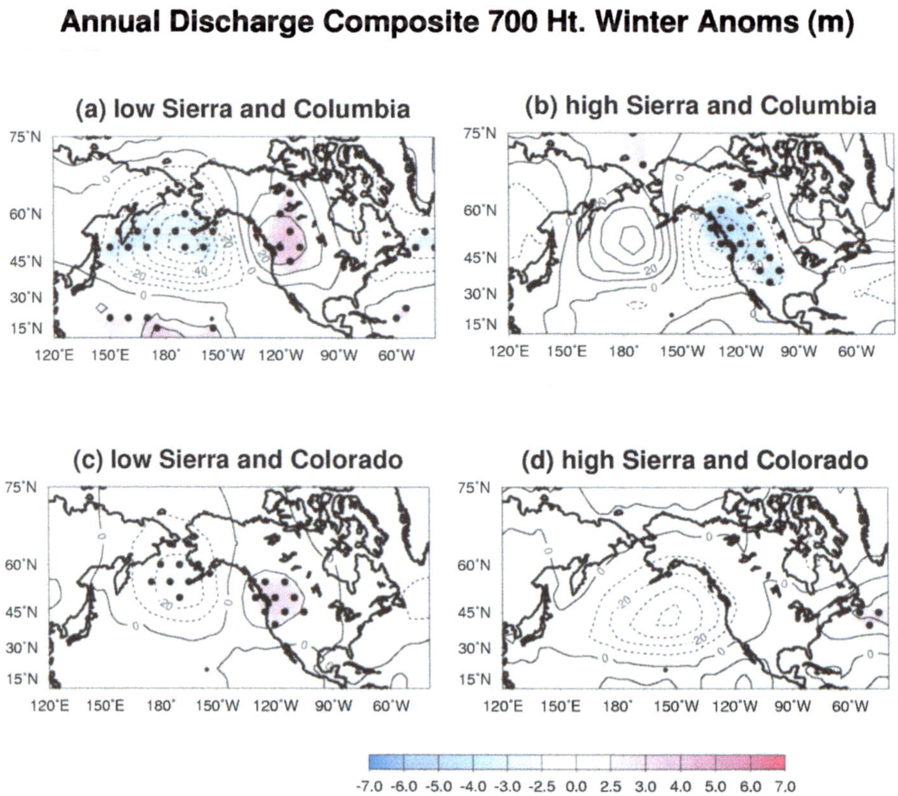


*Figure 11.* Average winter (December-January-February) 700 mb height anomalies during years when the Sierra and Columbia watersheds both had high (*a*) or low (*b*) annual discharges, and the Sierra and Colorado both had high (*c*) or low (*d*) annual discharges. Criteria for high and low discharges are as in Figure 4. Contours of positive/negative anomalies are solid/dashed. Grid cells at which average 700 mb height anomalies are significantly different from zero at 95% levels, using a *t*-test, are marked with heavy dots. *t*-values of positive/negative regions are shaded dark/light gray as in the key.

## 6. CONCLUSIONS

Climate variability and associated hydrologic variability have substantial impacts on California's hydropower and water supplies. It seems likely that these become even more important as the state's population and its needs for resources continue to grow. Climate variations both within and beyond the state boundaries have substantial influences on California resources. In order to characterize some of the transboundary hydroclimatic influences in California's water/hydropower setting, water-year river discharge totals from the west slope of the Sierra Nevada were compared to concurrent flows of the Columbia River (in the Pacific Northwest) and the Colorado River (in the southwestern United States).

Dry conditions in the three rivers have historically tended to cluster more in time than would be expected by chance. Dry conditions and wet conditions also tend to be more spatially extensive (among the three watersheds) than expected by chance. For example, years that are anomalously dry in two or more of the watersheds, or anomalously wet in two or more, occur with greater frequency than expected by chance. Dry/dry occurrences that pair the Sierra/Columbia watersheds or that pair the Sierra/Colorado watersheds are common. Co-occurrences of low flows or high flows in all three watersheds do not happen very often, but nonetheless are more frequent than expected by chance. Simultaneous low flows in all three basins occurred six times between 1906 and 1999, and, although it was not included in the present analysis, another of these massive regional dry events occurred in 2001. There is a modest tendency for opposing extremes to sometimes occur for the Columbia and Colorado Rivers, perhaps in response to PDO or ENSO episodes. Opposing extremes rarely occur for the Sierra and the Columbia or the Sierra and the Colorado watersheds.

Precipitation during winter, November through March, is critical in determining the status of the Columbia, Sierra, and (not as strongly) the Colorado water-year flow totals. Besides providing the regional water supply, the streamflows that are generated from this winter supply are strongly linked to the amount of hydroelectric power that each region produces and how much it is able to export or is compelled to import. In most years, the water year's supply is established by the end of February, owing to the dominance of winter storms in the annual precipitation cycle along the West Coast. The key factor that causes co-occurring discharge excesses and deficits in these watersheds is the broad scale of the winter atmospheric circulations, which form persistent patterns that activate or divert storms from the western states. ENSO and PDO have strong and consistent effects on the Columbia flows, but do not reliably produce high or low flows in the Sierra or Colorado Rivers. This is not to say that an organized form of the atmospheric circulation is not involved. Rather, each basin contains its own blend of circulation patterns that favor or disfavor ample yearly river flows for the region. Since most of these patterns have footprints that extend upstream over the North Pacific, it is important for progress in understanding and predicting these Pacific climate patterns to continue.

In addition to year-to-year and decade-to-decade variations in these systems, there are important seasonal regularities, which are driven at least partly by climate. These exist in the water and electric power generation and consumption systems in the western region. All three watersheds, were they unperturbed by humans, would have peak natural flows in spring to early summer, but all have been managed so that these spring–summer peaks have been substantially diminished. In California, releases from reservoirs are greatest in summer to satisfy irrigation needs and power demands then. California and the Southwest experience greatest peak demand for electric power during summer, presumably driven by air conditioning loads. In the Columbia, reservoir releases are greatest in winter to generate power because the Northwest region has greatest peak power demands then, probably to meet loads from heating and indoor appliances. These seasonally occurring regional contrasts are another complication that adds to or possibly mitigates impacts of unusual wet or dry (or cool or warm) climate spells that may persist for months to decades. The combined impacts of these regular and irregular climate influences will need to be incorporated into a truly comprehensive analysis of energy trading across the western region.

The present study has mostly examined the structure of excess or deficit annual aggregate discharge within and between these western watersheds. Not considered here is the potential added stress that may be imposed on water systems due to future climate changes. In addition to possible changes in precipitation, it is likely that the mountainous portions of these watersheds will experience major changes due to shifts in their snowmelt runoff timing due to climatic warming (Roos, 1991; Knowles and Cayan 2002).

## 7. ACKNOWLEDGMENTS

Funding was provided by the NOAA Office of Global Programs through the California Applications Center, by the U.S. Department of Energy, Office of Science (BER), Grant No. DE-FG03-01ER63255, and from the California Department of Water Resources, Environmental Services Office. We thank Jennifer Johns for word processing and illustrations and Emelia Bainto, Mary Tyree, and Larry Riddle for graphics. Thanks also to Henry Diaz and Jon Eischeid of the NOAA Climate Diagnostics Center for supplying daily precipitation data and Ross Miller of the California Energy Commission for supplying electrical energy data and information. Maurice Roos, Bruce McGurk, Tim Duane, Henry Diaz, and an anonymous reviewer gave very helpful comments and information.

## 8. REFERENCES

California Department of Water Resources. 1998. The California Water Plan Update, 1998, Bulletin 160-98, Volume 2.

Cayan, D.R. 1996. Interannual climate variability and snowpack in the western United States. *Journal of Climate* 9(5): 928–948.

Cayan, D.R., M.D. Dettinger, H.F. Diaz, and N.E. Graham. 1998. Decadal climate variability of precipitation over western North America. *Journal of Climate* 11(12): 3148–3166.

Dettinger, M.D., D.R. Cayan, H.F. Diaz, and D.M. Meko. 1998. North-south precipitation patterns in western North America on interannual-to-decadal time scales. *Journal of Climate* 11(12): 3095–3111.

Diaz, H.F., and C.A. Anderson. 1995. Precipitation trends and water consumption related to population in the southwestern United States: A reassessment. *Water Resources Research* 31(3): 713–720.

Eischeid. J.K., P. Pasteris, H.F. Diaz, M. Plantico, and N. Lott. 2000. Creating a serially complete, national daily time series of temperature and precipitation for the Western United States. *Journal of Applied Meteorology* 39: 1580–1591.

Fisher, J.V., and T.P. Duane. 2001. Trends in electricity consumption, peak demand, and generating capacity in California and the western grid 1977–2000. University of California Energy Institute, Program on Workable Energy Regulation (POWER), PWP #85.

Gershunov, A., T.P. Barnett, and D.R. Cayan. 1999. North Pacific interdecadal oscillation seen as factor in ENSO-related North American climate anomalies. *EOS* 80(3): 25–30.

Hamlet, A.F. 2003. The role of transboundary agreements in the Columbia River Basin: An integrated assessment in the context of historic development, climate, and evolving water policy. In, Diaz, H.F., and B. Morehouse (eds.), *Climate and Water: Transboundary Challenges in the Americas*. Dordrecht: Kluwer Academic Publishers (this volume).

Harding, B.L., T.B. Sangoyomi, and E.A. Payton. 1995. Impacts of a severe sustained drought on Colorado River water resources. *Water Resources Bulletin* 31(5): 815–824.

Karl, T.R., and R.W. Knight. 1985. *Atlas of Monthly and Seasonal Precipitation Departures from Normal (1895–1985) for the Contiguous United States. Historical Climatological Series* Vol. 3–12. National Climatic Data Center, Asheville, NC 219 pp.

Knowles, N. 2002. Natural and human influences on freshwater inflows and salinity in the San Francisco Estuary at monthly to interannual scales. *Water Resources Research*, in press.

Knowles, N., and D.R. Cayan. 2002. Potential effects of global warming on the Sacramento/San Joaquin watershed and the San Francisco estuary. *Geophysical Research Letters*, in press.

Knowles, N., D.R. Cayan, and D.H. Peterson. 2002. Seasonal to interdecadal variability of San Francisco estuary freshwater inflows and salinity. *Continental Shelf Research*, submitted.

Lord, W.B., J.F. Booker, D.M. Getches, B.L. Harding, D.S. Kenney, and R.A. Young. 1995. Managing the Colorado River in a severe sustained drought: An evaluation of institutional options. *Water Resources Bulletin* 31(5): 939–944.

Mann, M.E., and J.M. Lees. 1996. Robust estimation of background noise and signal detection in climatic time series. *Climatic Change* 33: 409–445.

Mantua, N.J., S.J. Hare, Y. Zhang, J.M. Wallace, and R.C. Francis. 1997. A Pacific interdecadal oscillation with impacts on salmon production. *Bulletin of the American Meteorological Society* 78: 1069–1079.

McCabe, G.J., Dettinger, M.D., and D.R. Cayan. 2002. Hydroclimatology of the 1950s drought. Submitted as a chapter in, Betancourt, J.L., and H.F. Diaz (eds.), *The 1950s Drought in the American Southwest: Hydrological, Ecological and Socio-economic Impacts*. Tucson : University of Arizona Press,.

Meko, D.M., C.W. Stockton, and W.R. Boggess. 1995. The tree-ring record of severe sustained drought. *Water Resources Bulletin* 31(5): 789–801.

Meko, D.M., M.D. Therrell, C.H. Baisan, and M.K. Hughes. 2001. Sacramento River flow reconstructed to A.D. 869 from tree rings. *Journal of the American Water Resources Association* 37(4): 1029–1040.

Namias, J. 1978. Multiple causes of the North American abnormal winter 1976–77. *Mon. Wea. Rev.* 106: 279–295.

National Research Council (NRC). 1999. *Improving American River Flood Frequency Analyses*. Washington, DC: National Academy Press, 120 pp.

Newcom, S. J.. 2002. The Colorado River: Coming to consensus. *Western Water*, March/April, 4–13, 17.

Peterson, D.H., D.R. Cayan, J. Dileo, M. Noble, and M.D. Dettinger. 1995. The role of climate in estuarine variability. *American Scientist* 83: 58–67.

Pulwarty, R.S., and K.T. Redmond. 1997. Climate and salmon restoration in the Columbia River Basin: The role and usability of seasonal forecasts. *Bulletin of the American Meteorological Society* 78(3): 381–397.

Roos, M. 1991. A trend of decreasing snowmelt runoff in northern California. *Proceedings of the 59th Western Snow Conference*, Juneau, Alaska, pp. 29–36.

Roos, M. 1994. Is the California drought over? *Proceedings of the Tenth Annual Pacific Climate (PACLIM) Workshop*, Asilomar, California, April 4–7, 1993, Technical Report 36 of Interagency Ecological Program for the Sacramento–San Joaquin Estuary, March 1994, pp. 123–128.

Slack, J.R., and J.M. Landwehr. 1992. Hydro-climatic data network (HCDN): A U.S. Geological Survey streamflow data set for the United States for the study of climate variations, 1874–1988. U.S. Geological Survey Open-File Rep. 92-129, 200 pp. (Available from U.S. Geological Survey, Books and Open File Reports Section, Federal Center, Box 25286, Denver, CO 80225.)

Stahle, D., M. Therrell, M.K. Cleaveland, D.R. Cayan, M.D. Dettinger, and N. Knowles. 2001. Ancient blue oaks reveal human impact on San Francisco Bay salinity. *EOS Transactions*, American Geophysical Union, 82(12).

12

# THE ROLE OF TRANSBOUNDARY AGREEMENTS IN THE COLUMBIA RIVER BASIN

## *An Integrated Assessment in the Context of Historic Development, Climate, and Evolving Water Policy*

Alan F. Hamlet

*Joint Institute for the Study of Atmosphere and Oceans, Climate Impacts Group, and Department of Civil and Environmental Engineering, University of Washington, Seattle, WA*

**Abstract** The historical development of the Columbia River Basin and its current reservoir operating policies has been strongly influenced by transboundary agreements between Canada and the United States, and particularly by the Columbia River Treaty (CRT) (1964) and adjunct agreements. Following this agreement, a number of major storage dams were built in Canada and the US, and an emphasis on flood control and winter hydropower production became the dominant water resources objectives in the main stem of the Columbia. The development of the basin for these purposes (and extensive irrigation in some sub-basins) has resulted in pronounced changes in the natural flow regime in the river, and corresponding ecological problems associated with degraded instream habitat that have yet to be resolved. The basin's operating policies are shown to more fully isolate human systems from climate variability than they do other uses of water in the basin, despite federal legislation calling for equal priority between hydropower and fish, and recent efforts to change the operating policies in the face of Endangered Species Act (ESA) listings of salmon and other endangered fish. Vulnerability of the current management system to low-flow conditions, inability to meet all objectives simultaneously in low-flow conditions, and the conflicting constraints of many existing agreements (among them the CRT), make changes to the basin's operating policies problematic.

Climate change, which is likely to reduce summer water availability due to changes in snowpack and the seasonality of natural streamflows, may exacerbate the existing weaknesses in the Columbia's operating policies and management framework, and is likely to create tensions between Canada and the United States in the context of maintaining summer instream flows. The Canadian snowfields in the basin are largely insulated from temperature-related changes in snowpack for the scenarios exam-



*Henry F. Diaz and Barbara J. Morehouse (eds.), Climate and Water*, 263-289.

ined, whereas some areas of the United States are much more strongly affected, particularly in the southern part of the basin. The CRT may be both an obstacle and a means to effective adaptation to climate change in the basin. For example, the CRT may add to the United States's vulnerability to summer low-flow conditions by inhibiting what was once natural flow across the border; however, the CRT could also potentially facilitate the transfer of water from Canadian storage in summer via changes in the existing seasonality of hydropower production.

# 1. INTRODUCTION

To understand the role that transboundary issues have played in shaping current water resources policy in the Columbia Basin, and their potential effect on future changes in the basin, a number of direct and indirect influences must be examined. This process of investigating a specific issue within the context of the whole has come to be called "integrated assessment." Using this general framework, the Columbia River Basin is examined here in the context of climate variability, historical development, and current water policy; and some important linkages between transboundary agreements and these key features are discussed. To examine the potential implications for the future, likely effects on regional snowpack and the seasonality of streamflows are quantified, and the impacts of climate change scenarios are examined. Some transboundary considerations inherent in existing water management institutions that may be affected by these potential future changes are discussed.

# 2. A BRIEF OVERVIEW OF THE COLUMBIA RIVER BASIN

The Columbia River Basin covers portions of seven western states (primarily Oregon, Washington, and Idaho), and about 30% of the basin is in British Columbia in Canada (Fig. 1). The Columbia River is one of the most developed water resources systems in the United States, with more than 200 large dams, and few feasible opportunities for further expansion. The basin's dams provide protection from spring flooding, supply about 75% of the electrical energy for the Pacific Northwest (PNW), provide irrigation water and power for pumping to supply agricultural needs in the arid interior of the basin, and also provide opportunities for river navigation.

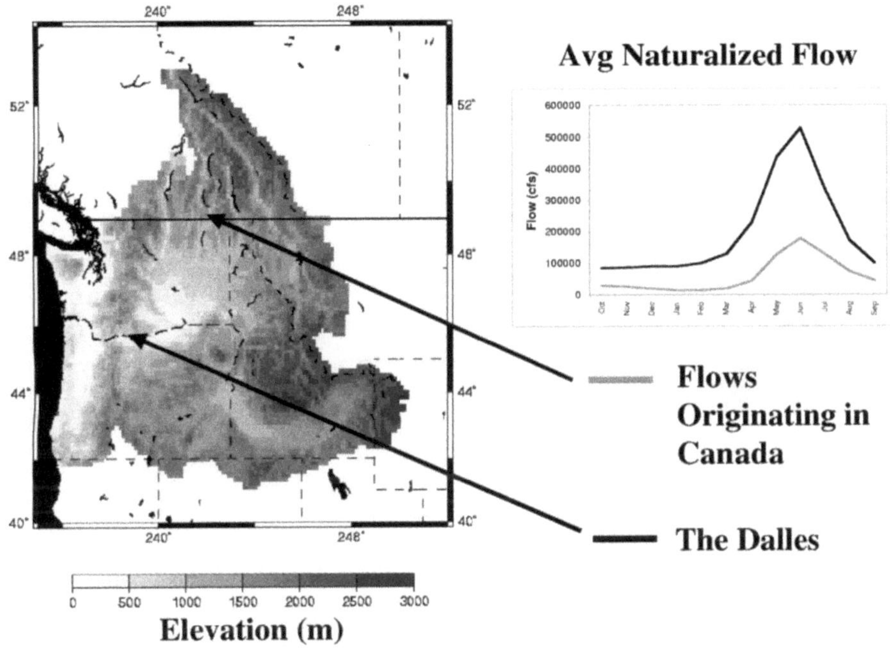


*Figure 1.* Topography of the Columbia River Basin, and composite mean monthly hydrographs for natural flows at The Dalles, Oregon, and flows originating in Canada.

More recently, the augmentation of regulated instream flow to protect endangered salmon, and recreation have become increasingly important uses of water in the region (Bonneville Power Administration [BPA] 1991). Figure 2 shows the major storage and run of river dams in the basin. While the figure may suggest a large amount of reservoir storage, in fact the Columbia reservoir system is able to store only about 30% of the river's annual flow. The PNW's water resources systems are able to function effectively with less man-made storage primarily because of natural storage of winter precipitation as snow in the mountains, which helps maintain streamflow throughout the PNW's dry summers. The Columbia's annual streamflows also vary relatively little from year to year in comparison with most other large water resources systems in the western United States, which also allows PNW water resources systems to function effectively with a smaller relative amount of reservoir storage than is typical in other regions.[1]

[1] The Colorado River, for example, can store about three times its annual flow in reservoirs (BPA 1991).

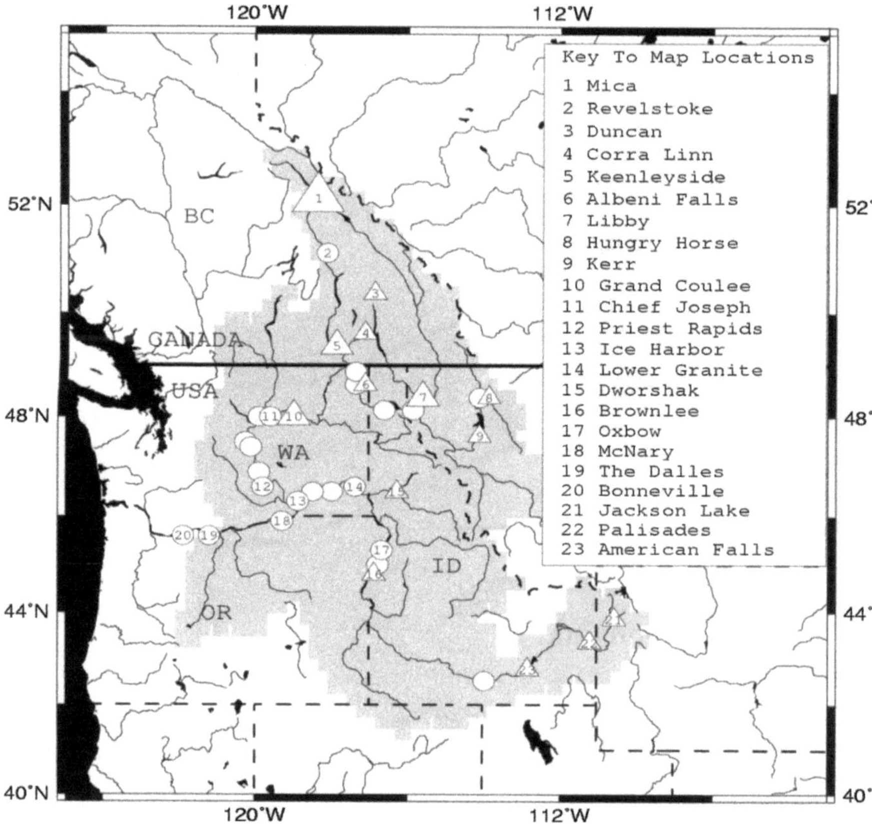


*Figure 2.* Major storage (triangles) and run-of-river (circles) reservoirs in the Columbia Basin (size of symbols for storage reservoirs reflect relative storage capacity). (Source: Hamlet and Lettenmaier 1999b.)

## 3. CYCLES Of PNW PRECIPITATION AND COLUMBIA RIVER STREAMFLOW

The PNW receives the dominant portion of the annual precipitation contributing to runoff from about October to March each year, and the Columbia Basin, which is at relatively high elevation overall (Fig. 1), accumulates much of this winter precipitation as snow, resulting in relatively low natural streamflows throughout the winter (Fig. 1 inset). As the weather warms in the spring, the winter accumulation of snow melts, creating a period of high natural flow in the spring and summer months that typically peaks in June (Fig. 1 inset, black trace). In the late summer there is typically a period of low flow until the fall rains begin. While annual river flows originating in Canada are consistent with the amount of basin area in Canada, about 50% of the late summer flows at The Dalles, Oregon, originate in Canada (Fig. 1 inset, gray trace). This effect is due primarily to the northern location and high elevation of the Canadian part of the basin, and (in some years) additional meltwater from glaciers.

## 4. THE COLUMBIA RIVER TREATY AND ASSOCIATED TRANSBOUNDARY AGREEMENTS

The Columbia River Treaty (CRT) between Canada and the United States is the most significant transboundary agreement affecting the Columbia Basin. This treaty authorized the construction of most of the major storage projects in the Canadian portion of the Columbia Basin, and created a formal agreement in which hydropower benefits for Canada and the United States and flood control benefits for the United States were combined. The treaty expires in 2024, and can be renegotiated by either party at any time. In the original agreement, half of the downstream power benefits resulting from releases of water from Canadian storage were granted to Canada, but Canada sold these rights to U.S. interests (the Columbia Storage Power Exchange, composed of 41 U.S. utilities) for 30 years from the time of completion of each project. Under the terms of this sale, Canada has recently regained control of the downstream power benefits for Duncan Dam (1998) and Keenleyside Dam (1999), and it will regain the benefits from Mica Dam in 2003. This is an important development because it means that the hydropower revenues that Canada receives from releases of water from Canadian storage can to a certain extent be managed to maximize the value of these downstream benefits by, for example, timing reservoir releases to coincide with favorable energy markets. Flood control operations and the associated evacuation of winter storage from Canadian and U.S. federal storage projects, however, will remain under the control of the U.S. Army Corps of Engineers. In addition, the CRT sets forth that flood control evacuation requirements will remain in force for the life of the Canadian storage projects, regardless of other changes in the CRT agreement. This condition also ensures that a certain amount of winter streamflow and hydropower capacity in the United States will be protected (with shared benefits to Canada and the United States under the CRT).

The CRT is supported by several other transboundary agreements, such as the Pacific Northwest Coordination Agreement (PNCA) and the Non-Treaty Storage Agreement (NTSA). The PNCA is an interstate and international agreement intended to facilitate mutually beneficial short-term energy and water trading within a group of hydropower projects operated as if they were owned by the same company. So, for example, releases from Canadian storage to augment U.S. hydropower can be requested by a member of the PNCA, and energy at a specified future time will be returned to Canada (this is called "wheeling" in the power industry). The NTSA is an agreement between the BPA (United States) and British Columbia (BC) Hydro (Canada) to extend the amount of storage available for coordinated use beyond that specified in the original CRT, with shared benefits to the BPA and BC Hydro (BPA 1991). This agreement also benefits U.S. interests by typically requiring less flood storage evacuation at Grand Coulee Dam.

## 5. CHANGES IN THE COLUMBIA'S NATURAL FLOW REGIME

The development of the Columbia has resulted in radical changes in the patterns of seasonal river flow. Figure 3a shows monthly average naturalized ("virgin") flows compared to modified flows[2], and simulated regulated flows resulting from the multiple uses of water for a constant 1990 level of development. Figure 3b shows the observed effects to regulated summer peak flows at The Dalles for 1858–2001. (Note the pronounced change in about 1975 following the completion of major dams and the unprecedented low flow resulting from the energy crisis in 2001.) Although irrigation plays a strong role in some sub-basins of the Columbia, the largest changes in the flow regime at The Dalles are due to the effects of hydropower production and flood control. The Columbia hydro system releases water from storage to produce energy in the winter, raising river flows above natural conditions, and captures spring and summer flows to prevent flooding and to refill the system storage each year, which reduces the intensity of the spring and summer peak flows in comparison with natural conditions.

These changes in the regulated flow regime occurred during a very short period of intense dam building from about 1965 to 1975 following the ratification of the CRT (Fig. 4), and are now believed to be one of the primary stresses on salmon populations in their freshwater habitat (Bisbal and McConnaha 1998; Cohen et al. 2000; Cone and Ridlington 1996; Mantua et al. 2001; Quinn and Adams 1996; Volkman 1997; Wood 1993).

---

[2] Modified flows are naturalized flows less net losses for irrigation and increased evaporation from larger pools behind dams.

A)

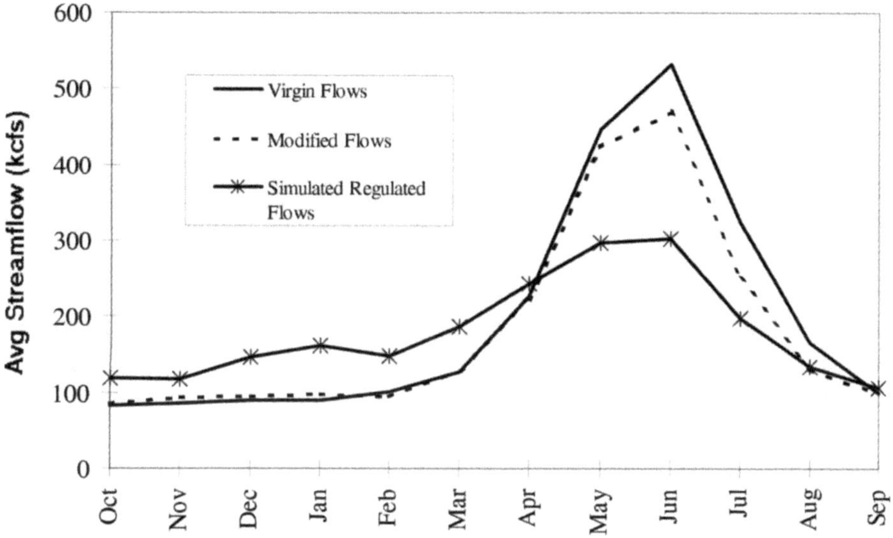


B)

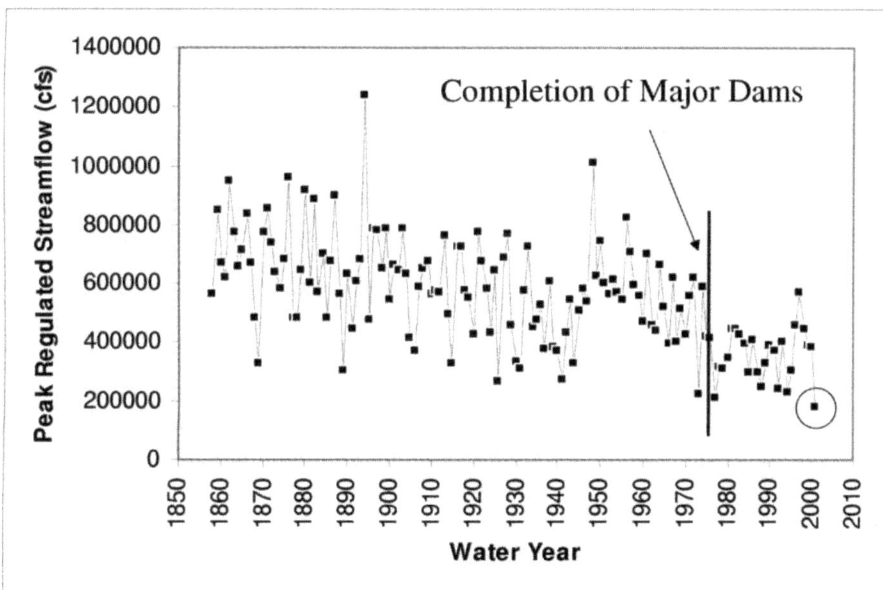


*Figure 3*. Effects of human development on flows at The Dalles, Oregon. (A) Mean monthly naturalized flow compared with estimated modified flows (see text for definition) and simulated regulated flows for 1990 level of development (Data 1931–89). (B) Observed regulated peak daily flows for 1858–2001. (Source: USGS.)

# 6. MAJOR MILESTONES IN COLUMBIA BASIN DEVELOPMENT

Figure 4 shows a timeline for a number of milestones in the development of the Columbia Basin. In 1942 the completion of Grand Coulee Dam, the first large, federal storage project on the main stem, effectively blocked all anadromous fish runs into the upper basin, and set the tone for the future development of the Columbia. Twenty years later, the Columbia River Treaty resulted in the construction of four new major dams (three of them in Canada) that have made possible the Columbia's current flood control operations, and increased the reservoir system's hydropower capacity substantially. In addition, the completion of Dworshak Dam on the Clearwater River (a tributary of the Snake River) in the United States in 1973 increased the total system storage and hydropower capacity to essentially their current levels.

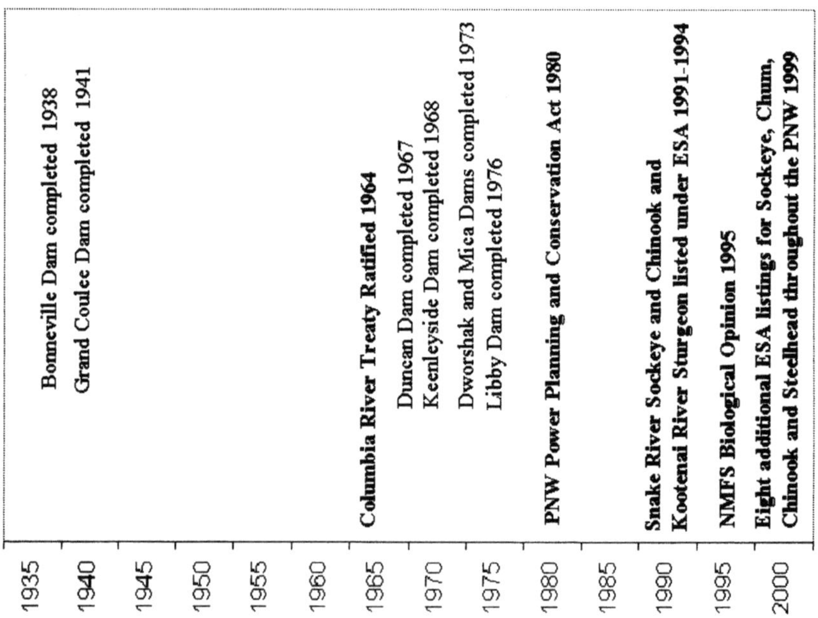


*Figure 4.* Timeline for major policy and development milestones for the Columbia River Basin (boldface type indicates policy change, regular type indicates development action).

As was mentioned above, prior to 1964, there was relatively little storage in the basin (primarily Lake Roosevelt behind Grand Coulee Dam and some irrigation projects in the Snake River Basin), and regulated flows at The Dalles were not strikingly different from the natural flows that occurred in the earlier part of the instrumental record (Volkman 1997). Following the completion of the last major storage dam (Libby Dam, completed in 1976), regulated flows were much like they are today, and problems with maintaining native salmon populations in the remaining freshwater habitat began to emerge.

In 1980, the Northwest Power Planning and Conservation Act established that hydropower and the maintenance of PNW salmon populations were of equal priority under the law in the Columbia Basin, and established the Northwest Power Planning Council in an attempt to coordinate basin operations towards this end. Increased use of hatcheries and an emphasis on engineering solutions to the salmon problem in the Columbia (e.g., the barging of salmon smolts to the ocean to avoid problems in the freshwater habitat) have avoided socioeconomic impacts to human systems, but these efforts have not been successful in reversing the decline of Columbia River salmon populations, particularly in the case of wild salmon species.[3] Poor ocean survival conditions for Washington and Oregon salmon populations beginning in about 1977 (Mantua et al. 1997) increased the pressure on Columbia salmon stocks, and the early 1990s brought Endangered Species Act (ESA) listings for Snake River sockeye and Kootenai River sturgeon. The National Marine Fisheries Service (NMFS) Biological Opinion (1995) resulted in changes in the Columbia's reservoir operating system intended to help endangered salmon, but 4 years later, eight additional ESA listings for PNW salmon species (not all of them in the Columbia) clearly demonstrated that the problems for PNW salmon have not been resolved (Mantua et al. 2001).

The Columbia River Treaty, with its focus on an "engineered river" for flood control and winter hydropower, marks a clear transition in the Columbia's history from a natural river to a managed water resources system. The human populations in the PNW have become very dependent upon maintaining the changes in the river's natural flow characteristics (particularly for flood control, energy production, and irrigation), whereas the natural ecosystems (such as salmon populations) are dependent on the natural

[3] Wild salmon must complete their full life cycle to successfully maintain their numbers, which means that they are much more dependent on favorable freshwater habitat for returning adult passage and spawning habitat than are hatchery fish, whose reproductive capacity can be maintained largely without suitable adult migration conditions or suitable spawning habitat. Hatchery fish also compete with wild salmon stocks both in the ocean and as returning adults (Mantua et al. 2001).

flow regime and riverine habitat for their well-being. Thus, it would appear to be impossible to decouple recent efforts to establish a more natural instream flow regime and improved riverine habitat from increased impacts to human systems, which are dependent on the current managed use of water resources. The ongoing emphasis on "engineering solutions" to the salmon problem, rather than restoration of instream habitat and a more natural flow regime, may be interpreted as societal reluctance to incur the socioeconomic impacts associated with a return to a more natural river.

## 7. IMPACTS OF CLIMATE VARIABILITY AND RESERVOIR OPERATING SYSTEM POLICY

Two global climate phenomena, the Pacific Decadal Oscillation (PDO) (Mantua et al. 1997) and the El Niño/Southern Oscillation (ENSO) (see, e.g., Battisti and Sarachik 1995) are well correlated with the variability of PNW climate and streamflow. Figure 5 shows the sea surface temperature (SST) signatures for the warm phase of PDO and ENSO, respectively (Mote et al. 1999). The primary difference between the SST signatures of the two phenomena is the characteristic period of variation, which for the PDO is on the order of 50 years, and for ENSO is on the order of 5–7 years. The SST signature of the PDO is also most pronounced in the North Pacific, whereas the SST signature of ENSO is most pronounced in the Tropical Pacific. Retrospective definitions of PDO and ENSO are constructed here from PDO epochs identified by Mantua et al. (1997), in which the PDO index is predominantly positive or negative, and from winter averages of the Niño3.4 index of Trenberth (1997). Average winter (December-January-February [DJF]) values of the Niño3.4 index more than 0.5 standard deviations above (below) the mean are defined here to indicate a warm (cool) phase of ENSO. All other years are defined as ENSO neutral.

The physical dynamics of the PDO are poorly understood in comparison with those of ENSO, and there is considerable debate in the atmospheric sciences community about whether the PDO is in fact a true oscillatory atmosphere/ocean phenomenon like ENSO, or whether the apparent decadal-scale variability in the North Pacific Ocean and the associated climate variations can be explained as (to give one alternative theory) a signature of atmospheric variability displaying red noise characteristics. Despite these uncertainties, the PDO is clearly associated with decadal-scale climate variability in the PNW in the twentieth century, although perhaps only in a diagnostic sense.

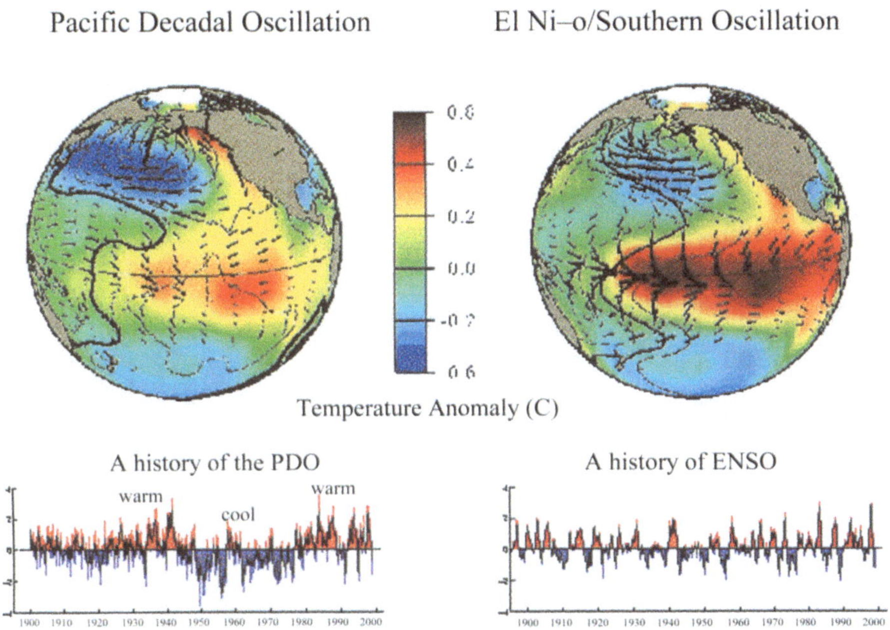


*Figure 5*. Mean SST anomalies for the warm phases of PDO and ENSO, and a time history of the PDO and Nino 3.4 indices. (Source: Mote et al. 1999.)

The warm phases of PDO and ENSO are associated with warmer and dryer winters in the PNW, whereas the cool phase of each phenomenon is associated with cooler, wetter winters (Mote et al. 1999). Although not well understood at a fundamental level, there are also apparent interactions between climatic variability associated with the PDO and those associated with ENSO (see, e.g., Gershunov and Barnett 1998). These interactions are apparent in the scatter plot of summer naturalized streamflow at The Dalles, Oregon, for the twentieth century shown in Figure 6.[4] The variation of summer streamflows at The Dalles (which are directly determined by winter climate via snowpack) is linked to decadal-scale variability in the Pacific associated with the PDO, superimposed with variations at shorter time scales associated with ENSO. Thus changes in the decadal average of summer streamflows are aligned in time with the observed shifts in the

[4] Naturalized flows for The Dalles were estimated for the period from 1900 through 1927 by Prof. David Jay and Pradeep Naik at the Oregon Graduate Center, and for the period from 1928 through 1998 by A.G. Crook and Co. (1993) and the Pacific Northwest River Forecast Center, supplied courtesy of the Bonneville Power Administration.

PDO, while ENSO affects the interannual variability about this shifting background mean (Hamlet and Lettenmaier 1999a).

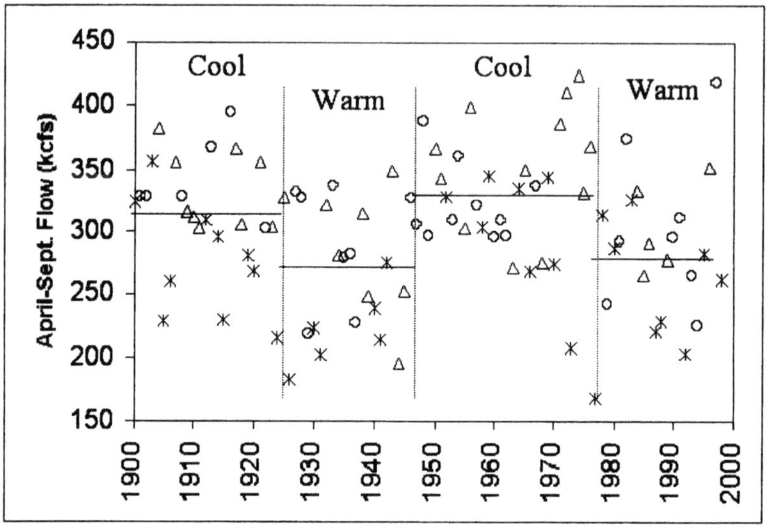


*Figure 6.* Effects of the PDO and ENSO on Columbia River summer streamflows. Asterisks are warm ENSO years, triangles are cool ENSO years, and circles are ENSO neutral. Warm and cool PDO epochs are identified, and the horizontal line shows the mean for each epoch. (Source: Mote et al. 1999.)

Note that for each warm and cool PDO "epoch," warm ENSO (El Niño) events, streamflow values tend to be in the bottom two-thirds of the distribution (within the epoch), and cool ENSO (La Niña) events tend to be focused in the upper two-thirds of the distribution. Extreme low flows are also more likely in warm PDO, warm ENSO years, and extreme high flows are more likely in cool PDO, cool ENSO years (Mote et al. 1999). These consistent patterns of variation over the twentieth century provide a useful way to characterize year-to-year climate variability in the PNW, and provide a useful framework to examine the impacts of climate variability on water resources.

While natural river flow ultimately determines the impacts of climate variability on Columbia Basin water resources, complex interactions between the quantity and timing of natural flow and the ability to meet various water resources objectives are best quantified by using a model in order to make a consistent assessment of the socioeconomic impacts. To accomplish this, the ColSim (Columbia Simulation) reservoir model is used to quantify the linkages between natural flow and socioeconomic impacts associated with water resources system performance. The monthly time step ColSim model is driven by naturalized flow (either observed or simulated) and simulates reservoir operations at the major storage and run-of-river projects (Fig. 2) for flood control, hydropower production, instream flow augmentation, irrigation diversions, and recreation (see Hamlet and Lettenmaier 1999b for more details on the model). Model outputs include simulated reservoir storage and regulated streamflow throughout the basin, and reliability of various system objectives listed above.

To evaluate the impacts of climate variability, a naturalized flow record from 1931 to 1989 used to drive the ColSim model was segregated into six climate categories based on combinations of PDO (warm and cool epochs) and ENSO (warm, cool, neutral) (Table 1), and the ability to meet a selected group of water resources objectives (Figure 7) was simulated in each of the six climate categories for two reservoir operating policies: (1) the status quo, and (2) a hypothetical management alternative giving high priority to fish flows. The results are shown in Figure 7 (status quo) and Figure 8 (fish alternative). In both cases the reliability[5] of system objectives impacted by low flows[6] generally increases with increasing natural flow (i.e., in moving from warm PDO/warm ENSO towards cool PDO/cool ENSO), whereas those objectives impacted by high flows[7] show the opposite effects. For the status quo operating policies, however, firm energy[8] is completely isolated from climate variability (by design), and other system objectives are variously affected by climate variability. For the fish-flow alternative, flows at McNary Dam on the main stem of the Columbia are almost completely isolated from climate variability (by design), and other objectives (including current levels of firm energy production) become more vulnerable to climate variability.

---

[5] Reliability is the simulated probability of successfully meeting a particular management objective in each month (e.g., if the objective is successfully met in 9 out of 10 months in the simulation, the reliability is reported as 90%).
[6] Irrigation, instream flow, energy production, recreation.
[7] Flood control and river navigation.
[8] Firm energy is associated with PNW winter energy requirements, and the status quo reservoir operating system is designed to meet these energy targets for all climate conditions likely to be encountered by the water resources system.

*Table 1.* Retrospective categories of PDO and ENSO for water years from 1900 through 1995.

| Climate Category | Warm PDO Warm ENSO | Warm PDO Neutral ENSO | Warm PDO Cool ENSO | Cool PDO Warm ENSO | Cool PDO Neutral ENSO | Cool PDO Cool ENSO |
|---|---|---|---|---|---|---|
| Number of Events | 14 | 15 | 12 | 18 | 16 | 21 |
| Water Years in Category | 1926 | 1927 | 1925 | 1900 | 1901 | 1904 |
| | 1930 | 1928 | 1932 | 1903 | 1902 | 1907 |
| | 1931 | 1929 | 1934 | 1905 | 1908 | 1909 |
| | 1940 | 1933 | 1938 | 1906 | 1913 | 1910 |
| | 1941 | 1935 | 1939 | 1912 | 1916 | 1911 |
| | 1942 | 1936 | 1943 | 1914 | 1922 | 1917 |
| | 1977 | 1937 | 1944 | 1915 | 1947 | 1918 |
| | 1978 | 1946 | 1945 | 1919 | 1948 | 1921 |
| | 1980 | 1979 | 1984 | 1920 | 1949 | 1923 |
| | 1983 | 1981 | 1985 | 1924 | 1953 | 1950 |
| | 1987 | 1982 | 1986 | 1952 | 1954 | 1951 |
| | 1988 | 1990 | 1989 | 1958 | 1957 | 1955 |
| | 1992 | 1991 | | 1959 | 1960 | 1956 |
| | 1995 | 1993 | | 1964 | 1961 | 1963 |
| | | 1994 | | 1966 | 1962 | 1965 |
| | | | | 1969 | 1967 | 1968 |
| | | | | 1970 | | 1971 |
| | | | | 1973 | | 1972 |
| | | | | | | 1974 |
| | | | | | | 1975 |
| | | | | | | 1976 |

Source: Hamlet and Lettenmaier (2000).

These simple experiments demonstrate several important characteristics of the Columbia Basin's water resources system and its operating policies. First, the relative priority between energy production and instream flow for fish is clearly weighted towards reliability of energy production for the status quo operating policies, despite the legislative mandate that salmon and hydropower are equal in priority in the Columbia Basin.[9] Second, it is apparent that the Columbia Basin's reservoir operating policies cannot be optimized to isolate all system objectives from climate variability simultaneously. Thus, any changes in the reservoir operating plan, while they may effectively redistribute the impacts of climate variability among various us-

[9] Pacific Northwest Electrical Power Planning and Conservation Act, 1980.

ers and uses of water, cannot resolve the inherent conflicts over water (Miles et al. 2000). Miles et al. (2000) have also made the point that the historical

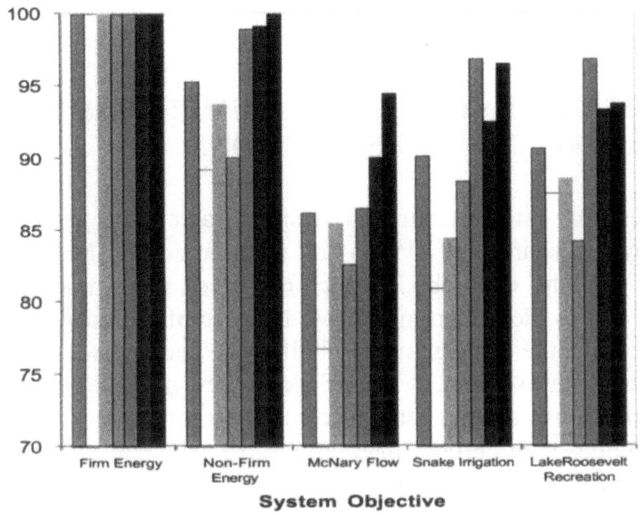


*Figure 7.* Effects of natural variability on Columbia Basin water resources system objectives for the status quo operating policies

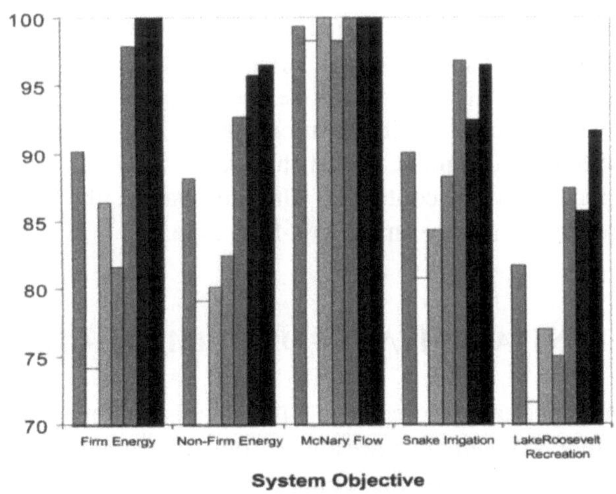


*Figure 8.* Effects of natural variability on Columbia Basin water resources system objectives for a hypothetical fish-flow operating policy.

development of these conflicts in the Columbia Basin is therefore a major obstacle to future change, because attempts to alter the status quo will inevitably polarize potential "winners" and "losers" in the legal and political arenas.

## 8. ASSESSMENT OF THE COLUMBIA BASIN'S MANAGEMENT INSTITUTIONS

Research on the Columbia Basin's water management institutions reported by Callahan et al. (1999) and summarized by Miles et al. (2000) concluded that the Columbia's water resources were more vulnerable to conditions of low flow than high flow, because of the nature of the institutions governing the management of different uses of water in the basin. Flood control, for example, is one of the highest priorities in the status quo reservoir operating system, and is centrally coordinated on a basin-wide scale (including well-defined interactions between Canada and the United States) by the U.S. Army Corps of Engineers. On the low-flow side of the spectrum, literally hundreds of individual agencies must interact within a poorly defined framework to attempt to protect various conflicting objectives. So, for example, in low-flow years, irrigation and fisheries managers in the Yakima River Basin (a tributary of the Columbia) must attempt to balance impacts to farmers and riverine habitat for salmon without any central management authority designed to coordinate these activities (Gray 1999). The impacts of the very low streamflows in water year 2001 highlight the inability of the PNW water resources systems to cope effectively with low flows (Mapes 2000a,b; Bernton 2001). This disparity in the ability of the Columbia management system to adapt effectively to high and low flows is important in the context of climate change, the likely impacts of which are focused primarily on the low-flow side.

## 9. POTENTIAL IMPACTS OF CLIMATE CHANGE

Many uncertainties regarding the details of the effects of climate change on the PNW remain unresolved, but the potential vulnerability of PNW water resources to rising regional temperatures simulated by global climate models[10] is now reasonably well understood. Impacts associated

[10] The Climate Impacts Group has examined scenarios from eight different global climate models. While each model presents a somewhat different picture with regard to the seasonal

with increased regional temperature stem primarily from reductions in spring snowpack and subsequent summer streamflows upon which most PNW water resources systems depend. These effects have been shown to be largely independent of uncertainties associated with the range of precipitation changes predicted by various global climate models, although higher or lower precipitation can increase or decrease impacts associated with streamflow timing shifts (Gleick 2000; Hamlet and Lettenmaier 1999b; Mote et al. 1999; Cohen et al. 2000).

Figure 9 shows the effects on Columbia Basin snow extent from March through June for the current climate compared to warming scenarios for the 2040s and 2090s compared to a modern base. As the climate warms in winter, more of the low-elevation areas, particularly in the southern part of the basin, have lost their snow by April. The spring snow extent in the northern part of the basin in Canada, by comparison, is largely unaffected. These changes in seasonal snow accumulation affect streamflows in a fairly straightforward manner. Warmer winter temperatures mean that more precipitation falls as rain in the winter and runs off, with corresponding increases in winter streamflow at The Dalles. Decreases in spring snowpack mean that less water is available to contribute to spring and summer streamflows. Furthermore, this reduced snowpack typically melts earlier, which effectively increases the length of the summer season, potentially increasing total evapotranspiration and decreasing late summer soil moistures and the associated base flows in the river prior to the onset of fall rains (Hamlet and Lettenmaier 1999b; Nijssen et al. 2001). The net result for the Columbia Basin is more streamflow in winter, and less in summer for all scenarios by 2040 (Fig. 10) (Hamlet and Lettenmaier 1999b). Note that the streamflow timing shifts in the northern parts of the basin at Corra Linn Dam (see Fig. 2) are much less pronounced than they are at Ice Harbor Dam on the Snake River, which is at a lower elevation in the southern part of the basin. Summer low-flow conditions below a "drought" threshold equal to water year 1992[11] were also shown to occur twice to three times as often than for the status quo, despite the fact that annual precipitation increased significantly in some of the scenarios examined.

---

details of regional warming, all showed increases in winter temperatures in the PNW associated with increases in greenhouse gases (Mote et al. 1999; Nijssen et al. 2001).

[11] Drought thresholds are always somewhat arbitrary, but water year 1992 produced significant drought impacts across the PNW, and the Bonneville Power Administration, to give one example, suffered losses of about $275 million, due to energy shortfalls from the Columbia hydropower system.

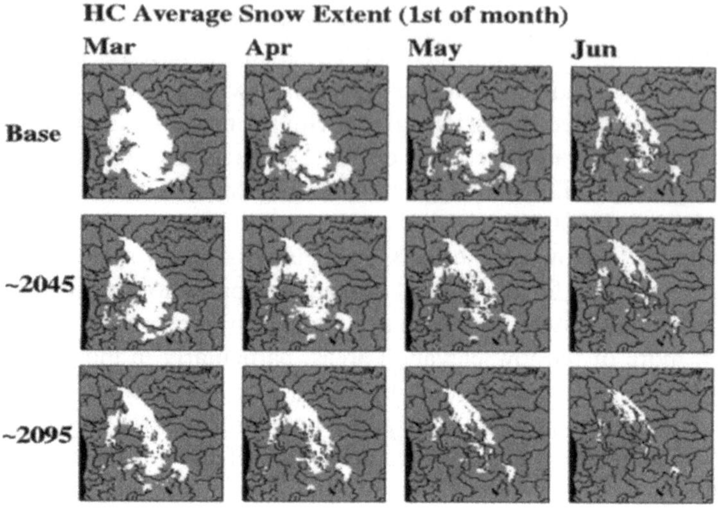


*Figure 9.* Simulated effects of regional warming on long-term mean Columbia Basin snow extent in the U.K. Meteorological Office's Hadley Center model HadCM2 (HC). Shown are warming scenarios for different months at different times in the future compared to a modern base scenario. (Source: Hamlet and Lettenmaier 1999b.)

Not surprisingly, investigation of the impacts on water resources using the ColSim reservoir model (Hamlet and Lettenmaier 1999b) showed that these changes in streamflow timing would result in the greatest impacts on those system objectives most affected by summer low flows (e.g., irrigation, summer non-firm energy production, instream flow, and recreation). Figure 11 shows the impacts to selected water management objectives associated with several climate change scenarios (each from a different climate model) at different times in the twenty-first century. Note that the reliability of firm energy, which is affected by the combined energy capacity of the reservoir system associated with carryover reservoir storage and streamflows in winter, is altered significantly only by the very dry ECHAM4 (Max Planck Institute in Hamburg, Germany) model scenarios, whereas most of the system objectives affected by low summer streamflows are reduced for all but one of the scenarios shown, due to the relatively consistent timing shifts in annual streamflows (Hamlet and Lettenmaier 1999b).

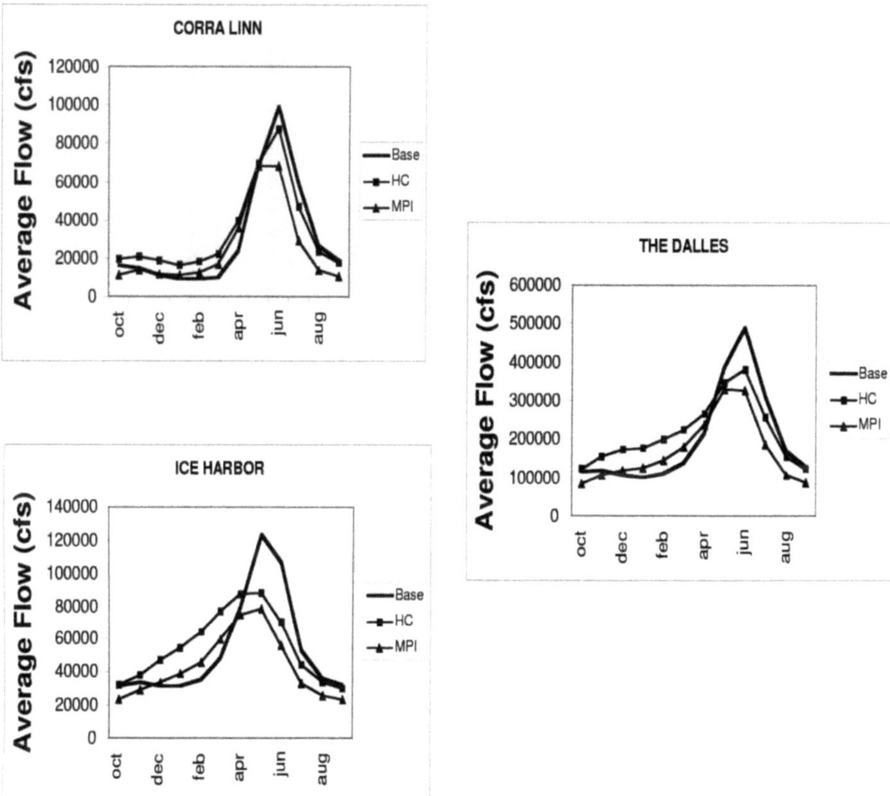


*Figure 10.* Changes in Columbia Basin natural streamflows associated with two warming scenarios for the 2040s (MPI is the Max Planck Institute's ECHAM 4 model scenario, HC is the Hadley Centre's HadCM2 scenario). (Source: Hamlet and Lettenmaier 1999b.)

Because Columbia Basin management institutions have difficulty in coping with climate variability that produces low-flow conditions, PNW water resources systems are also likely to be vulnerable to climate change because of the likelihood of decreased flows in summer on average, and increased frequency of extreme low flows. Projected increases in PNW population and corresponding increases in the demand for energy and water over the next 40 years, while quantitatively uncertain, seem likely to exacerbate rather than mitigate these potential vulnerabilities (Cohen et al. 2000; Miles et al. 2000; Mote et al. 1999).

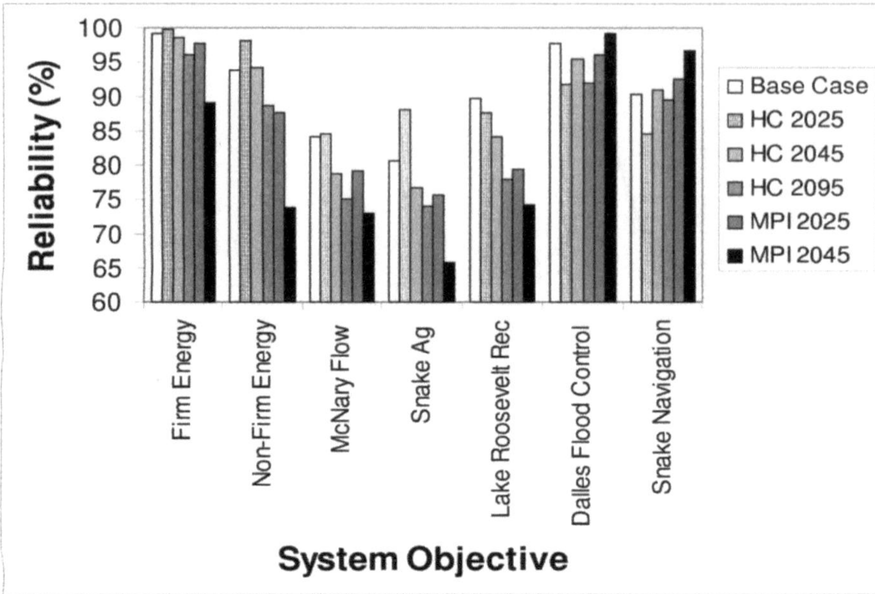


*Figure 11.* Reliability of water resources system objectives for a base case (current climate) and several climate warming scenarios derived from the Hadley Center's HadCM2 and MPI's ECHAM4 global climate models. Snake Ag: Snake River Basin agricultural water needs; Snake Navigation: Instream flows associated with desirable navigation conditions on the Snake River, Lake Roosevelt Rec: reliability of high lake elevations at Lake Roosevelt in summer suitable for recreation (Source: Hamlet and Lettenmaier 1999b.)

Although not examined explicitly in the investigations described above, regional warming may also significantly affect the long-term glacial mass balance in the Canadian ice fields, which may in turn reduce late summer instream flow on longer time scales.[12]

[12] Cycles of glacial accumulation and ablation are not easily analyzed from the relatively short records available, because of nontrivial linkages to antecedent climate conditions at decadal (or longer) time scales. Glaciers in the PNW have clearly been decreasing in mass over the past 20 years or so, for example, but it is not clear whether this is simply a cyclical effect resulting in a short-term trend towards the long-term mean state (melting high levels of glacial accumulation that occurred in the cool and wet conditions that occurred from 1948 through 1976), or whether these changes reflect significant adjustments in the long-term mean associated with global warming (see, e.g., Hodge et al. 1998).

## 10. TRANSBOUNDARY ISSUES BETWEEN CANADA AND THE UNITED STATES

Transboundary issues between Canada and the United States are embedded in some of the primary water resources challenges facing the Columbia River Basin with regard to the maintenance of instream flow in summer. For the status quo, the agencies that manage the Columbia River in the United States are struggling in the midst of conflicts over available water. Such conflicts present few attractive options for increasing instream flow in summer that do not threaten other uses of water with important economic benefits to the region, such as winter hydropower production and irrigated agriculture.

The Columbia River Treaty resulted in increased reservoir storage capacity and has facilitated mutually beneficial use of Canadian and U.S. dams for winter hydropower production and flood control over the past 35 years. The treaty may also have increased the vulnerability of the United States to changes in climate in the context of maintaining summer instream flows. Despite the fact that roughly 50% of the natural flows at The Dalles in late summer originate in Canada, within the framework of the current management institutions and international agreements Canada has limited incentives to provide impounded water in summer for instream flow in the United States to protect endangered salmon. To begin with, salmon runs in the Columbia main stem are completely blocked from the Canadian portion of the basin by Grand Coulee Dam in the United States, so Canada currently has no anadromous fish runs in the Columbia Basin. Furthermore, construction of dams in Canada has resulted in important resident lake fisheries in the upper basin, which have management objectives that conflict with U.S. needs for instream flow in summer. Drawing the Canadian lake systems down in summer to provide instream flow in the United States in summer is likely to negatively impact resident fisheries and recreation opportunities in Canada.

Climate change, with the potential to alter seasonal patterns of snowpack and streamflow, is likely to accentuate these conflicts between the interests of Canada and the United States by increasing the likelihood of low flows in summer (Miles et al. 2000). Assessment of the impacts of climate change on the hydrology of the Columbia Basin demonstrates that a disproportionate amount of the available storage (both snowpack and reservoir storage) in summer will be in Canada for a warmer regional climate (Fig. 9; Hamlet and Lettenmaier 1999b). In conjunction with more pronounced changes in the timing of streamflows in the lower basin, this suggests diminishing U.S. access to important summer water supplies needed to maintain instream flows (even at current levels) in summer in the lower basin.

In this context, the Columbia River Treaty may constitute both an obstacle, and a means, to the transfer of water between Canada and the United States. On the one hand, the treaty is an obstacle for the reasons outlined above and because there is no explicit provision within the existing agreement for the transfer of water in summer between Canada and the United States for the purpose of maintaining instream flow. On the other hand, greater emphasis on marketing of hydropower to California (or other markets outside the PNW) in summer could provide a mutually beneficial transfer of water for both Canada and the United States; this kind of interchange is at least consistent with the economic basis of the Columbia River Treaty, and it has some precedent in current practice. Furthermore, because downstream power benefits are now reverting back to Canada, there is a clear economic incentive for Canada to release water when it is most valuable for energy production. Deregulated energy markets and potential reductions in late summer flows under climate change may increase seasonal price differences (prices could be lower than they are now in winter, higher than they are now in summer) in such a way that economic incentives to release water in late summer will strengthen with time. One obstacle to such changes is associated with the continued need to meet winter energy demand in the PNW. If winter hydropower capacity is decreased, more energy production from thermal sources may be needed to replace this lost capacity, which may in turn increase energy costs in the PNW. One could argue, however, that this fundamental issue will need to be addressed in any case due to current disparities between supply and demand in the PNW and expected increases in PNW population. Furthermore, evolving energy technologies (e.g., fuel cells, second generation nuclear plants) may make alternate methods of energy production more attractive in the future than they are now.

## 11. POTENTIAL CHANGES IN WATER POLICY TO ADAPT TO CLIMATE CHANGE

The seminal water resources problem in the context of adapting to climate change in the Columbia Basin is associated with the need to transfer water from winter to summer to compensate for the timing shifts in natural streamflow that are likely to occur. This need is exactly opposed to current practice in the Columbia Basin as a whole, which emphasizes the use of natural storage in mountain snowpack, and which operates the dams to refill reservoirs in the summer and to make releases from storage in the winter. It is possible, however, that changes in the objectives of reservoir operating policies may reduce some of the vulnerability to summer low flows associated with climate change. Alignment of annual hydropower production with instream flow needs in summer, which was discussed previously, may provide such an opportunity. In sub-basins whose storage is used primarily for irrigation (e.g., the Snake River Basin), however, transfers of water from winter storage to summer instream flow cannot be accomplished without impacts to the water supply available for irrigation later in the summer. This suggests potentially greater difficulty in avoiding climate change impacts associated with simultaneously attempting to protect current levels of irrigation reliability and summer instream flows. One potential solution that has been proposed for areas with significant irrigation demand is to allow a market-based system of voluntary water transfers (both short-term and permanent) to reduce irrigation demand, providing more water for other uses such as instream flow enhancement or summer hydropower production (see e.g., Whittlesey et al. 1986).

## 12. SUMMARY AND CONCLUSIONS

The snowmelt-dominated Columbia River is most strongly influenced by winter climate, which largely determines annual precipitation volumes that contribute to significant runoff production. Storage of winter precipitation as snowpack, reservoir storage equivalent to about 30% of mean annual flow), and relatively small interannual fluctuations in streamflow have made it possible to meet various water resources objectives during the dry Pacific Northwest summers. A disproportionate amount of late summer streamflow originates in Canada under natural conditions. About 30% of the basin is in Canada, but 50% of the natural flow in late summer originates in Canada on average.

Development of the Columbia Basin from about 1940 to 1975 created an extensive flood control and hydropower dam system, with an emphasis on the construction of major storage projects in Canada from 1965 to 1975, and cooperative agreements between the United States and Canada (the Columbia River Treaty and adjuncts) designed to simultaneously provide spring flood storage and winter hydropower production. Irrigation has also played a strong role in the development of some parts of the region (most notably the Snake River Basin), although it accounts for a relatively small amount (about 6%) of the annual flow in the Columbia Basin as a whole. The construction and operation of the Columbia River dam system has provided huge economic benefits to the Pacific Northwest and Canada, and has had major impacts on the natural flow regime of the river, upon which many ecosystems depend.

Following the period of rapid water resources development of the basin, which ended in about 1975, conflicts between the preservation of native salmon populations and the hydropower system began to emerge, and changing social values put increasing pressure on the region's water management agencies to improve habitat for salmon in the river. The Pacific Northwest Power Planning and Conservation Act of 1980, for example, established that hydropower and Columbia River salmon populations were equal in priority under the law. In the context of instream flow, it is apparent that the goals of this legislation have not been met, although many engineering solutions have attempted to improve salmon survival in the river without significantly altering the patterns of regulated flow upon which the region has become dependent for flood control, energy production, and irrigation. In addition to these problems, with increasing pressure on the river it has become impossible to isolate all water resources objectives in the basin from climate variability. Conflicts over summer water supplies in even moderately low-flow years create unavoidable impacts to some objectives. Pressure from increasing human populations in concert with potential changes in climate associated with increasing greenhouse gases threaten to create escalating conflicts over summer water supplies in the region.

Given the dependence of main stem streamflows in the lower basin in late summer on inflows originating in Canada, a number of transboundary tensions are likely to develop between Canada and the United States. Canada, with no anadromous fish runs in the Columbia River, has no direct incentive to preserve instream flow for salmon in the lower basin, and in fact such objectives may conflict with maintenance of healthy lake ecosystems in the upper basin. In addition, much of the snowpack and associated summer water supplies in the basin may originate in Canada in a warmer climate, while existing treaty agreements have no explicit provisions for transfer of water from Canada to the United States for the purposes of maintaining instream flow. Building Canadian storage has therefore arguably increased the vulnerability of the United States to future conflicts over summer water supplies, since natural flows from Canada are now impounded, and do not flow across the international boundary unimpeded.

Despite these problems, the agreements within the Columbia River Treaty and other associated transboundary agreements present some potential opportunities to facilitate the mutually beneficial release of water from Canadian storage in summer to help meet instream flow objectives in the United States. The growing summer demand for energy outside the Pacific Northwest, for example, may provide opportunities to market Canadian hydropower at high prices at times when fish flows are needed in the lower basin. While there is currently no explicit provision for these kinds of water transfers in the existing treaties, the economic framework of the Columbia River Treaty, at least, appears to be consistent with these kinds of summer water transfers. Deregulated energy markets and potential increases in summer energy prices associated with climate change (increasing summer demand, decreasing summer water supply) may increase economic incentives to release water from Canadian storage in summer over time.

## 13. ACKNOWLEDGMENTS

This publication (submission #895) is supported by a grant to the Center for Science in the Earth System funded through The Joint Institute for the Study of the Atmosphere and Ocean (JISAO) under NOAA Cooperative Agreement No. NA17RJ1232. The author would also like to thank and acknowledge his colleagues at the Climate Impacts Group whose research is referenced herein.

## 14. REFERENCES

Battisti, D.S., and E. Sarachik. 1995. Understanding and predicting ENSO. *Reviews of Geophysics* 33: 1367–1376.

Bisbal, G.A., and W.E. McConnaha. 1998. Consideration of ocean conditions in the management of salmon. *Canadian Journal of Fisheries and Aquatic Sciences* 55: 2178–2186.

Bernton, H. 2001. Klamath Basin farmers, communities reel at cutoff of irrigation water in favor of endangered fish. *Seattle Times*, July 1.

Bonneville Power Administration (BPA), U.S. Army Corps of Engineers (USACOE), North Pacific Div., U.S. Bureau of Reclamation (USBR), Pacific Northwest Region. 1991. *The Columbia River System: The Inside Story*. Report DOE/BP-1689 published for the Columbia River System Review by the USACOE and the USBR, September.

Callahan, B., E. Miles, and D. Fluharty. 1999. Policy implications of climate forecasts for water resources management in the Pacific Northwest. *Policy Sciences* 32: 269–293.

Cohen, S.J., K. Miller, A. Hamlet, and W. Avis. 2000. Climate change and resource management in the Columbia River Basin. *Water International* 25: 253–272.

Cone, J., and S. Ridlington. 1996. *The Northwest Salmon Crisis: A Documentary History*. Corvallis: Oregon State University Press.

Crook, A.G. 1993. *1990 Level Modified Streamflow, 1928–1989*. Report of A.G. Crook Company to Bonneville Power Administration, Contract # DE-AC79-92BP21985

Gershunov, A., and T.P. Barnett. 1998. Interdecadal modulation of ENSO teleconnections. *Bulletin of the American Meteorological Society* 79: 2715–2725.

Gleick, P.H. 2000. *Water: The Potential Consequences of Climate Variability and Change for the Water Resources of the United States*. Report of the Water Sector Assessment Team of the National Assessment of the Potential Consequences of Climate Variability and Change, U.S. Climate Research Program, September.

Gray, K.N. 1999. The impacts of drought on Yakima Valley irrigated agriculture and Seattle municipal and industrial water supply. M.M.A. Thesis, School of Marine Affairs, University of Washington, Seattle, Washington.

Hamlet, A.F., and D.P. Lettenmaier. 1999a. Columbia River streamflow forecasting based on ENSO and PDO climate signals. *American Society of Civil Engineers, Journal of Water Resources Planning and Management* 125: 333–341.

Hamlet, A.F., and D.P. Lettenmaier. 1999b. Effects of climate change on hydrology and water resources in the Columbia River Basin. *Journal of the American Water Resources Association* 35: 1597–1623.

Hamlet, A.F., and D.P. Lettenmaier. 2000. Long-range climate forecasting and its use for water management in the Pacific Northwest region of North America. *Journal of Hydroinformatics* 2.3: 163–182.

Hodge, S.M., D.C. Trabant, R.M. Krimmel, T.A. Heinrichs, R.S. March, and E.G. Josberger. 1998. Climate variations and changes in mass of three glaciers in western North America. *Journal of Climate* 11: 2161–2179.

Mantua, N., S. Hare, Y. Zhang, J.M. Wallace, and R. Francis. 1997. A Pacific interdecadal climate oscillation with impacts on salmon production. *Bulletin of the American Meteorological Society* 78: 1069–1079.

Mantua, N.J., R.C. Francis, P.W. Mote, and D. Fluharty. 2001. Climate and the Pacific Northwest Salmon Crisis: A Case of Discordant Harmony. In, Amy Snover (ed.), *Rhythms of Change, An Integrated Assessment of Climate Impacts on the Pacific Northwest.* Boston: Massachusetts Institute of Technology Press (in review).

Mapes, L. 2001a. Water woes to force closure of taps at Baker River dam. *Seattle Times*, February 24.

Mapes, L. 2001b. Chinook salmon pay price as Skagit runs low on water. *Seattle Times*, February 27.

Miles, E.L., A.K. Snover, A. Hamlet, B. Callahan, and D. Fluharty. 2000. Pacific Northwest regional assessment: The impacts of climate variability and climate change on the water resources of the Columbia River Basin. *Journal of the American Water Resources Association* 36: 399–420.

Mote, P. (ed.), and co-authors. 1999. *Impacts of Climate Variability and Change: Pacific Northwest.* Report to the PNW regional assessment for the National Assessment of the Impacts of Climate Variability and Change, Climate Impacts Group, University of Washington, November.

National Marine Fisheries Service (NMFS). 1995. Biological opinion: Re-initiation of consultation on 1994–1998 operation of the federal Columbia River power system and juvenile transportation program in 1995 and future years. Endangered Species Act, Section 7 Consultation. National Marine Fisheries Service, Northwest Region, March 2, 1995.

Nijssen, B., G.M. O'Donnell, A.F. Hamlet, and D.P. Lettenmaier. 2001. Hydrologic sensitivity of global rivers to climate change. *Climatic Change* (in press).

Quinn, T.P., and D.J. Adams. 1996. Environmental changes affecting the migratory timing of American shad and sockeye salmon. *Ecology* 77: 1151–1162.

Trenberth, K.E. 1997. The definition of El Niño. *Bulletin of the American Meteorological Society* 78: 2771–2777.

Volkman, J.M. 1997. *A River in Common: The Columbia River, the Salmon Ecosystem, and Water Policy*. Report to the Western Water Policy Review Advisory Commission.

Whittlesey, N., J. Hamilton and P. Halverson. 1986. *An Economic Study of the Potential for Water Markets in Idaho*. Report to the Snake River Technical Studies Committee, joint report of the Idaho and Washington Water Resources Research Institutes, Moscow, December.

Wood, C.A. 1993. Implementation and evaluation of the water budget. *Fisheries* 18: 6–17.

# 13

# CLIMATE DOESN'T STOP AT THE BORDER

*U.S.-Mexico Climatic Regions and Causes of Variability**

Andrew C. Comrie

*Department of Geography and Regional Development, University of Arizona, Tucson, Arizona, U.S.A.*

**Abstract** Climate studies frequently make use of data stored in national data centers, and by national origin. Of course, rainfall or temperature patterns and the climatic processes that cause them are distributed across the earth system independently of political boundaries. Yet, much of the observation-based research into the nature and causes of climate variability uses data sets defined by climatologically arbitrary political boundaries. This is true of many projects and published papers covering the U.S.-Mexico border.

This chapter reports on results and implications of a precipitation regionalization study for the entire U.S.-Mexico border region. The analysis is based on seasonality and variability of monthly precipitation at 309 stations in the southwestern United States and northern Mexico for the period 1961 through 1990. Principal components analysis (PCA) methods were used to divide this climatologically complex study area into physically meaningful regions for subsequent analysis. Nine consistent and largely contiguous regions were obtained in several solutions, including regions for the North American monsoon, the low deserts, and the California Mediterranean region, and for summer precipitation regimes adjoining the Gulf of Mexico. The monsoon region was also analyzed to identify four monsoon subregions. The applicability of the regionalization is illustrated via an analysis of relationships between monsoon precipitation variability and atmospheric circulation.

---

*This chapter is from a paper by Comrie and Glenn (1998) that originally appeared in the journal *Climate Research*. It is reproduced here in slightly adapted form by permission of Inter-Research Science Publishers.



*Henry F. Diaz and Barbara J. Morehouse (eds.), Climate and Water*, 291-316.

# 1. INTRODUCTION

Precipitation is the critical climate variable in the arid borderlands of the United States and Mexico. Although climate is of course contiguous across the political boundary, historically, broad-scale analyses of precipitation across this border region have most often treated the U.S. and Mexican sides separately, partly for reasons of data availability. The region is physically and climatically complex, making simple comparisons of precipitation awkward. Yet, there are several well-recognized climatic controls on precipitation across the region that reflect its subtropical location. The subtropical anticyclones and associated upper-level ridges centered over the region and adjacent Atlantic and Pacific Oceans exert a dominant influence year-round, modulated by the role of the mid-latitude westerly circulation in winter and the North American summer monsoon (Carleton 1987; Cavazos and Hastenrath 1990; Burnett 1994; Comrie 1996; Woodhouse 1997; Adams and Comrie 1997; Higgins et al. 1998). Subregionally, physiography (Fig. 1) and atmospheric circulation combine to create a spatially complex precipitation climatology (Fig. 2), with strong seasonal contrasts and areas of considerable interannual variability.

For many analytical purposes (e.g., examination of temporal trends), it is convenient to divide the spatial continuum of a precipitation climatology into a manageable number of quasi-homogeneous areas. Climate regionalization is a useful technique that enables generalization about areas on the basis of a spatially and temporally varying parameter such as precipitation. Eigenvector-based techniques are widely used for delimiting climate regions (Bärring 1988; Mallants and Feyen 1990; Lyons and Bonell 1994; Fernàndez Mills et al. 1994; Fernàndez Mills 1995), and such approaches have been successful in discriminating precipitation regions based on similar seasonality and long-term variability characteristics (Kutzbach 1967; Richman and Lamb 1985, 1987; Ogallo 1989; Eklundh and Pilesjö 1990; White et al. 1991; Bunkers et al. 1996; Cahalan et al. 1996). The work of White et al. (1991) reviews this approach and compares a range of rotation algorithms for eigenvector-based climate regionalization. They found oblique rotations to be generally the most stable for climate regionalization, while orthogonally rotated and unrotated solutions were less stable.

Principal components analysis (PCA) was used to identify precipitation regions for the U.S.-Mexico border region based on the above characteristics. It was decided explicitly to avoid relying on the absolute precipitation totals in order to exclude the otherwise overwhelming elevation effects. This is a region of complex terrain in which local precipitation amounts are governed by topographic controls at finer scales than the atmospheric controls of overall seasonality and variability. Consequently, this chapter aims to identify spatially cohesive precipitation regions based on common patterns of seasonality and variability rather than on simple precipitation totals. There are no studies that explicitly regionalize precipitation in this area, although several authors have examined related topics such as harmonic analysis of precipitation, predominant seasonal air masses, and synoptic climatology (Horn and Bryson 1960; Mitchell 1976; Barry et al. 1981; Comrie 1996).

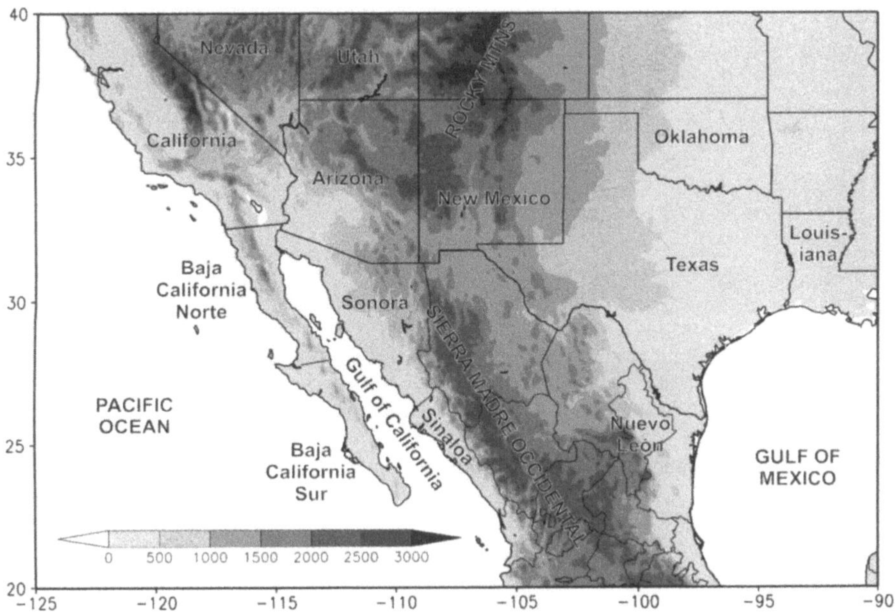


*Figure 1.* Terrain elevation (m) of the study area with major place names and physical features mentioned in the text.

Specifically, this investigation applies the general recommendations of White et al. (1991) in a broad-scale precipitation regionalization over the large domain of the border region. It examines the performance of the method itself, evaluates the various defined regions, and compares two techniques for locating region boundaries. Because the value of any climate classification lies in its usefulness, the applicability of the regionalization is illustrated via an analysis of precipitation variability and 500 millibar (mb) circulation for the North American monsoon region. Thus, the broader goals behind the study are first to simply identify and describe these regions as they exist across both sides of the political boundary, noting their characteristics and controls, and second, to provide appropriate spatial units for analyses of precipitation variability in this and future work.

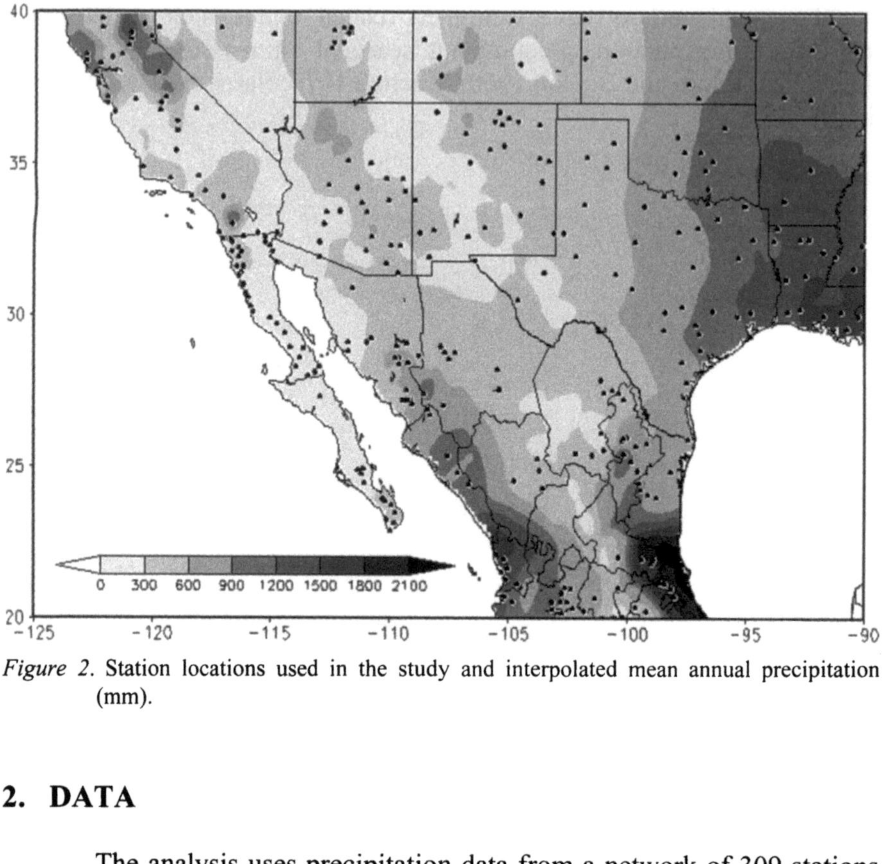


*Figure 2.* Station locations used in the study and interpolated mean annual precipitation (mm).

## 2. DATA

The analysis uses precipitation data from a network of 309 stations throughout the study region, which covers land areas between 20° and 40°N and 125° and 90°W (see Figs. 1 and 2). The domain was chosen to be somewhat larger than the perceived "edges" of the immediate border region so that precipitation regions extending beyond this zone might be resolved with respect to neighboring areas. The data set comprises monthly precipitation totals from 1961 through 1990 (n = 360) for stations across the study area, extracted from the Global Historical Climatology Network (GHCN) (Vose et al. 1992) archived at the National Climatic Data Center. Several other data sources were evaluated, but the quality control of the GHCN data (systematic screening for missing data, outliers, station moves, internal consistency, etc.) made them the optimal choice. A dense pattern of points was chosen, as precipitation is highly variable (White et al. 1991), and a regular spacing of stations was attempted because some studies have shown that a PCA performed on unevenly spaced points can alter the loading patterns significantly (Karl et al. 1982). To do this, all stations were ranked by the proportion of missing data for each, resulting in 173 stations in the United States that had at least 96.6% data present. For Mexico, 136 stations were used that had at least 75.0% data present. While many of the Mexican stations had better records than this, some of the lower-ranking stations were needed for better spatial coverage. Also, much of the missing Mexican data were for the period after the 1985 Mexico City earthquake, during which many records were lost. No estimates of these missing data were made, as they tended to be missing in chronological blocks for adjacent stations. The lower data percentages were of some concern, as correlation matrices generated from data containing missing values can yield negative eigenvalues (Rummel 1970). This problem was circumvented to some degree by pairwise (rather than listwise) deletion across the matrix. For the application to precipitation variability in the monsoon region, data for 500 mb geopotential height for 1961–90, obtained from the National Centers for Environmental Prediction (NCEP)/National Center for Atmospheric Research (NCAR) reanalysis project (Kalnay et al. 1996), were used.

## 3. METHODS

For the analysis, S-mode PCA (multiple stations over time) was applied using a correlation matrix with pairwise deletion as above. Use of the correlation matrix, as opposed to the covariance matrix, allowed dry stations in the deserts to be directly compared to relatively wet stations in the mountains. Thus, stations with the same seasonal timing of monthly rainfall were correlated and grouped together, presumably because of the same atmospheric controls, even if absolute precipitation amounts differed simply because of the elevation effect. Naturally, the types of patterns that emerge from such an analysis depend on the aspect of precipitation that is being regionalized, such as annual totals, seasonal totals, or some other measure of temporal variability.

The selection of the number of principal components (PCs) to retain was relatively clear in this case, and based on scree tests (Cattell 1966), eigenvalues > 1.0, and the eigenvalue separation test (North et al. 1982); nine PCs (i.e., regions) explaining 65% of the variance in the data were retained. White et al. (1991), in their comparison of rotation techniques, found that oblique rotations generally produced the best results for climate regionalization. Therefore, the retained components were subjected to oblique rotation (direct oblimin) to enhance interpretability, and to avoid Buell (1975, 1979) patterns. The value of the parameter $\gamma$ controls obliquity. Negative values create less oblique solutions (greater orthogonality); positive values result in more obliquity (greater collinearity) (Clarkson and Jennrich 1988). Initially, $\gamma = 0$ was used (intermediate obliquity), and in a later analysis $\gamma = -0.2$ was used. For this regionalization method, no clustering algorithm is applied after the PCA. Instead, the rotated loading values ($\lambda$) for each PC are plotted on a map. An interpolation algorithm can produce isolines, and regions can be produced from them by using the $\lambda = 0.4$ loading contours as boundaries (Richman and Lamb 1985, 1987; White et al. 1991).

An alternative maximum-loading approach was also evaluated, in which each station is assigned to the component upon which it loads most highly. Regions can then be drawn around stations assigned to the same PC. Both of these methods were used and compared. The choice of 9 retained components was confirmed via favorable comparison of mapped results between this and several alternative solutions that were obtained, as well as to analyses of the U.S. and Mexican subsets alone.

Thus, in using oblique rotation, the general recommendations of White et al. (1991) for their work in Pennsylvania were followed in applying this regionalization method to the U.S.-Mexico border region. However, in this case there are several distinctions: The domain is an order of magnitude larger, comprising more than the 10 border-region states versus the state of Pennsylvania; the precipitation patterns are more complex with strong elevation and rain-shadow effects; the seasonal regimes are also more complex and are not limited to classic frontal mid-latitude systems; and the spatiotemporal data coverage, while the best available, is less than ideal.

Once the initial results were examined, four subsequent analyses were performed. The first and second of these were carried out as checks for robustness of the regions: an analysis of precipitation data transformed closer to normal, and a change in the obliquity of the rotation algorithm. The third analysis was a further subregionalization of the monsoon region to provide an additional level of detail for this large area. This analysis formed the basis for the fourth analysis, an examination of interannual monsoon precipitation variability, in which a compositing analysis of the 500 mb circulation was performed based on anomalous wet and dry summers in each of the monsoon subregions.

# 4. RESULTS AND DISCUSSION

## 4.1. Loadings Maps

Figure 3 shows the rotated loadings patterns for the nine PCs, with PC1 representing the most common variance in the rotated solution, and PCs 2 through 9 representing successively less common variances. The nine regions are each distinct, physically reasonable, and intuitively satisfying with respect to the climatology of the area. All regions have a clear core, with almost no confusing secondary focus areas. In Figure 4, the upper panel (a) shows the regions delineated by using the $\lambda = 0.4$ contour rule (–0.4 for the nominally negative loadings in PC5), and the lower panel (b) shows the regions delineated by using the maximum loading rule described above.

The results are essentially the same for both mapping methods; for the contour rule, overlapping lines indicating varying transition gradients across region boundaries, while the maximum loading rule is categorical. Depending on the application, these may be positive or negative attributes. Note that a small number of stations at the northern limit do not fall within any 0.4 contours and might be labeled as unclassified. However, the value of 0.4 is convenient but arbitrary, so all stations could be classified by using 0.35, or far fewer stations could be classified by using the 0.5 contour. All stations are classified under the maximum loading rule, and although a minimum loading threshold could be used, it was not employed here. Because stations on the periphery of the study area by definition fall within climatic regions underrepresented in the data set, concern is less with them than with those in the center, which were all well classified by both rules. Because of applications to some future work, and for consistency in this chapter, subsequent results are shown using the maximum loading rule.

## 4.2. Regions

To facilitate discussion of the regions, Figure 5 presents the mean monthly precipitation across all stations in each of the nine regions. Determining the spatial form of these nine seasonality regimes was the primary purpose of the PCA. Recall that use of the correlation matrix focuses on the shape of the seasonal cycles over the study period and not the amount of precipitation. Region 1 is the monsoon region, which stretches along the Sierra Madre Occidental in northwestern Mexico and follows the Continental Divide through eastern Arizona and most of New Mexico in the United States. This region is characterized by the strong mid- to late summer precipitation maximum with considerably less precipitation during the rest of the year. The western boundary appears to skirt the escarpment along the Gulf of California and then snake around the southeastern highlands and the Mogollon Rim of Arizona, closely mirroring the patterns of thunderstorm activity during the monsoon (Adams and Comrie 1997) and the first harmonic of precipitation (Horn and Bryson 1960). The northern and eastern boundaries likewise provide a reasonable match with precipitation-based definitions of the North American monsoon, and they include the extreme southern and western portions of Colorado and Texas, respectively, running south along the Mexican Altiplano (highlands). The narrow southern boundary of the monsoon region was not resolved within the domain, and it most likely blends into the Intertropical Convergence Zone (ITCZ)–related rainfall of central and southern Mexico. This region covers a large latitude range, and it is central to the climate of the entire borderlands, so an additional analysis was performed to identify subregions, described later in the section on monsoon subregions.

Region 2 is a desert region to the west of the monsoon region, and it encompasses areas surrounding the Mojave Desert and the lower Colorado River Valley, including southeastern California, southeastern Nevada, western Arizona, far northwestern Sonora, and Baja California Norte. It is dry essentially year-round (with many areas receiving less than 300 mm/yr), with marginally more precipitation in winter. The monsoon precipitation peak is notably absent, and the region is at the southernmost limit of midlatitude frontal systems that supply the limited winter precipitation. The coastal ranges of southern California and the Sierra Nevada appear to create the boundary with Region 5 over California, which has a distinct wet winter and dry summer Mediterranean regime caused by a classic pattern of dominant westerlies in winter and the dominant North Pacific anticyclone and associated West Coast ridge in summer. Region 7 covers Baja California Sur and a narrow coastal strip of the Mexican mainland just across the Gulf of California. The summer rainfall peak and low winter precipitation partly relate this region to the monsoon, but the rains peak in September rather than in July and August, possibly because of increased tropical storm activity (Adams and Comrie 1997; Stensrud et al. 1997).

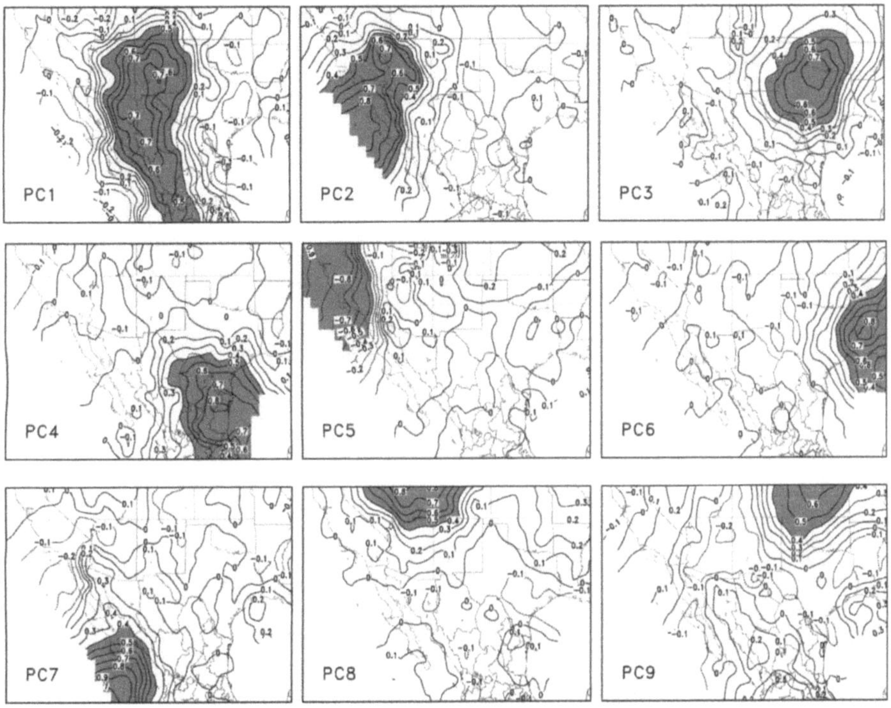


*Figure 3.* Rotated loadings for the original nine principal components (PC1-PC9).

To the east of the monsoon region, Region 3 covers much of central Texas and Oklahoma, and it has dual early/late summer precipitation peaks in May and September. This seasonal regime is similar to that for Region 4, covering northeastern Mexico and adjacent coastal areas in Texas, where the early summer peak is in June and the September peak is relatively a little higher (Region 3 likely experiences greater frontal influence early and late in the summer half-year than Region 4). Region 4 corresponds to the regime examined by Cavazos (1997). Both regimes are strongly controlled by the North Atlantic subtropical anticyclone and its interaction with regional air masses and the continental elevated mixed layer (Cavazos and Hastenrath 1990; Lanicci and Warner 1991; Cavazos 1997), and both experience the midsummer dry spells known in Mexico as *caniculas*. The two maxima in Regions 3 and 4 are partly associated with tropical storms and hurricanes over the Caribbean/Atlantic Basin. Regions 6, 8, and 9 are peripheral to the border region, but they serve to reinforce the boundaries of the core border regions. Region 6 over eastern Texas, Louisiana, and the southeastern United States receives year-round precipitation, while Regions 8 and 9 appear to be related to regimes beyond the study area over the Intermountain West and the Midwest United States.

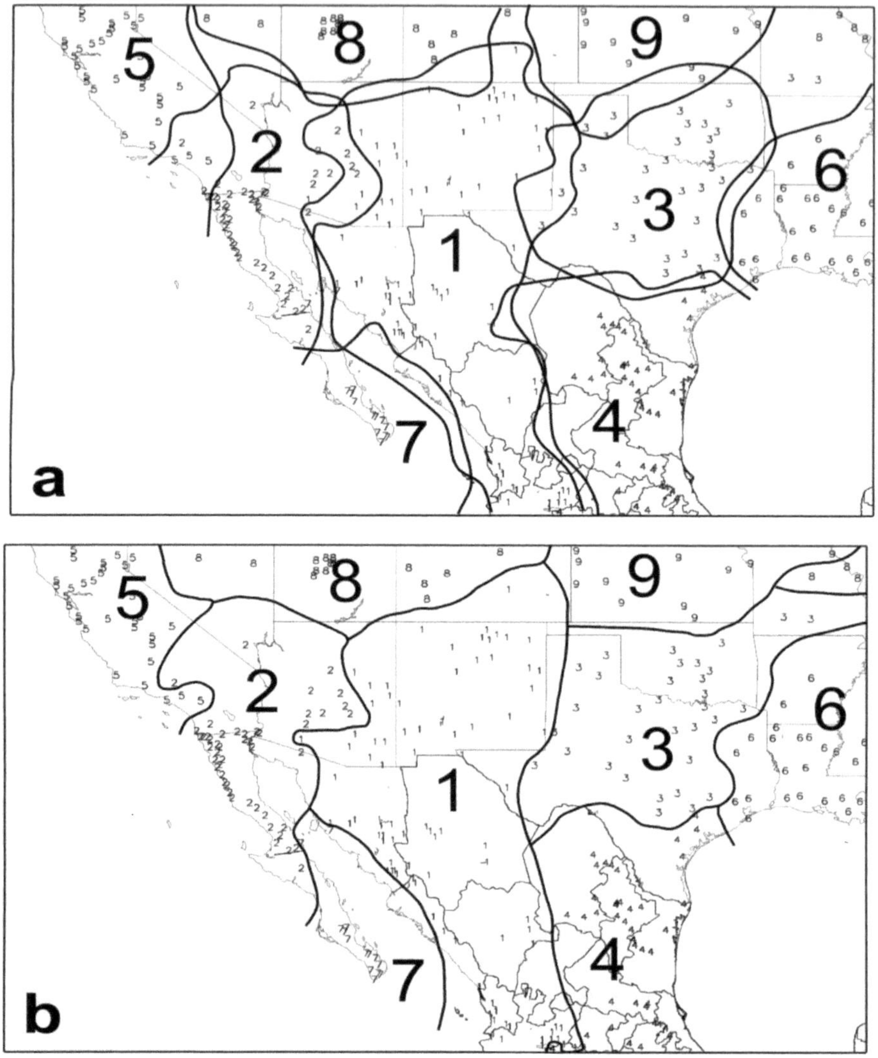


*Figure 4.* The nine regions delineated by using (a) the $\lambda = 0.4$ contour rule and (b) the maximum loading rule.

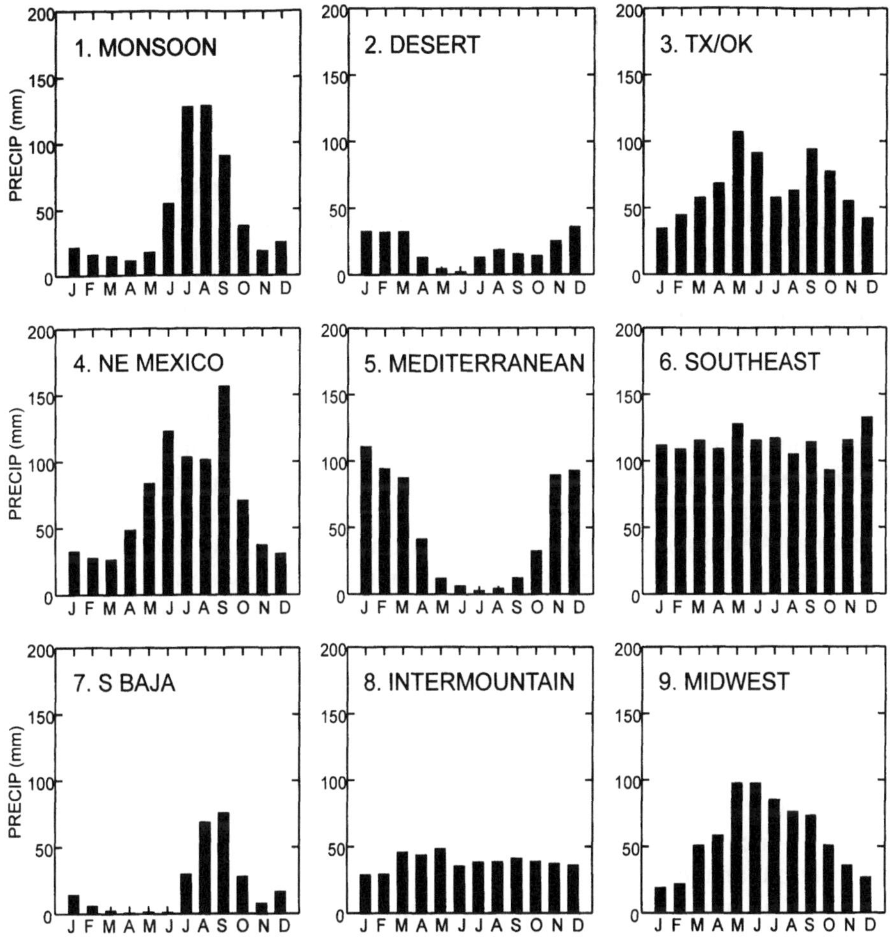


*Figure 5.* Mean monthly precipitation of all stations in each of the regions.

Overall, the regions correspond well with the respective climatological controls on seasonality. Region 2, the desert region, is perhaps the least definite because of the intrinsically flat precipitation distribution with very little seasonal signal from low precipitation amounts (e.g., the Pacific coast of Baja California Norte might arguably be more Mediterranean in nature). Climatic controls in this region are likely a marginal mix of those for the neighboring monsoon and Mediterranean regions. Nonetheless, most regions are quite distinct, and because common seasonality within regions implies common controls, the set of regions provides justifiable, cohesive climatic units for study.

### 4.3. Transformed Data

Pearson's correlation coefficient used in the PCA input matrix can be sensitive to non-normality (White et al. 1991). In practice, PCA seems quite robust to moderate departures from normality of input data, but given the unusual precipitation distributions in some parts of the study area (especially stations with many zero precipitation months in the record), it seemed prudent to evaluate the regionalization by using data transformed somewhat closer to normal. Frequency distributions across stations were examined, and experiments were made with several transformations, including the gamma distribution that is bounded on the left by zero and positively skewed, and is therefore used in some precipitation studies. However, a simple square root transformation is also commonly used, and it appeared most useful for monthly precipitation ($p$) when considered across all stations (i.e., $p^{0.5}$). Figure 6 shows frequency distributions for four example stations before and after the $p^{0.5}$ transform (Tucson, Arizona; Albuquerque, New Mexico; Monterrey, Nuevo León; and Cabo San Lucas at the tip of Baja California Sur), and following the transformation all are moderately normal except the most extreme example (Cabo San Lucas). With the transformed data as input, the analyses were performed exactly as before. Remarkably, perhaps, the appropriate number of PCs to retain was again 9, this time explaining 64.8% of the variance. The rotation (oblique, $\gamma = 0$) and mapping (maximum loading rule) procedures were carried out as before. Figure 7 shows the resulting nine regions, in which the PC numbering has changed to reflect the new explained variances of each, but where the overall regions are quite similar to those for the original analysis.

The monsoon region is largely unchanged, except for the elimination of some stations in northeastern New Mexico and the appearance of a southern boundary. The latter adjoins a small, new peripheral central Mexico region. The desert, Mediterranean, and southern Baja regions are identical to those for the initial analysis. The boundaries of the northeastern Mexico/southeastern Texas region and the southeastern United States region are the same except for a few stations where the two regions meet. The core of the Texas/Oklahoma region is much the same, but along the northern edge of the study area there are several changes in boundaries, with an additional split region. Thus overall, apart from edge effects, the central border precipitation regions are the same for data transformed by using $p^{0.5}$.

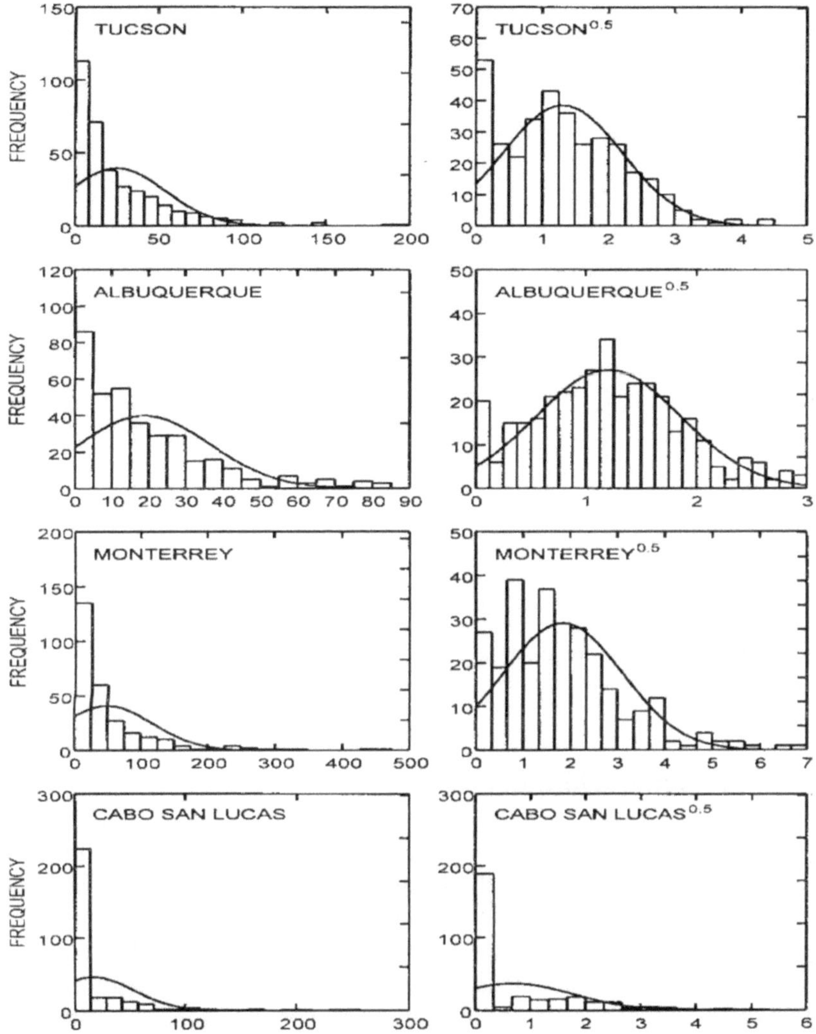


*Figure 6.* Frequency distributions for four example stations, showing original precipitation data (left) and square root transform ($p^{0.5}$) data (right) with normal curves based on sample means and standard deviations (units in millimeters).

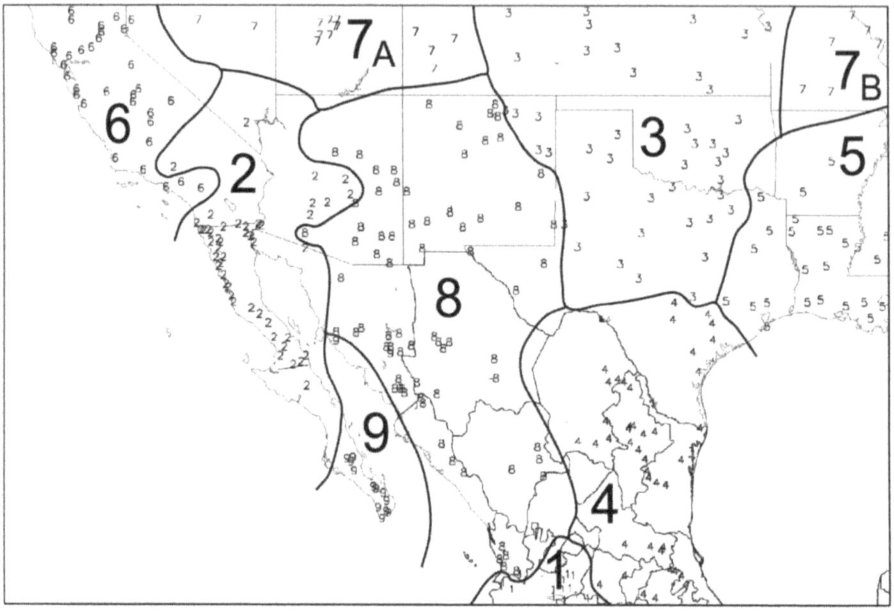


*Figure 7.* The nine regions from the square root transform ($p^{0.5}$) analysis, delineated by using the maximum loading rule.

## 4.4. Changing Obliquity

One of the main points made by White et al. (1991) in their comparison of PCA rotations was the greater stability of obliquely rotated solutions for regionalization (and hence the use of oblique rotation in this study). The direct oblimin algorithm (see Methods section) was used because of its availability in a statistical package, but White et al. (1991) mentioned that it may result in slightly over-oblique solutions compared to some other algorithms. Their application of oblique rotation used $\gamma = 0$, the standard case, as was done above.

To evaluate a less oblique regionalization solution (approximating other algorithms), therefore, direct oblimin rotation was performed with a less oblique $\gamma = -0.2$ on the original (untransformed) PCA results, and results were mapped by using the maximum loading rule. Figure 8 shows the nine regions for the $\gamma = -0.2$ solution, which again are consistent with previous patterns and show very similar numbering. Apart from a few individual station changes, the most notable difference in the monsoon region is the movement of the boundary with southern Baja, where the mainland coastal stations along the Gulf of California join the latter region. A new central Mexico region appears, as for the transformed data, with some concomitant minor changes in the boundary with the northeastern Mexico region. However, the new region is one of two disconnected parts, the other part being the original Midwest region (which added a few stations in the Texas panhandle and northeastern New Mexico). Other than these small differences, the southeastern United States, Texas/Oklahoma, desert, Mediterranean, and Intermountain West regions are basically unchanged, although the latter region has a smaller disconnected area (one station) over Missouri than before. Again, for the core border regions of principal interest, the results for changed obliquity are very similar to those for the initial analysis.

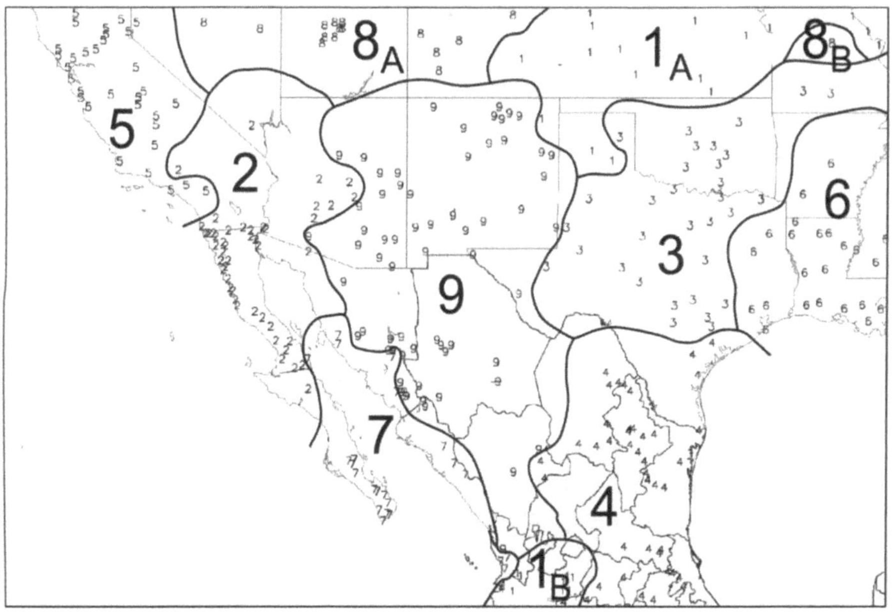


*Figure 8.* The nine regions from the $\gamma = -0.2$ oblique rotation analysis, delineated by using the maximum loading rule.

## 4.5. Monsoon Subregions

As was mentioned earlier, the monsoon region covers a broad area over a large range of latitude from central Mexico to Colorado, as well as straddling the Continental Divide. To examine the region more closely, the overall analysis methodology for the 89 stations in Region 1 (i.e., stations with maximum loading on PC1 in the initial analysis) was replicated. Following S-mode PCA of the monthly precipitation data, 4 components were retained, explaining 72% of the variance. Again, the 4 PCs were subjected to direct oblimin rotation ($\gamma = 0$), and the results were mapped.

Figure 9 shows the four subregions, and Figure 10 illustrates the mean monthly precipitation for each subregion. Subregion 1 lies within Mexico, Subregions 2 and 4 straddle the international border and lie to the west and east of the Continental Divide, respectively, while Subregion 3 covers northeastern New Mexico and southern Colorado. The Mexican subregion (1) has the highest rainfall, as expected, with the summer rains initiating in June, peaking in July and August, and dissipating in October. Not all of this rainfall is due to the monsoon; some is due to other tropical systems in the Caribbean/Atlantic Basin mentioned earlier. This region is also affected by late-summer and early-fall tropical storms off the west coast of Mexico. The other subregions receive less rainfall each month over a slightly shorter season, being on the northern edge of the monsoon circulation and farther from moisture sources. Subregion 2 has a longer dry period in the early summer, with monsoon rainfall principally in July and August and a small winter precipitation signal. In Subregion 4 to the east, there is some rainfall in May and June (early summer), but the peak summer rainfall falls between July and September. In Subregion 3, the early summer lacks the typical pre-monsoon dry period (perhaps a result of being the northernmost subregion and under a more westerly circulation), but the mid- and late summer does have the characteristic July and August rainfall peak. These timing differences across western Texas, southern New Mexico, eastern Arizona, and neighboring parts of Mexico match well with current understanding of the onset of the North American monsoon, which begins generally in the eastern portions of the greater monsoon region and transitions westward (Adams and Comrie 1997).

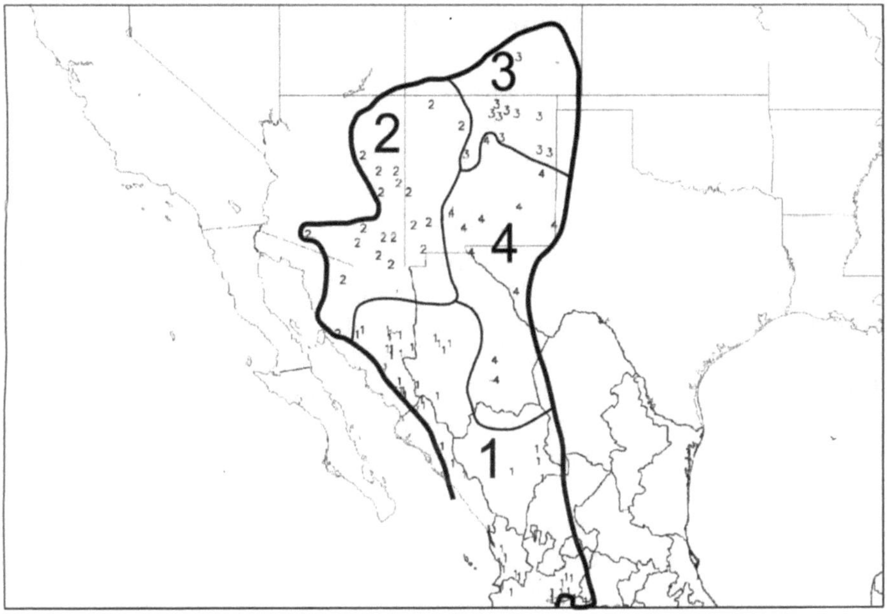


*Figure 9.* The four monsoon subregions delineated by using the maximum loading rule.

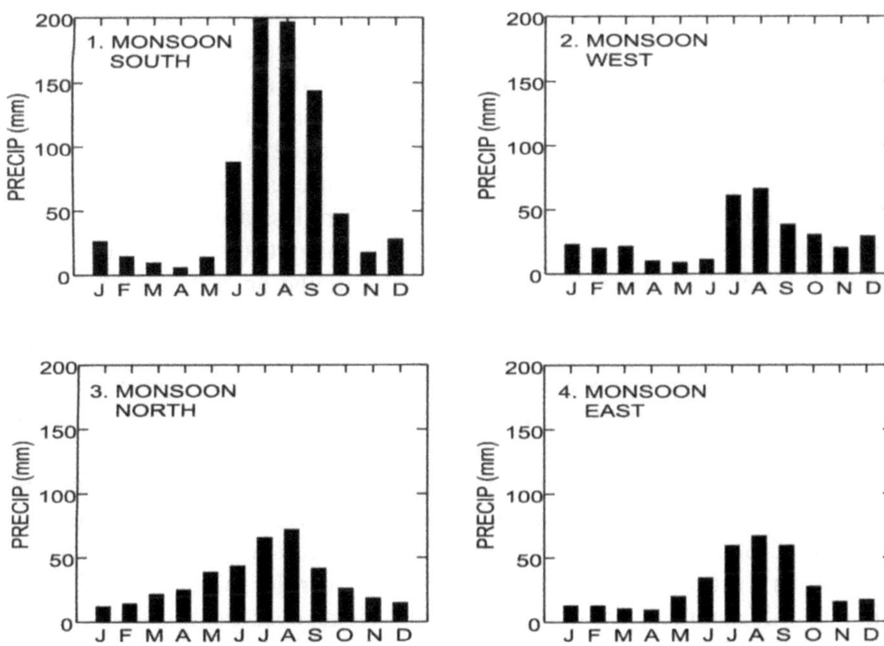


*Figure 10.* Mean monthly precipitation of all stations in each of the four monsoon subregions.

## 4.6. Monsoon Subregional Precipitation Variability

The North American monsoon is central, geographically and intellectually, to understanding the climates of the southwestern United States and northern Mexico. It is known that important seasonal and interannual controls on monsoon variability occur at both the synoptic and mesoscales (Higgins et al. 1998; Higgins and Shi 1999, 2001), although the role of each and their interactions are not well explored. However, there is an ongoing paucity of mesoscale climate data, precluding analysis at that scale. At the synoptic scale, most studies have focused on the monsoon margin that falls in the United States (Adams and Comrie 1997), often at the 500 mb level. Therefore, to extend such analyses over the entire binational monsoon region, and to illustrate application of the regionalization results, relationships between precipitation variability in the monsoon subregions and the 500 mb circulation were examined.

Anomalies for total summer (July, August, and September) precipitation during the study period for each of the monsoon subregions were calculated. The standardized anomaly time series are shown in Figure 11, representing interannual variability that ranges up to ±100 mm across all subregions. Because the mean summer precipitation in Subregion 1 (542 mm) is greater than those in the other subregions (166 to 183 mm), this represents roughly a 20% variability in Subregion 1 and as much as 50% variability in Subregions 2 through 4. While there are a number of coincident wet and dry summers among certain subregions, precipitation variability across the subregions can vary considerably (correlations range between –0.3 and 0.3).

The driest and wettest seven summers in each anomaly time series (representing roughly the 25th and 75th percentiles, arbitrarily) were selected, and the matching daily 500 mb pressure heights ($n = 644$) were composited. Figure 12 shows the mean 500 mb circulation patterns associated with the wet and dry summers for each subregion. By subtracting the mean of the remaining "middle" 15 summers ($n = 1380$), matching wet and dry anomaly fields and their significance were also calculated. Figure 13 illustrates these 500 mb anomalies for wet and dry summers, with shaded areas indicating values significant at $\alpha \leq 0.05$.

In monsoon Subregion 1 to the south over Mexico, wet summers exhibit anomalous ridging over northern monsoon regions and over the central United States, while in dry summers the high intensifies, leading to an anomalous west-to-east trending ridge directly over the subregion. In Subregion 2, wet summers show an anomalous ridge over the western United States as the high expands over the subregion and areas to the northwest, but in dry summers the high expands over large areas to the west southwest and east, with relatively strong anomalous ridging west of the Baja peninsula and over the eastern United States. Subregion 3 experiences fairly broad expansion of the anticyclone in wet summers, focused on an anomalous ridge to the northwest similar to that for Subregion 2, while in dry summers the anticyclone intensifies and causes a relatively strong west-to-east aligned anomalous ridge right over the subregion. Subregion 4 along the east of the monsoon region is the most different from the other subregions. In wet summers the entire anticyclone shifts east, leading to an anomalous ridge-trough-ridge pattern (with the trough partially over the subregion), but in dry summers the ridge expands south and east of its typical position with an anomalous ridge moving over the subregion and an anomalous trough to the northwest.

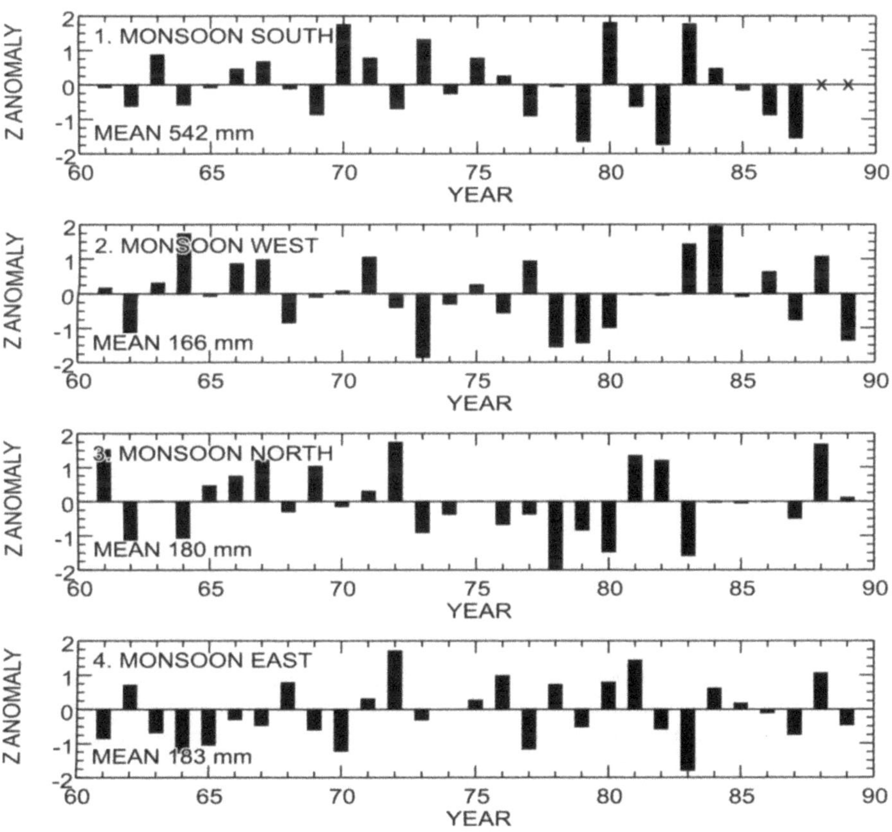


*Figure 11.* Standardized anomalies of July-August-September precipitation totals in the four monsoon subregions (X denotes missing data).

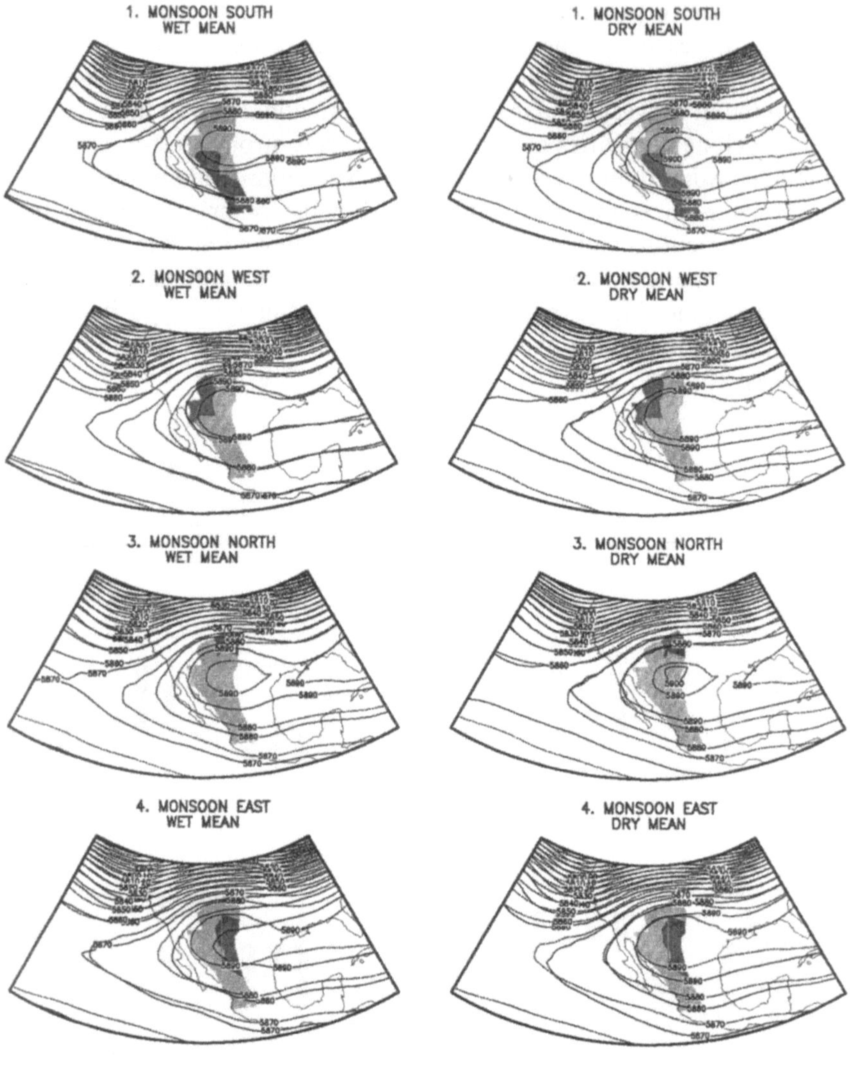


*Figure 12.* Composite 500 mb heights (m) for the seven wettest and driest summers (solid lines) in the four monsoon subregions, as compared to remaining 15 "middle" years (dashed lines), with shaded areas indicating individual subregions (dark) and the whole monsoon region (light).

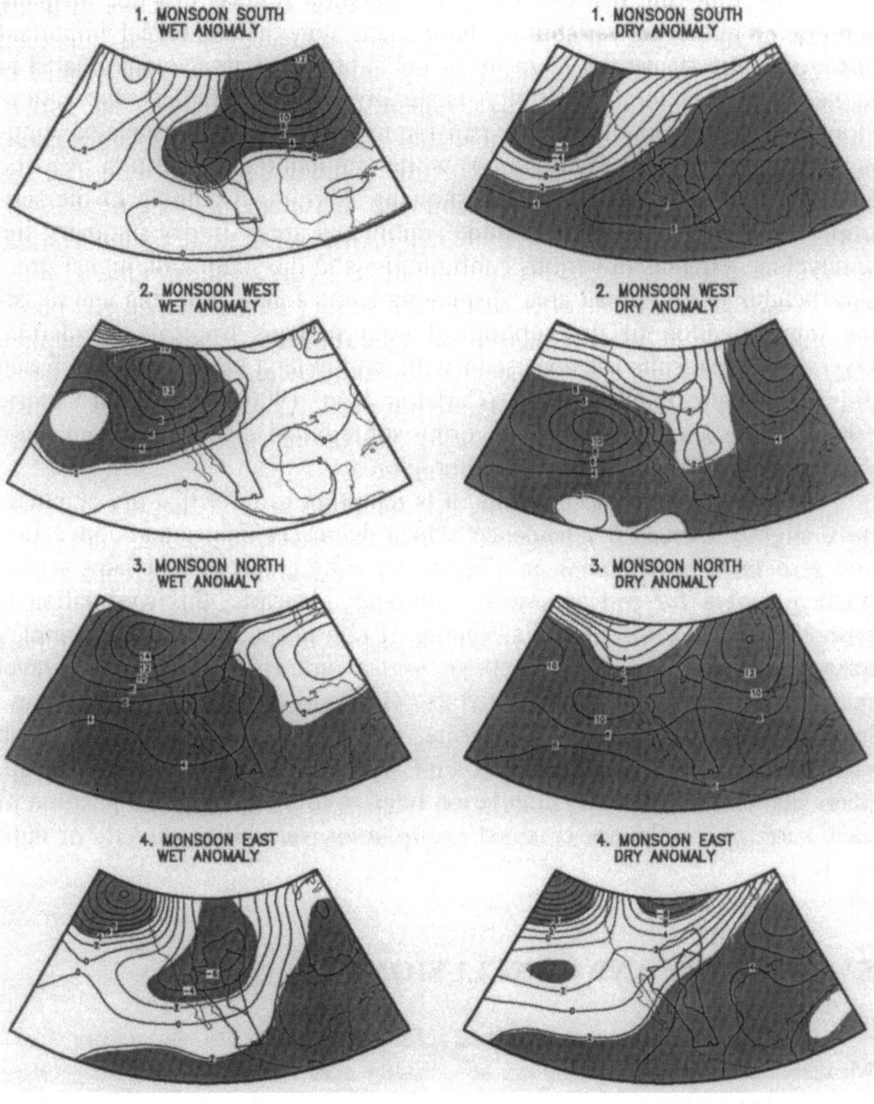


Figure 13. 500 mb height anomalies (m) for the data in Figure 12. Shaded areas indicate significant differences for $\alpha \leq 0.05$. The monsoon region and appropriate subregion are outlined on each map.

Recognizing that the 500 mb circulation reflects just one of many sources of monsoon variability, these maps nonetheless reveal important changes in the shape and intensity of the subtropical anticyclone related to monsoon precipitation variability. Generally, in wet summers, the anticyclone undergoes some form of expansion to the north with associated continental ridging (except Subregion 4) while maintaining its position over the southern part of the study region, showing meridional bulging of the subtropical circulation into mid-latitude continental areas. In dry summers, the anticyclone expands in various configurations to the south with higher pressure heights over a broad area, displaying some zonal stretching and possible intensification of the subtropical (and perhaps tropical) circulation. Overall, these results are consistent with, and extend earlier findings based on, Arizona precipitation alone (Carleton et al. 1990; Adams and Comrie 1997). There are, however, noteworthy subregional differences from these general patterns such as those for Subregion 4.

In explaining these results, it is tempting to infer that dry summers have slightly increased subsidence, which decreases monsoonal convection and associated precipitation in a particular subregion (with perhaps a contrasting analog for wet summers). Although plausible, this speculation is probably too generalized and sweeping, if one keeps in mind the complex terrain and the absence of climatological information on lower-level mesoscale monsoon dynamics such as "Gulf surges" of moisture and low-level jets along the Gulf of California. Yet, significantly different 500 mb circulation patterns are associated with wet and dry monsoons, and it appears that shifts in 500 mb circulation relative to the geographic position of each subregion influence seasonal precipitation variability, directly or indirectly.

## 5. SUMMARY AND CONCLUSION

This study has determined climate regions for the entire U.S.-Mexico border region based on seasonality and variability of precipitation (rather than precipitation totals). In doing so, it has demonstrated the applicability of the White et al. (1991) regionalization method to a large, diverse area at the subcontinental scale. The method performs well and quite consistently under various analyses and transformations, giving good confidence in the results. The resulting contiguous regions, subregions, and associated boundaries in the core of the study area make physical sense, and they are robust to transformed input data and changes in rotation. Certainly, the results convey a congruous picture of precipitation regimes across the borderlands.

In some parts of the study area, improving the sparse spatial distribution of stations might offer some greater definition of regional boundaries (e.g., Mojave Desert region, southwestern Texas/northern Mexico). In areas where the data are good and behave coherently (including some of the sparser areas), the two techniques for delineating region boundaries map closely to one another and, as expected, they diverge where the data do not behave coherently. Yet, overall, the regions are remarkably consistent among analyses, and they provide a climatologically satisfying and useful division of border region precipitation climatology that relates well to the driving atmospheric processes.

The contiguous groups of stations identified in this study were created to enable analyses of regional variability, secular trends, and other signals of global climate change. To illustrate the utility of the regionalization, the regionalization results for the monsoon region were applied to an examination of relationships between summer monsoon precipitation variability and 500 mb circulation. The results show significantly different patterns in the overlying pressure height fields associated with anomalous wet and dry conditions in each of four monsoon subregions.

Currently, the regionalization results are being used for a follow-on study of summer and winter precipitation variability across the entire border region, and they are also being employed in an integrated assessment project for the southwest United States. Further work on monsoon variability should elucidate in more detail the potential controlling mechanisms, including among others the role of the Hadley circulation and the ITCZ in the shape and strength of the subtropical circulation over the region.

More generally, the results of this study give some indications about the use of this methodology for further applications in other regions. The success of the basic oblimin regionalization over Pennsylvania (White et al. 1991), a quite different region, suggests that it has wide applicability to different climatological regimes. Yet it is easy to blindly apply such a recommendation when in reality there is no "right" answer. A range of statistics always provides the researcher with greater insight. Therefore, for most studies it is also recommended that several data and rotational transformations be examined to supply a measure of reliability to the regionalization results.

Finally, this chapter highlights in particular the importance of examining precipitation variability and controlling processes across political boundaries. More generally, it is not uncommon to avoid the difficulties of obtaining and analyzing physical and ecological data across such boundaries. However, in doing so it is possible to be unaware of, or even ignore, both the scale and extent of atmospheric and other natural factors playing out across a socially constructed divide.

## 6. ACKNOWLEDGMENTS

I wish to thank Erik Glenn for his early work on this project and the original research paper, and I am grateful to Inter-Research Science Publishers for granting permission to reproduce it here.

## 7. REFERENCES

Adams, D.K., and A.C. Comrie. 1997. The North American monsoon. *Bulletin of the American Meteorological Society* 78(10): 2197–2213.

Bärring, L. 1988. Regionalization of daily rainfall in Kenya by means of common factor analysis. *Journal of Climatology* 8: 371–389.

Barry, R.G., G. Kiladis, and R.S. Bradley. 1981. Synoptic climatology of the western United States in relation to climatic fluctuations during the twentieth century. *Journal of Climatology* 1: 97–113.

Buell, C.E. 1975. The topography of empirical orthogonal functions. Fourth Conference on Probability and Statistics in Atmospheric Science, Tallahassee, Florida, American Meteorological Society, pp. 188–193.

Buell, C.E. 1979. On the physical interpretation of empirical orthogonal functions. Sixth Conference on Probability and Statistics in Atmospheric Science, Banff, Alberta, Canada, American Meteorological Society, pp. 112–117.

Bunkers, M.J., J.R. Miller, and A.T. DeGaetano. 1996. Definition of climate regions in the northern plains using an objective cluster modification technique. *Journal of Climate* 9: 130–146.

Burnett, A.W. 1994. Regional-scale troughing over the southwestern United States: Temporal climatology, teleconnections, and climate impact. *Physical Geography* 15: 80–98.

Cahalan, R.F., L.E. Wharton, and M.-L. Wu. 1996. Empirical orthogonal functions of monthly precipitation and temperature over the United States and homogeneous stochastic models. *Journal of Geophysical Research* 101(D21): 26309–26318.

Carleton, A.M. 1987. Summer circulation climate of the American Southwest: 1945–1984. *Annals of the Association of American Geographers* 77: 619–634.

Carleton, A.M., D.A. Carpenter, and P.J. Weber. 1990. Mechanisms of interannual variability of the Southwest United States summer rainfall maximum. *Journal of Climate* 3: 999–1015.

Cattell, R.B. 1966. The scree test for the number of factors. *Multivariate Behavioral Research* 1: 245–276.

Cavazos, T. 1997. Downscaling large-scale circulation to local winter rainfall in northeastern Mexico. *International Journal of Climatology* 17: 1069–1082.

Cavazos, T., and S. Hastenrath. 1990. Convection and rainfall over Mexico and their modulation by the Southern Oscillation. *International Journal of Climatology* 10: 377–386.

Clarkson, D.B., and R.I. Jennrich. 1988. Quartic rotation criteria and algorithms. *Psychometrika* 53: 251–259.

Comrie, A.C. 1996. An all-season synoptic climatology of air pollution in the U.S.-Mexico border region. *Professional Geographer* 48(3): 237–251.

Comrie, A.C., and E.C. Glenn. 1998. Principal components–based regionalization of precipitation regimes across the southwest United States and northern Mexico, with an application to monsoon precipitation variability. *Climate Research* 10: 201–215.

Eklundh, L., and P. Pilesjö. 1990. Regionalization and spatial estimation of Ethiopian mean annual rainfall. *International Journal of Climatology* 10: 473–494.

Fernàndez Mills, G. 1995. Principal component analysis of precipitation and rainfall regionalization in Spain. *Theoretical and Applied Climatology* 50: 169–183.

Fernàndez Mills, G., X. Lana, and C. Serra. 1994. Catalonian precipitation patterns: Principal component analysis and automated regionalization. *Theoretical and Applied Climatology* 49: 201–212.

Higgins, R.W., and W. Shi. 1999. Dominant factors responsible for interannual variability of the Southwest Monsoon. *Journal of Climate* 13: 759–776.

Higgins, R.W., and W. Shi. 2001. Intercomparison of the principal modes of interannual and intraseasonal variability of the North American monsoon system. *Journal of Climate* 14: 403–417.

Higgins, R.W., K.C. Mo, and Y. Yao. 1998. Interannual variability of the United States summer precipitation regime with emphasis on the southwestern monsoon. *Journal of Climate* 11: 2582–2606.

Horn, L.H., and R.A. Bryson. 1960. Harmonic analysis of the annual march of precipitation over the United States. *Annals of the Association of American Geographers* 50: 157–171.

Kalnay, E., M. Kanamitsu, R. Kistler, W. Collins, D. Deaven, L. Gandin, M. Iredell, S. Saha, G. White, J. Woollen, Y. Zhu, A. Leetmaa, R. Reynolds, M. Chelliah, W. Ebisuzaki, W. Higgins, J. Janowiak, K.C. Mo, C. Ropelewski, J. Wang, R. Jenne, and D. Joseph. 1996. The NCEP/NCAR 40-year reanalysis project. *Bulletin of the American Meteorological Society* 77: 437–471.

Karl, T.R., A.J. Koscielny, and H.F. Diaz. 1982. Potential errors in the application of principal component (eigenvector) analysis to geophysical data. *Journal of Applied Meteorology* 21: 1183–1186.

Kutzbach, J.E. 1967. Empirical eigenvectors of sea-level surface temperature and precipitation complexes over North America. *Journal of Applied Meteorology* 6: 791–802.

Lanicci, J.M., and T.T. Warner. 1991. A synoptic climatology of the elevated mixed-layer inversion over the southern Great Plains in spring. Part 1: Structure, dynamics, and seasonal evolution. *Weather Forecasting* 6: 181–197.

Lyons, W.F., and M. Bonell. 1994. Regionalization of daily mesoscale rainfall in the tropical wet/dry climate of the Townsville area of north-east Queensland during the 1988–1989 wet season. *International Journal of Climatology* 14: 135–163.

Mallants, D., and J. Feyen. 1990. Defining homogeneous precipitation regions by means of principal components analysis. *Journal of Applied Meteorology* 29: 892–901.

Mitchell, V.L. 1976. The regionalization of climate in the western United States. *Journal of Applied Meteorology* 15: 920–927.

North, G.R., T.L. Bell, R.F. Cahalan, and F.J. Moeng. 1982. Sampling errors in the estimation of empirical orthogonal functions. *Monthly Weather Reviews* 110: 699–706.

Ogallo, L.J. 1989. The spatial and temporal patterns of the east African seasonal rainfall derived from principal component analysis. *International Journal of Climatology* 9: 145–167.

Richman, M., and P.J. Lamb. 1985. Climatic pattern analysis of 3- and 7-day summer rainfall in the central United States: Some methodological considerations and a regionalization. *Journal of Climate and Applied Meteorology* 24: 1325–1343.

Richman, M., and P.J. Lamb. 1987. Pattern analysis of growing season precipitation in southern Canada. *Atmosphere-Ocean* 25(2): 137–158.

Rummel, R.J. 1970. *Applied Factor Analysis*. Evanston, Illinois: Northwestern University Press, 617 pp.

Stensrud, D.J., R.L. Gall, and M.K. Nordquist. 1997. Surges over the Gulf of California during the Mexican monsoon. *Monthly Weather Review* 125: 417–437.

Vose, R.S., R.L. Schmoyer, P.M. Steurer, T.C. Peterson, R. Heim, T.R. Karl, and J. Eischeid. 1992. The Global Historical Climatology Network: Long-term monthly temperature, precipitation, sea level pressure, and station pressure data. ORNL/CDIAC-53, NDP-041, Carbon Dioxide Information Analysis Center, Oak Ridge National Laboratory, Oak Ridge, Tennessee.

White, D., M. Richman, and B. Yarnal. 1991. Climate regionalization and rotation of principal components. *International Journal of Climatology* 11: 1–25.

Woodhouse, C.A. 1997. Winter climate and atmospheric circulation patterns in the Sonoran Desert Region, USA. *International Journal of Climatology* 17: 859–873.

14

# CLIMATE AND CLIMATE VARIABILITY IN THE ARENAL RIVER BASIN OF COSTA RICA

Jorge A. Amador,* Rafael E. Chacón,** and Sadí Laporte**

**Centro de Investigaciones Geofisicas y Escuela de Física, Universidad de Costa Rica, P.O. Box 2060, San José, Costa Rica*

***Centro de Servicio de Estudios Básicos de Ingeniería (Area de Hidrología), Instituto Costarricense de Electricidad y Centro de Investigaciones Geofisicas, Universidad de Costa Rica, San Jose, Costa Rica*

**Abstract** This work examines some of the effects of climate and climate variability in the Arenal River Basin of Costa Rica, the site of the largest hydropower complex in the country. The Arenal system, which drains part of the north-central portion of Costa Rica, covers a total area of approximately 493 $km^2$; it is managed mainly for electric power generation and produces nearly a quarter of the electricity in Costa Rica. Monthly, pentad (5-day means), and daily precipitation data are used to study signals associated with climate and shorter-term atmospheric disruptions in the basin. Although the study area is relatively small, strong spatial and temporal contrasts are found in the precipitation patterns there. A clear distinction in the seasonal distribution of precipitation is observed over short distances (~30–40 km), between the northwestern (NW) lowlands of the basin compared to the southeastern (SE) sector. The former region exhibits a bimodal precipitation distribution, with maxima in June and September–October, and relative minima in July and December–April. The July minimum suggests a weak midsummer drought, or "veranillo," signal. The latter region has practically no dry season, with the highest precipitation values occurring during the second part of the calendar year. As is determined by principal component analysis (PCA) of anomalies in monthly precipitation data, the main disruption of the normal pattern of precipitation appears to be related to the El Niño/Southern Oscillation (ENSO) signal in the northwest region, whereas the southeast sector shows a positive correlation with Caribbean low-level wind changes. Some of the latter changes are associated with warm or cold ENSO episodes, which seem to modulate wind intensities of the low-level jet over the Caribbean. Precipitation effects in the basin for selected extreme cases, such as those of Hurricane Mitch and other so-called "temporales," are also analyzed. The importance of these systems as fundamental components of the basin's hydrological cycle is well established. ENSO-



*Henry F. Diaz and Barbara J. Morehouse (eds.), Climate and Water*, 317-349.

related variability of the regional summer circulation (such as that of the low-level jet in the western Caribbean), and the appearance of cases of extremely strong trade winds during the winter circulation, are also important forcing mechanisms for precipitation variability in the basin. In these cases, the interaction between the basin's complex topography, and changes in the flow pattern and intensity, seem to be of fundamental importance for precipitation variance. Some socioeconomic impacts of precipitation variability, as well as a discussion about the potential use of climate variability information for water management in the basin, are presented.

# 1. INTRODUCTION

Costa Rica's economy used to rely mainly on agriculture. During the last decade, however, some high-technology industries have been established, increasing the demand for qualified professionals, and stimulating the adaptation and improvement of higher education and training programs at universities and technological institutes. Tourism has become an important source of national income; therefore, building and access to facilities in natural and ecologically rich environments have expanded substantially. Also, public health services and telecommunications are widely spread throughout the country and its more than 3 million inhabitants. As might be expected from this socioeconomic situation, there has been an urgent requirement for providing various forms of energy to users, especially electricity. Lately, the demand for electricity has increased dramatically in Costa Rica, where electricity generation is provided mainly by hydropower (75%). Thermal, geothermal, and wind energy generation (18%, 5%, and 2%, respectively) are the other sources of energy used to supply electricity to industry and residential users.

Climate variability on various time scales, from interannual to seasonal, and shorter-term meteorological phenomena have affected Costa Rica's economy and jeopardized the society in many different ways during the past. After the severe impacts of El Niño and La Niña during the last decade, especially those associated with the 1997–98 El Niño event, and perhaps after the impact of Hurricane Mitch in 1998, the general public realized the relevance and importance of timely meteorological and climate information in the planning, management, and utilization of water resources in the country.

A relatively small country, Costa Rica also possesses a very complex topography (Fig. 1), with distinct annual rainfall distributions, even on small spatial scales. Large mean annual precipitation totals and suitable sloping terrain conditions have made hydropower one of the most important natural resources of the country. The Costa Rica Institute of Electricity (Instituto Costarricense de Electricidad, ICE), a technical body of the Costa Rican government created in 1949, has been given the responsibility of building infrastructure, and planning and managing water resources for hydroelectric production. Costa Rica, as has been shown in the last few years, is very sensitive to water management policies and to changes in water availability. Climate variability, and in some cases related extreme events, can represent significant challenges to the long-term well-being of a country that is seeking to develop both economically and socially. Agriculture is still, of course, a major activity in many rural areas of the country and a major provider of national income; however, major industries (e.g., the electronics industry), and large commercial factories that also contribute important percentages to the national income, are located in urban areas and depend on a reliable supply of electricity.

This chapter deals with the climate setting and with major climate disruptions affecting the largest hydroelectric system in Costa Rica, the Arenal-Corobicí-Sandillal (ARCOSA) complex in the Arenal River Basin. The study is intended to focus on some regional climate systems affecting the Arenal Basin's own mesoscale climate, and on how changes in these physical mechanisms are related to disruptions in precipitation patterns; these changes in turn may have significant effects on reservoir inflows and water levels, and thus on electricity production. Despite the importance of the basin for hydropower generation, relatively few studies have been published regarding meteorological and climatic conditions in the reservoir and their relationships with regional or global-scale phenomena. Fernández et al. (1986) discussed some of the mesoscale characteristics of the region, and focused on the effects on the reservoir of several meteorological parameters around the basin, such as increases in locally measured wind velocities. More recently, Amador et al. (2000a,b) performed an extensive study of disruptions in precipitation patterns due to some selected El Niño/Southern Oscillation (ENSO)events, and extreme meteorological events that affected the basin, especially during the last decade. Some of the major findings of these works will be discussed below.

There has been general agreement that the skill in ENSO prediction has improved considerably in the last few years (Hoerling and Kumar 2000). There is a need, therefore, to identify and understand major deviations in regional precipitation patterns in order to improve the application of prediction schemes associated with ENSO events. It is known that El Niño has an important impact on Costa Rica's climate, especially along the Pacific slope (Fernández and Ramírez 1991; Waylen et al. 1996a).

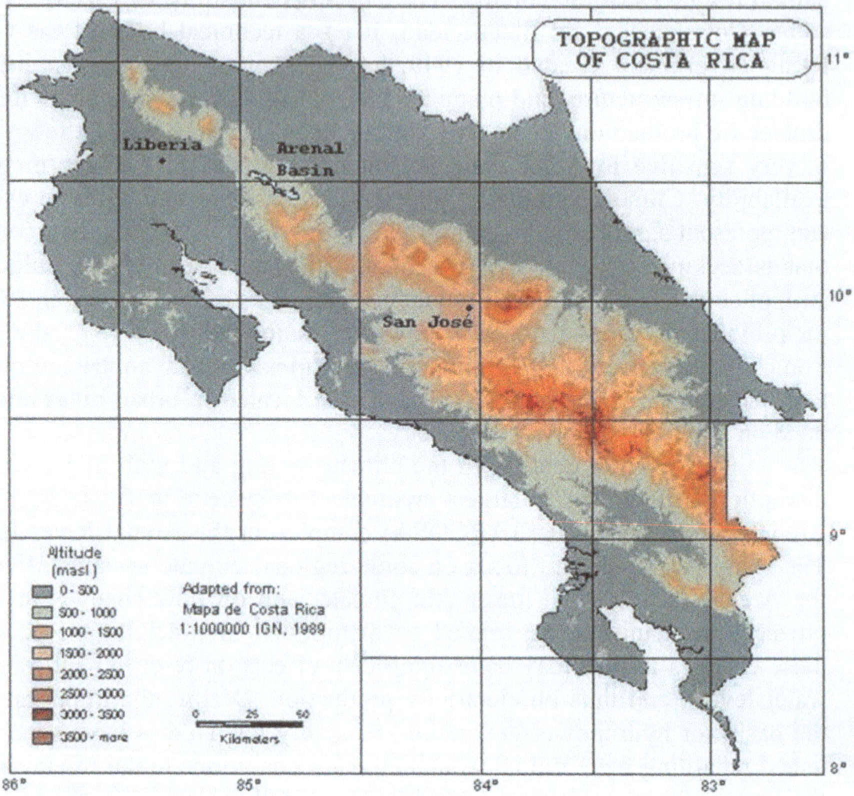


*Figure 1.* Topographic map of Costa Rica showing the approximate location of the Arenal Basin with respect to the town of Liberia in the Guanacaste Province, and San José, the capital city of the country. Altitude is given in meters above sea level (masl).

Here, we study precipitation variations in the Arenal Basin for both the 1997–98 El Niño and the 1998–2000 La Niña events, in the context of regional and global-scale phenomena. The study of any form of variability related to ENSO or non-ENSO signals requires, however, an adequate description of the "normal" climate conditions in the basin. Therefore, the approach taken here would allow decision makers to take advantage of any additional information about the above-mentioned phenomena when they are planning and operating power generation facilities.

In this context, the seasonal precipitation cycle is examined through analysis of the relative contribution of different precipitation systems during the course of the year, and the associated rainfall variability in the basin.

The main characteristics of both summer (June-July-August, JJA) and winter (December-January-February, DJF) circulations are analyzed and related to relevant features of the temporal and spatial distribution of sea surface temperatures (SSTs) in the adjacent oceans. In the latter context, the midsummer drought (Magaña et al. 1999), and the low-level jet over the Caribbean during the boreal summer (Amador 1998; Amador and Magaña 1999; Amador et al. 2000c), constitute two important physical mechanisms associated with the seasonal cycle leading to some unique characteristics in the distribution of mean monthly precipitation within the Arenal Basin.

Finally, in order to show the wide range of climatic variability that can modulate the spatial and temporal precipitation distribution in the basin, the anomalous contributions to precipitation in some case studies for the 1997–98 El Niño and the 1998–2000 La Niña events are presented. A goal of this study is to contribute to a better understanding of climate and its variability in this important region for hydropower generation in Costa Rica, with the goal of providing governmental and private institutions with improved climate information for decision making. By improving the understanding of the elements that control regional climate and its variability, in relation to both ENSO and non-ENSO signals, more accurate and tailored climate predictions could be developed to fulfill some of the needs of this particular socioeconomic sector.

In the following, we first present information about the more important hydroelectric projects in Costa Rica; we then follow with an analysis of general regional climate patterns and those that pertain specifically to Costa Rica. We then consider the hydroclimatology of the Arenal River Basin and the large-scale teleconnection patterns that affect the general climate of the region.

## 2. THE ARENAL BASIN PROJECT

### 2.1. Brief Historical Account

The Arenal Dam and Reservoir are located at the southern tip of the Guanacaste Mountains, in the Guanacaste Province (Fig. 1), some 190 km northwest of San José, the capital city of Costa Rica. The Arenal system is located in a northwest-southeast trending chain of volcanic cones that is part of Central America's range of active volcanoes. The bedrock at the dam site, referred to as the "Aguacate formation," is characterized by a complex of lava flow rock, tuff, agglomerate, and volcanic breccia, with occasional thin interbeds of sandstone and shale. Lava flows are known to have blocked the most important local river, the Arenal River, in the geologic past. The lava flows resulted in deposits, up to 40 m thick, of fluvio-lacustrine river channel and old floodplain sediments overlying the local bedrock at all but the highest elevations at the dam site. In recent geological times, most of the dam site has been covered by a mantle of unconsolidated pyroclastic deposits varying up to 15 m in thickness, and consisting of alternating, and often discontinuous, layers of volcanic ejecta, which probably originated from the nearby Arenal Volcano (Wahler and Associates 1975).

The basin covers an area of approximately 493 $km^2$, about one-quarter of the Arenal Conservation Area. The reservoir comprises about 17% (84 $km^2$) of the basin's area, and the current water surface area is about three times that of the original lake. The Arenal River used to drain the lake towards the Caribbean, as a tributary of the San Carlos and San Juan Rivers, the latter being part of the border between Costa Rica and Nicaragua. After the first field studies a few decades ago, the region showed extraordinary potential for electricity generation; consequently, the ICE initiated the Arenal project back in 1967. In contrast to the original Arenal River drainage course towards the Caribbean Sea, once the project was completed, the reservoir was left draining water towards the Gulf of Nicoya in the Pacific Ocean. Other projected hydroelectric plants, such as Corobicí and Sandillal, were planned to make use of the water resource in a sort of cascade system.

Planning of water use and reservoir system management included from the beginning not only the generation of hydropower, but also the use of water for irrigation projects in the Guanacaste Province, one of the driest regions in Costa Rica. This area receives tremendous socioeconomical benefits from the Arenal-Tempisque irrigation project, especially those sectors related to crop activities. The amount of irrigated land dedicated to agriculture in 1995 was of the order of 16,000 hectares (ha). Positive changes in productivity after irrigation have been very important for social well-being and farming, since rice and sugar yields have increased from 3 to 8 metric tons per hectare, and from 8 to 14 metric tons per hectare, respectively.

## 2.2. The Arenal-Corobicí-Sandillal Hydroelectrical Complex

The ARCOSA complex consists of three power generating plants located at different elevations, namely Arenal, Corobicí, and Sandillal, which have a total installed generating capacity of 362 megawatts (MW) of power. Figure 2 shows the relative locations of the Arenal, Corobicí, and Sandillal plants with respect to the reservoir. Corobicí alone contributes 174 MW to the system. This capacity has recently been surpassed by the plant at Angostura (177 MW), located on the Reventazón River on the Caribbean slope of Costa Rica and inaugurated in December 2000. The Arenal plant has a generating capacity of 157 MW. Altogether, the ARCOSA plants contribute about one-quarter of the total annual electrical energy that is consumed in the country, hence their great importance in the national electrical system. The Arenal reservoir has a storage capacity of 2.4 billion $m^3$ at its maximum operation level of 546 meters above sea level (masl), and has the potential to support electricity demands of 1700 GWh.

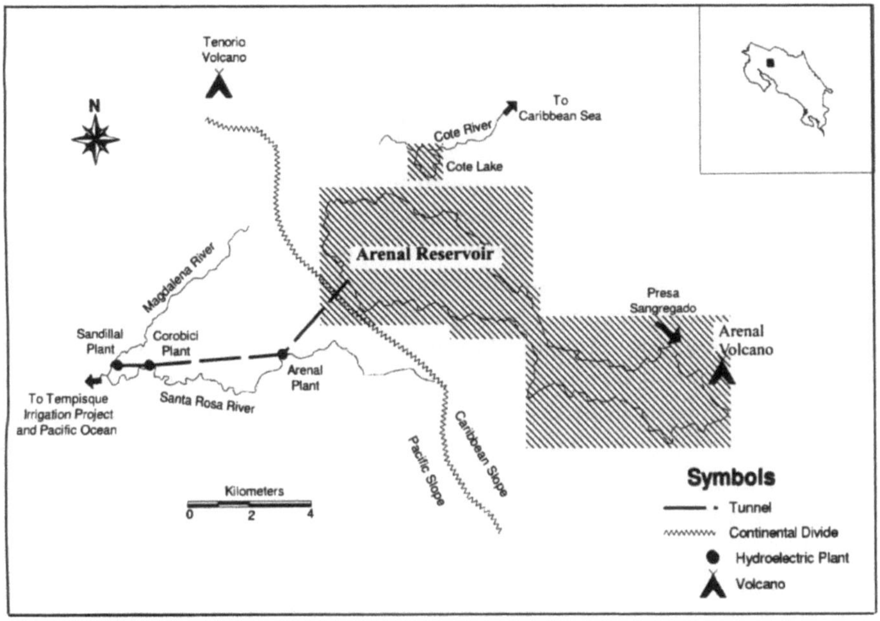


*Figure 2.* Relative locations of the Arenal, Corobicí, and Sandillal hydroelectric plants with respect to the Arenal Reservoir.

## 3. REGIONAL CLIMATE SYSTEMS

As has been discussed recently by Magaña et al. (1999), the annual distribution of rainfall in the Caribbean side of Central America contrasts dramatically with that in the Pacific side. The latter slope exhibits a bimodal distribution, with maxima in June and in September-October, and a relative minimum during July-August. This reduction in precipitation is the so-called midsummer drought, known locally as the "veranillo" or "canícula." The existence of such a reduction in the annual distribution of rainfall has been shown to be part of the seasonal cycle of precipitation over much of Central America (Magaña et al. 1999). According to these authors, the "veranillo" is related to changes in the intensity of convective activity over the northeastern Pacific "warm pool." During the midsummer drought period, the trade winds over the Caribbean strengthen, due in part to the dynamical response of the low-level atmosphere to the magnitude of the convective forcing in the Intertropical Convergence Zone (ITCZ), which in turn is associated with SST distribution. The starting date, intensity, and duration of the midsummer drought vary from year to year, and the drought constitutes an important climate disruption for sectors of the economy that are affected by water availability along much of the Pacific side of Central America. From December through March, the Pacific slope of Central America enjoys warm and mostly dry conditions.

The most interesting dynamical feature over the Caribbean Sea during summer is the low-level jet that develops during June, reaches a maximum in July, and weakens in September (Amador 1998; Amador and Magaña 1999). This intense easterly current over the Caribbean is part of the summer trade wind regime, and it constitutes the most important factor in determining regional climate during this season. The long-term mean (1968–96) of the vector wind at 925 hPa (approximately 800 m height) for July, based on data from the National Centers for Environmental Prediction/National Center for Atmospheric Research (NCEP/NCAR) Reanalysis Project (Kalnay et al. 1996), is shown in Figure 3. As can be seen from the wind field, values in excess of 14 m/s dominate the central Caribbean Sea near 14°N, 75°W. The origin of this strong jet has not been fully established; however, its barotropically unstable nature (Molinari et al. 1997; Amador 1998; Amador and Magaña 1999) suggests that it is through the interaction with transients, such as the easterly waves, that the jet may obtain part of its kinetic energy. Another possibility may be that this strong flow feeds energy to the easterly waves, increasing their meridional amplitude to the north of its mean position. Also, from early summer to November, easterly waves and tropical cyclones constitute major elements among the tropical systems that produce rain over the region. Portig (1976) and Hastenrath (1991), among others, have documented in some detail the role of these systems in the annual distribution of precipitation in Central America.

In the Caribbean region, precipitation during the winter months is closely related to mid-latitude air intrusions (Schultz et al. 1997, 1998) and to low-level cloud systems traveling from the east (Velásquez 2000). Generally speaking, weather conditions during the winter months along the Caribbean slope of Central America are wetter and more humid than conditions prevailing along the Pacific side during the same season.

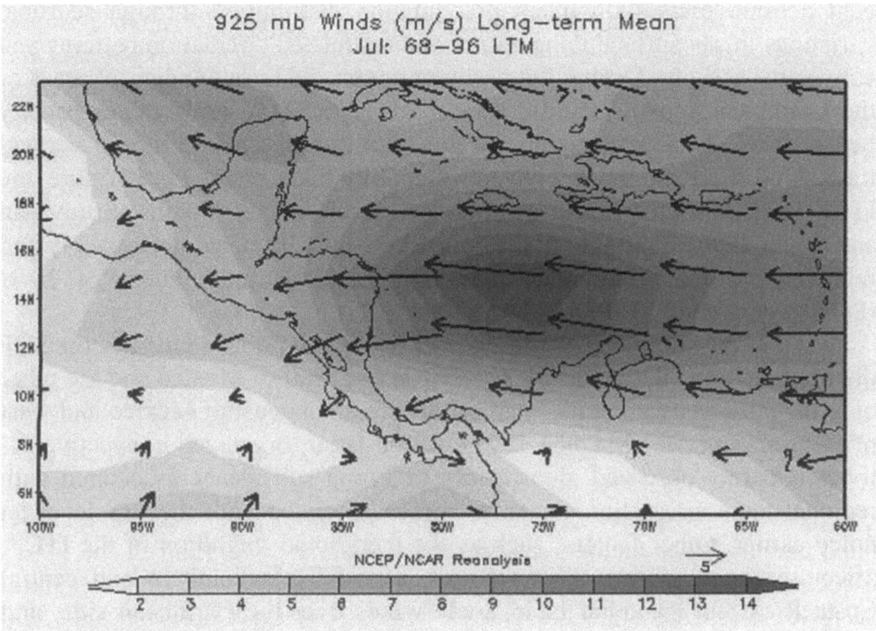


*Figure 3.* Long-term 925 hPa (~800 m) mean wind vector (m/s) showing the location of the low-level jet over the Caribbean during July based on NCEP/NCAR Reanalysis data for the period 1968–96.

Regarding the beginning and ending of rainy spells, several studies have shown that SST fluctuations in the tropical Atlantic and Pacific Oceans are related to variations in the duration and timing of the rainy season in Central America (e.g., Enfield and Alfaro 1999). Tropical north Atlantic SSTs, for which Alfaro (2000) defined warm and cold episodes, show the largest influence over the region when compared with other parameters that, have a positive correlation with rainfall in Central America (Alfaro and Cid 1999). In contrast, the Niño3 region was found to have a lower negative correlation with precipitation in the region, influencing only the Pacific slope of Central America.

## 3.1. The Climate of Costa Rica

When one deals with the elements that control climate in Costa Rica, several factors should be considered. One of these elements, as could be inferred from the previous section, is the large influence of the two adjacent oceanic masses on mean precipitation distribution through seasonal variations in sea surface temperature and associated circulation patterns and convective activity. During the northern winter, SSTs over adjacent areas of the Caribbean and the Pacific are relatively uniform, with values usually below 28°C. As a consequence of this, and the presence of strong vertical trade wind shear, no major convective activity takes place. Furthermore, the ITCZ is at its southernmost position during the winter months (Srinivasan and Smith 1996). Throughout the course of winter, trade winds intensify and relatively cold air intrusions frequently reach latitudes south of 10°N (Schultz et al. 1997, 1998).

During summer, the presence of two warm pools dominates the SST distribution—one pool over the Caribbean and Gulf of Mexico and the other over the northeastern Pacific, just off the southern coast of Mexico and west of Central America (Magaña 1999). In the latter, organized convective activity is barely observed, due mainly to strong subsidence associated with regional-scale circulations, such as those associated with the low-level jet noted earlier. Other factors, such as the meridional migration of the ITCZ, affect seasonal rainfall characteristics, especially in southern and central Costa Rica. On the other hand, trade winds over the Caribbean side, and southwesterlies on the Pacific side, are important mechanisms for moisture convergence and rainfall production. The interactions between the main flow and topographic barriers constitute a predominant process that contributes to precipitation all year round.

In countries with very irregular topography such as Costa Rica, precipitation exhibits very high temporal and spatial variability. Figure 4 shows the mean annual distribution of precipitation over Costa Rica (main map), and the seasonal precipitation patterns at some selected stations (inserted figures) for different periods. The average annual isohyetal contours were hand drawn by using 347 stations with precipitation data for 1970–89. Over some mountain basins, such as those of central Costa Rica for which precipitation data are scarce, isohyets were depicted by utilizing vertical precipitation profiles from neighboring basins that are known to have similar general topographic and climatic characteristics. Precipitation values estimated by means of this method were constrained in such a way that they did not exceed the mean runoff of any particular basin.

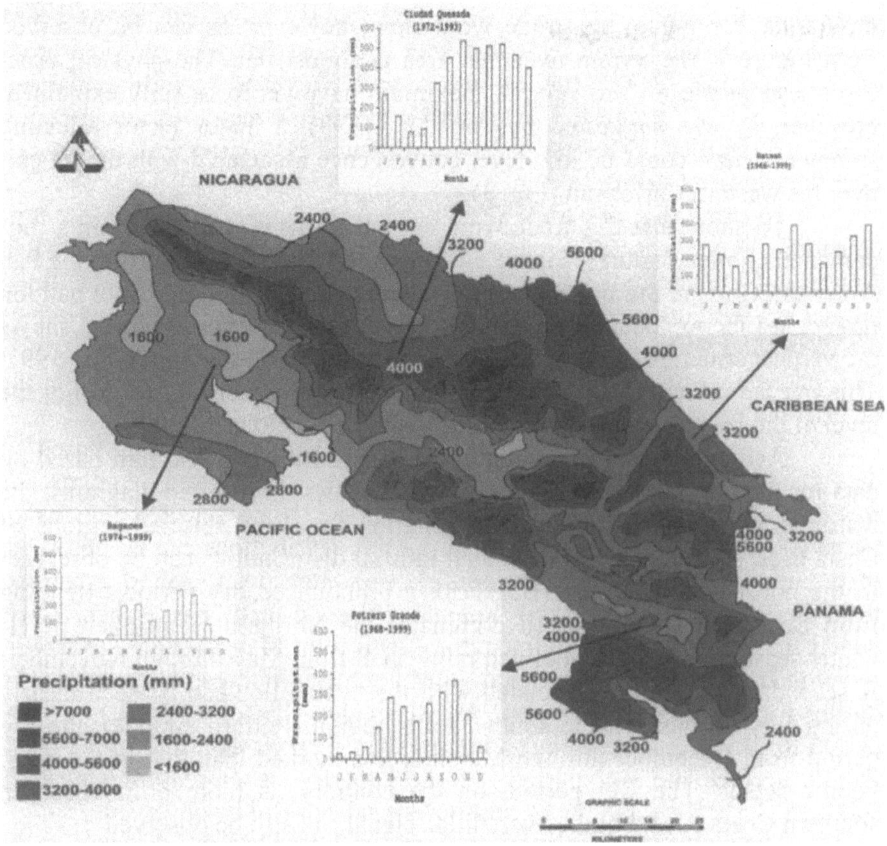


*Figure 4.* Mean annual distribution of precipitation over Costa Rica (main map) for the period 1970–89, and seasonal rainfall distributions at some selected stations (inserted figures) for selected periods. Precipitation contours are in millimeters per year and intervals are shown in gray scale.

From the map in Figure 4, it can be seen that precipitation in the Pacific side varies from about 1,400 mm in the northwestern region in the Guanacaste Province, to over 7,000 mm in some isolated high-elevation areas of Costa Rica. On the Caribbean side, precipitation ranges from about 1,600 mm in the eastern region of the Central Valley and Cartago (to the east of San José), to 9,000 mm in the central Caribbean slope of the northwest-southeast trending cordillera (see also Fig. 1). Note the relative rainfall maximum (>5,000 mm) in the Caribbean side, near the border between Costa Rica and Nicaragua. This maximum, in fact, extends well along the Nicaragua coast to the north, reaching southeastern Honduras (Amador and Magaña 1999). At first sight, it seems that this maximum may be due to orographic forcing of the trade wind flow; however, as can be observed from Figure 1, the terrain over that area is almost flat. The physical processes responsible for this rainfall maximum have yet to be fully explained; however, as was suggested by Amador (1998), a major factor affecting summer rainfall could be low-level convergence associated with the jet exit over the western Caribbean (Fig. 3).

To show that this trade wind current forms part of the basin's fundamental climate features, Figure 5 is presented. Based on data from the Pan American Climate Studies Sounding Network (PACS-SONET) pilot balloon project at Liberia, located some 50 km west of the reservoir, Figure 5 shows the vertical structure of the jet downstream of the basin region for July 1997. This intense current attains its maximum values of about 10 m/s, just at the level of the basin's main topographical features (1,500 to 2,000 m).

Average annual precipitation for Costa Rica is 3,300 mm based on data for the period 1970–89. Figure 4 also shows, in inserted diagrams, the distributions of mean monthly precipitation for some selected regions of Costa Rica. Different types of annual rainfall distributions can be observed. In the north Pacific region (Bagaces), a pronounced dry period extending from December through April dominates the Northern Hemisphere (NH) winter season, with a well-defined rainy spell from May through November. A secondary minimum associated with the "veranillo" can be seen clearly during July–August. In the south Pacific region (Potrero Grande), the dry period from December through March is less marked than that in the north Pacific sector. The wet period, on the contrary, is more pronounced in southern Costa Rica, but the "veranillo" signal can still be observed.

The contrast between Pacific and Caribbean rainfall distributions is noteworthy. At the eastern Caribbean region (Bataán), rainfall shows two minima, one in March and the other one in September–October. The maximum rainfall occurs in July, probably associated with the Caribbean low-level jet, which develops during this period. It may appear that orography plays an important role in the occurrence of this maximum; however, Bataán is located in an almost flat terrain, some 50 km to the east of the mountain range. A secondary maximum is observed in December as a result of the intensification of the trade winds, and the southward displacement of air masses, during the Northern Hemisphere winter. These relatively cold northerly winds, which appear especially during December–February are often referred to locally as "los nortes" (Portig 1976). Finally, the northern region of Costa Rica (Ciudad Quesada) shows a distribution with a minimum in March–April. The maximum rainfall tends to occur in July, although precipitation is relatively heavy during most of the wet season .

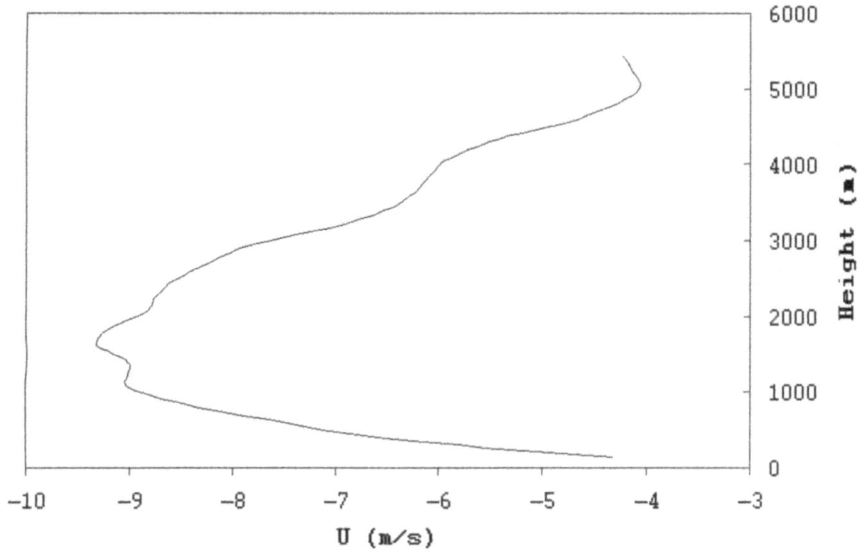


*Figure 5.* Vertical structure of the mean zonal component (meters per second) of the low-level jet, based on (PACS-SONET pilot balloon data obtained at Liberia, Costa Rica, for July 1997.

## 3.2. Hydroclimate of the Arenal River Basin

The average rainfall pattern over the Arenal Basin is illustrated in Figure 6. The distribution for the 1970–99 period was estimated by using 25 gauging stations (6 conventional, 19 automatic) located within the basin and in adjacent regions. To ensure a proper rainfall analysis, the station precipitation records were verified by means of the double accumulation curve method, and those sites having errors of a systematic type were corrected. From Figure 6, it can be seen that annual rainfall varies from a maximum of 5,000 mm in the southeastern (SE) region to 2,000 mm in the northwestern (NW) lower side of the basin. A secondary maximum of 4,000 mm is located in the highlands of the northwestern region. The above-mentioned rainfall maxima are due partly to the interaction of the trade wind regime with the mountain ranges having a southeast to northwest orientation. Rainfall tends to diminish towards the western basin sectors in the Guanacaste Province, comprising the driest region of Costa Rica, with just over 1,400 mm per year (see Fig. 4).

Although the study area is relatively small, strong spatial and temporal contrasts are found in precipitation patterns (Fig. 7). A clear distinction in the seasonal distribution of precipitation is observed over short distances (~30–40 km), between the northwest lower part of the basin (represented by Naranjos Agrios rainfall data, hereafter NA) as compared to the southeast part (represented by Caño Negro data, hereafter CN). As is shown in Figure 7a, which uses pentad data, the northwest region exhibits a bimodal precipitation distribution with maxima around May and September–October, and a relative minimum in late June to late July. This minimum suggests a weak midsummer drought, or "veranillo," signal. The southeast region (Fig. 7b) has only a relatively short dry season, with the highest precipitation occurring during the second part of the calendar year. In this distribution, a rainfall peak is noticeable during the summer months, which indicates the importance of the interaction of the trade winds, associated with the Caribbean low-level jet, with the basin topographic features. The analysis also suggests that changes in the intensity of the low-level jet could be an important mechanism for precipitation variability during the summer months, which in turn may affect inflows to the reservoir.

Additional hydroclimatological information for the Arenal Basin is presented in Figure 8. The monthly mean distribution (for the indicated periods) of extreme maximum and minimum air temperatures for two transects are shown in Figure 8a. Nueva Tronadora (NT) is located in the northwest lowlands, and Presa Sangregado (PS) is situated in the SE part of the basin. As discussed earlier based on Figure 7, the above two regions show the largest contrast in monthly precipitation distribution within the basin. As was expected, changes in radiation and cloudiness are responsible for extreme surface temperature behavior. Note that absolute maximum temperature ($T_{max}$) corresponds well with reduced periods of precipitation, from March to April approximately, in both regions of the basin (see Fig. 7). As rainfall increases during April–May, as a consequence of the onset of the rainy season and the ITCZ northward migration, maximum temperature drops at both sites to a relative minimum in July, suggesting a weak or masked "veranillo" signal. Since the presence of the midsummer drought (Fig. 7a) implies a relative reduction of precipitation and cloudiness, $T_{max}$ should show a relative maximum during the midsummer drought, as opposed to what is observed in the northwest sector of the basin. A plausible explanation for above-noted behavior of $T_{max}$ could be the presence of strong winds associated with the low-level jet during the summer season, which may act to lower temperature somewhat. Minimum temperatures

($T_{min}$) are consistent with the development of the dry and wet seasons, but they do not show any significant changes that could be associated with the "veranillo," as Magaña et al. (1999) showed in their study. During July–August an increase in $T_{min}$ is indicated for station PS that could be related to greater atmospheric moisture content associated with an increase in cloudiness and rainfall, due to the interaction of the strong summer winds with topography in the southeast sector of the basin.

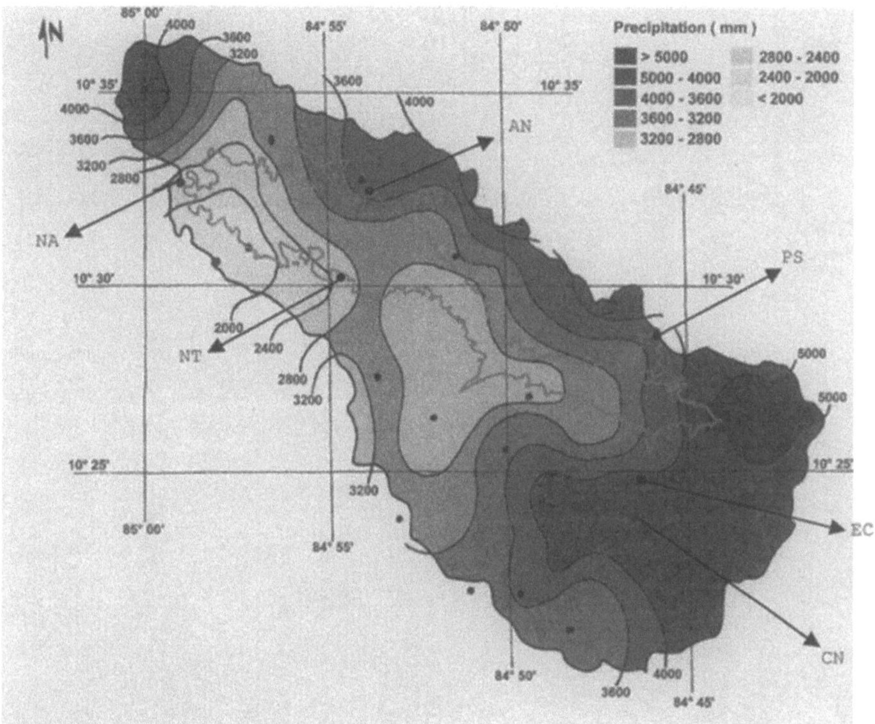


*Figure 6.* Mean annual distribution of precipitation in the Arenal Basin for the period 1970–99 showing, approximately, the hydrometeorological station network used in this study (black dots). The stations for which data are explicitly discussed are NA (Naranjos Agrios), NT (Nueva Tronadora), CN (Caño Negro), EC (El Cairo), PS (Presa Sangregado), and AN (Nuevo Arenal). The solid green line shows approximately the boundaries of the Arenal Reservoir. Isohyets are contoured at 400 mm intervals from 2,000 to 4,000 mm, and at 1,000 mm intervals above 4,000 mm.

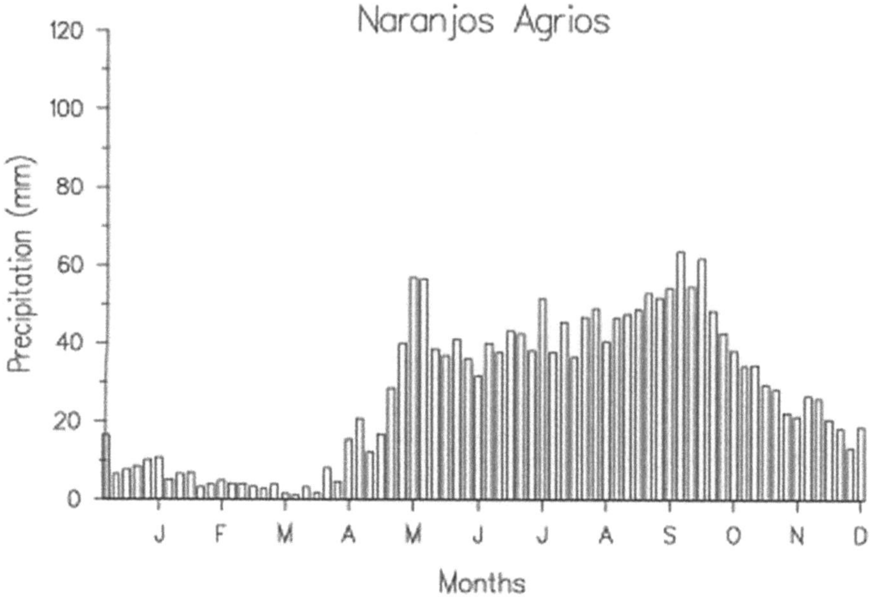


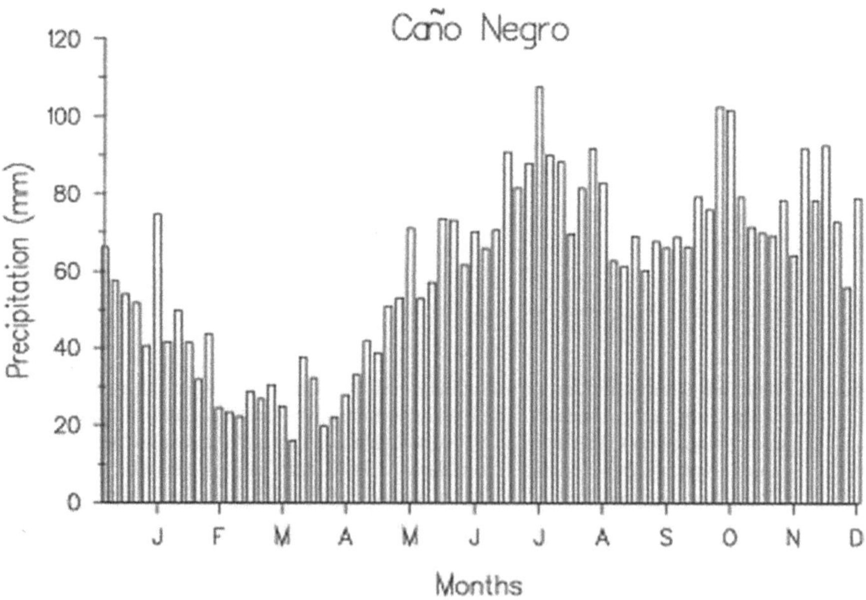


*Figure 7.* Mean pentad precipitation distributions for (a) Naranjos Agrios for the period 1974–2000, and (b) Caño Negro for the period 1966–2000. Precipitation units are millimeters per pentad.

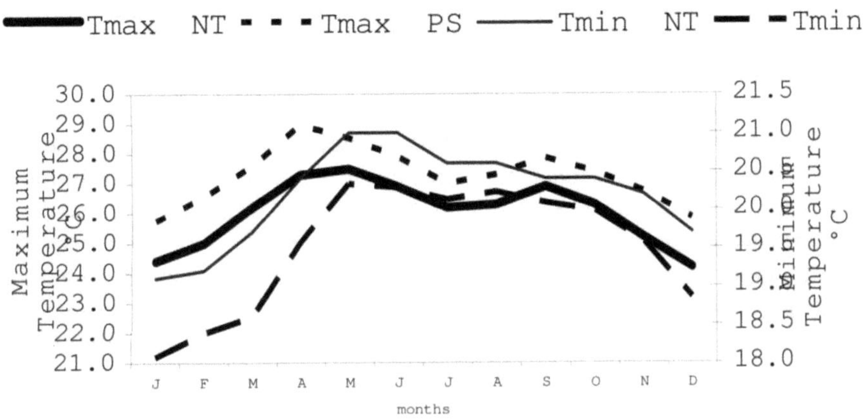


a)

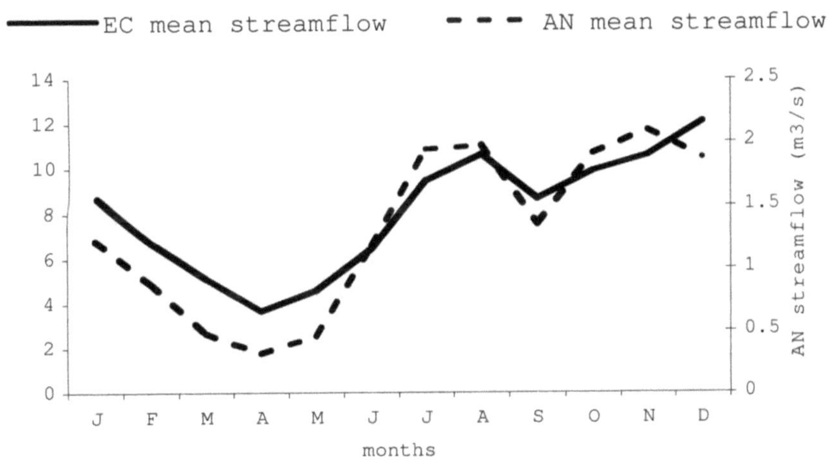


b)

*Figure 8.* Mean monthly distributions of (a) maximum and minimum temperatures ($T_{max}$ and $T_{min}$, respectively, both in °C) at Nueva Tronadora (NT) for 1978–99 (left scale), and at Presa Sangregado (PS) for 1983–99 (right scale), and (b) streamflow in cubic meters per second at El Cairo (EC) for 1975–2000 (left scale), and at Nuevo Arenal (NA) for 1978–2000 (right scale).

As noted earlier, the increase in mean precipitation during the summer months over the eastern part of the basin appears to be associated with the development of the low-level jet in the Caribbean by means of the interaction of the mean wind and the easternmost topographical features of the basin. In Figure 8b, mean streamflow for two stations (El Cairo, EC, for 1975–2000, and Nuevo Arenal, AN, for 1978–2000) is presented. The presence of a maximum in this parameter during July–August, corresponding to the period of maximum wind speed associated with the low-level jet over the Caribbean, can be seen clearly in Figure 8b. Two minima are found in this parameter at both stations, one in April and the other one in September; the former corresponds to the dry spell along the Pacific slope of Costa Rica and the latter occurs just after the drop in trade wind intensity in September (Amador 1998). After June–July, the intense interaction of the strong low-level trade winds associated with the low-level easterly jet with topography, produces an important contribution to mean precipitation in the southeast sector of the basin; consequently, there is a substantial seasonal increase in reservoir water levels. Both the decrease in streamflow and the reduction in the intensity of the low-level jet suggest that an index associated with this intense low-level flow over the Caribbean should be included in any scheme for predicting precipitation variability in the basin.

## 4. PHYSICAL MECHANISMS OF CLIMATE VARIABILITY IN THE ARENAL RIVER BASIN

Empirical orthogonal function (EOF) analysis of monthly rainfall anomalies was performed for two different subsets of station data for the basin. The subsets of stations were subjectively chosen following the two main seasonal precipitation distributions observed in the basin (see Fig. 7). In the northwest lower terrain and relatively drier region, Naranjos Agrios (10°32' N, 84°59' W), Nueva Tronadora (10°30' N, 84°55' W), Toma de Arenal (10°30' N, 84°59' W), La Tejona (10°31' N, 84°59' W), Dos Bocas (10°33' N, 84°55' W), and Aguacate (10°33' N, 84°57' W) were used for EOF analysis for the period 1979–98. In the southeast, Presa Sangregado (10°29' N, 84°46' W), Pueblo Nuevo (10°26' N, 84°47' W), Pastor (10°25' N, 84°45' W), Pajuila (10°30' N, 84°47' W), Jilguero (10°27' N, 84°43' W), and Caño Negro (10°24' N, 84°46' W) were utilized also for the same period (use Fig. 6 to approximately locate the stations used). The first principal components (PC) of precipitation anomalies for the northwest and southwest regions explains 72% and 83% of the variance in this variable, respectively.

Cross-correlation analysis of the dominant mode for both regions was carried out with corresponding anomalies from several global and regional indices, such as the SST index for the Niño3.4 region, North Atlantic

Oscillation (NAO), Southern Oscillation Index (SOI), and 925 and 700 hPa mean wind speeds. The latter parameter was averaged over the area 10°–20°N, 60°–80°W, as a simple index of the mean intensity of the low-level trade wind regime. Broadly speaking, the correlations are relatively small for both regions, indicating the presence of other dominant mechanisms for rainfall variability. The following aspects can, however, be identified from the EOF analysis. From the estimates of the correlations for lags up to 6 months that are significant at a 99% confidence level, rainfall changes in the northwest drier region have a maximum correlation coefficient of –0.23 for a lag of 2 months with the El Niño3.4 index. In other words, changes in SST over the central Pacific associated with warm (cold) ENSO events lead by nearly 2 months to a tendency for abnormally dry (wet) conditions in the Pacific side of the basin. Regarding the southeast region of the basin (Caribbean side), precipitation anomalies do not seem to be significantly correlated to SOI or SSTs in the Pacific.

Waylen et al. (1996b), in their study of time and space variability of annual precipitation in Costa Rica in relation to SOI, noted that there is a marked difference in response in those areas draining towards the Pacific and those draining towards the Caribbean, which, in a broad sense, agrees well with the results of this study. George et al. (1998) also found that annual discharges for rivers within the Pacific watershed were positively associated with SOI values, whereas in the Caribbean, discharges showed less clear and coherent patterns of association. They also suggested that this difference in response could be related to the elevations of the river basins. The complexity of the response in relation to topography has been pointed out by Waylen et al. (1996b), in reference to the differences in dominant precipitation generating mechanisms within the basins. For the southeast region, however, 925 hPa wind speed changes averaged over the area 10°–20°N, 60°–80°W, are positively correlated (0.24) at a 99% confidence level, with disruptions of rainfall from the normal pattern at zero lag. Stronger (weaker) than normal winds are related to wetter (drier) than average conditions over the Caribbean side of the basin. In order to gain some insight into the nature of the prevailing mechanisms affecting precipitation, the seasonality of the above results is analyzed below by using a finer temporal precipitation resolution for the phases of El Niño 1997–98 (approximately from March 1997 through June 1998) and La Niña 1998–2000 (from July 1998 through July 2000).

Figures 9a and 9b present the normalized pentad precipitation anomalies for Naranjos Agrios and Caño Negro, respectively, for the 1997–98 El Niño, and Figures 9c and 9d present them for Naranjos Agrios and Caño Negro, respectively, for the 1998–2000 La Niña. Pentad precipitation anomalies were estimated as departures from the mean values for the period of analysis and were normalized by the corresponding standard deviation. For both stations, during El Niño 1997–98, relatively long subperiods of negative anomalies can be clearly identified. From the annual perspective, the impact of this event was to produce weak below average conditions as a whole, implying a deficit in water availability for electricity generation. These results are consistent with those of Amador et al. (2000a) for the El Niño events of 1992–93 and 1994–95, although the physical mechanisms responsible for this kind of response have yet to be proposed. Previous results from principal component analysis discussed earlier suggest the importance of identifying the dominant regional climate variability modes to explain changes in precipitation within the basin.

We have noted that precipitation anomalies are associated with changes in trade wind intensity and in the low-level jet, which are also associated with ENSO episodes. Figure 10 shows the composite anomaly of the 925 hPa wind vector associated with the low-level jet over the Caribbean during several El Niño (Fig. 10a) and La Niña (Fig. 10b) events, for the summer months (June through August). Composite anomalies of the wind vector were estimated by using at least nine episodes for each of the two ENSO phases. El Niño and La Niña events were defined following Kiladis and Diaz (1989), and by using a procedure similar to that proposed by Trenberth (1997).

El Niño (La Niña) summers are characterized by stronger (weaker) than normal winds over the central Caribbean associated with the low-level jet core region. From Figure 9b, a marked period of positive precipitation anomalies is observed at Caño Negro during July–August 1997, which is consistent with the idea discussed earlier—that changes in wind intensities are associated with variations in the low-level jet, which in turn are related to precipitation anomalies over the basin through a wind–topography interaction mechanism. Note that in Naranjos Agrios (Fig. 9a), July–August 1997 exhibit mainly negative precipitation anomalies, as expected from a strong descending wind flow over the northwest sector of the basin. An El Niño event during summer implies a stronger low-level jet, which is reflected in positive precipitation anomalies in the Arenal Basin; this is especially so in the southeast sector, which dominates the contribution of precipitation to the reservoir water level this time of year.

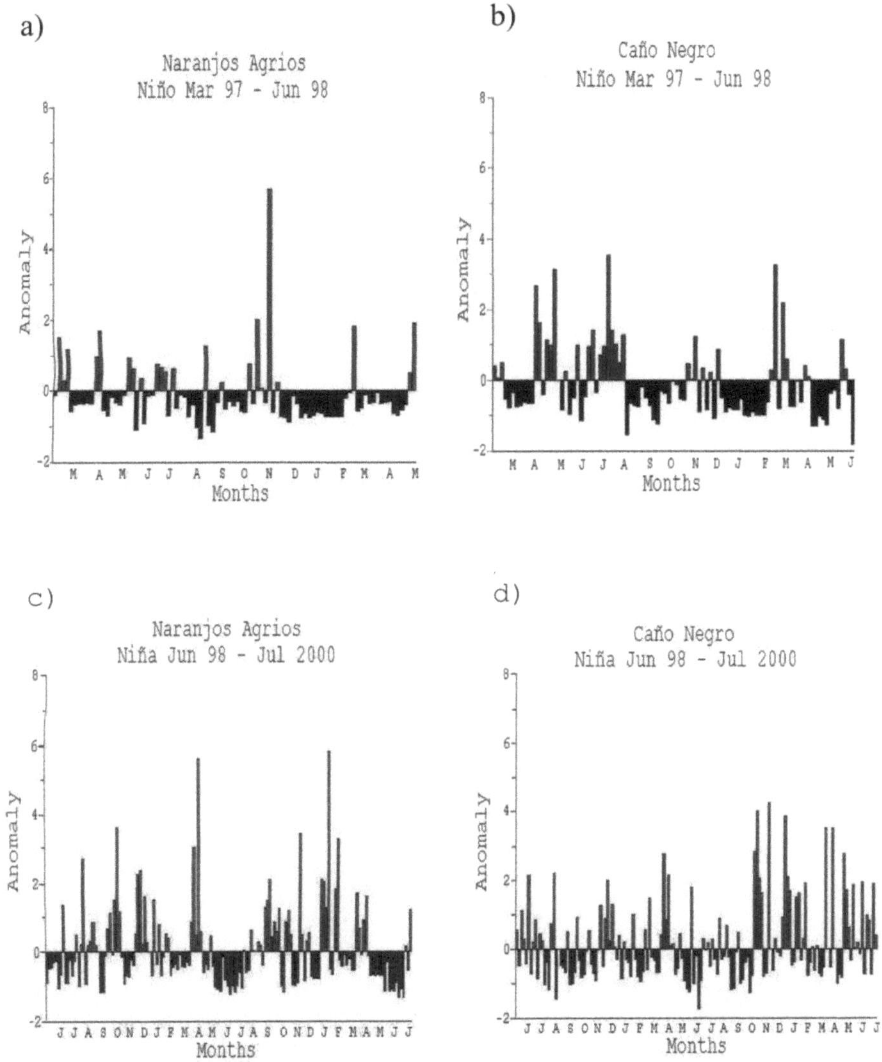


*Figure 9.* Normalized precipitation anomalies (standard deviations) by pentad for (a) El Niño 1997–98 at Naranjos Agrios, (b) as in (a) but for Caño Negro, (c) La Niña 1998–2000 at Naranjos Agrios, and (d) as in (c) but for Caño Negro.

(a)

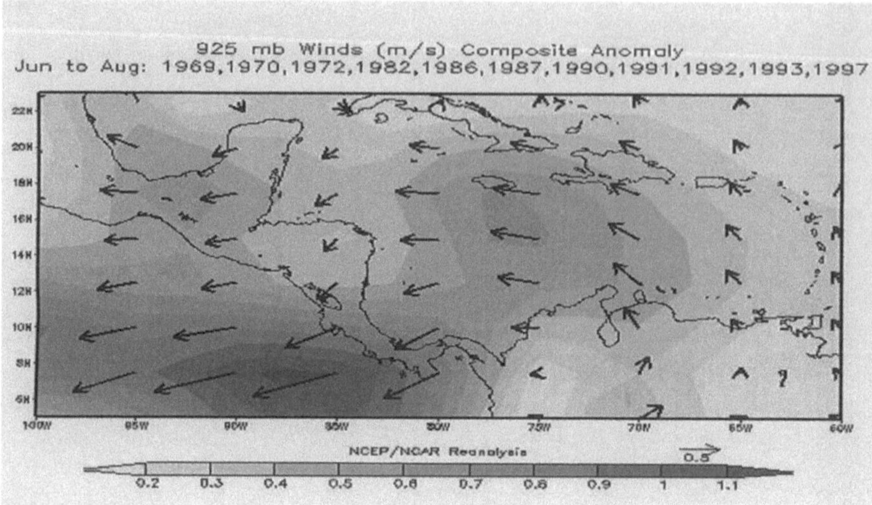


(b)

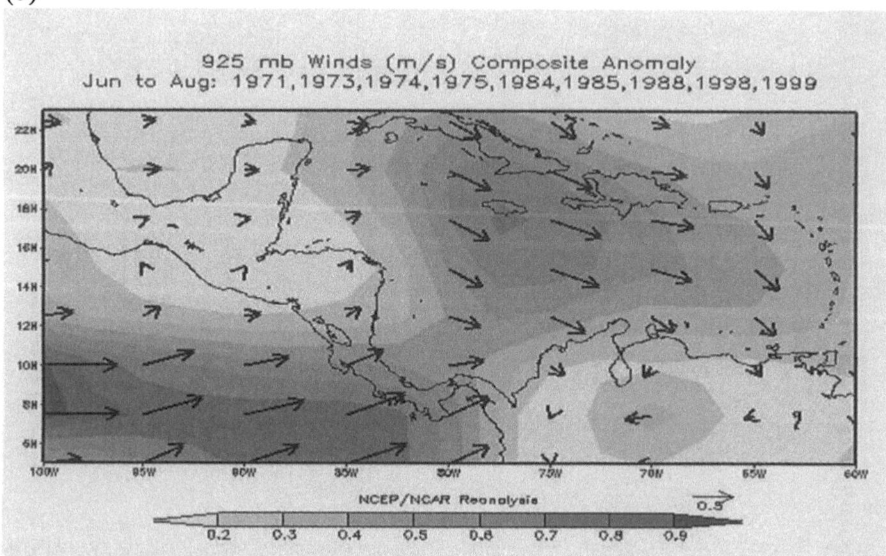


*Figure 10.* Composite anomaly of the 925 hPa wind vector (meters per second) for (a) El Niño summers (June through August), and (b) La Niña summers, based on NCEP/NCAR Reanalysis data. El Niño and La Niña events follow basically the definition provided by Kiladis and Diaz (1989) and the procedure proposed by Trenberth (1997).

Kiladis and Diaz (1989) detected a relatively weak trend toward drier than average conditions in the Caribbean basin from July through October of a canonical warm ENSO phase. Their result is consistent with the one found here, at least for the western Caribbean. Conditions are unfavorable for convection due to increased vertical wind shear related to the strengthening of the trades during El Niño summer months, (Amador et al. 2000c). Dryness is also associated with negative SST anomalies that develop, at least in part because of enhanced evaporation. Once the trade winds weaken in September–October, the SSTs start to recover slowly towards the end of the rainy season, again favoring convective activity before the start of the winter months. Negative anomalies during the El Niño winter of 1998 are discussed below in conjunction with results shown in Figure 11.

Positive anomalies for other periods at Naranjos Agrios and Caño Negro during the 1997–98 El Niño correspond to the so-called "temporales" and mid-latitude air intrusions. Some of these cases are discussed in more detail later in the chapter. Figures 9c and 9d show for the Naranjos Agrios and Caño Negro regions, respectively, the pentad precipitation anomalies during the most recent La Niña event (1998–2000). Contrary to what is indicated in Figures 9a and 9b, the 1998–2000 cold phase of ENSO presents both positive and negative anomalies with no apparent relationship with global or regional-scale systems. A more detailed analysis, however, reveals relatively drier than normal conditions at Naranjos Agrios and Caño Negro during the summer months (July–August 1998 and July–August 1999). Figure 10b indicates that the low-level jet during summer of a cold ENSO phase is exhibits weaker than normal wind velocities, which is consistent with the negative summer anomalies found at both stations in Figures 9c and 9d, under the assumption of weaker flow-topography interactions. Also, one could argue that weaker (stronger) trades associated with the low-level jet entail decreased (increased) influx of atmospheric moisture from the Caribbean into Costa Rica, and as a consequence into the Arenal Basin. Changes in the strength of the low-level jet lead to favorable (unfavorable) conditions for the generation of rainfall. A similar mechanism that partially explains hydrological anomalies in some areas of the Pacific slope of Colombia is also associated with a low-level westerly jet that penetrates inland, especially from August through November, with maximum winds of about 6 m/s in October (Poveda et al. 1998; Poveda and Mesa 2000).

(a)

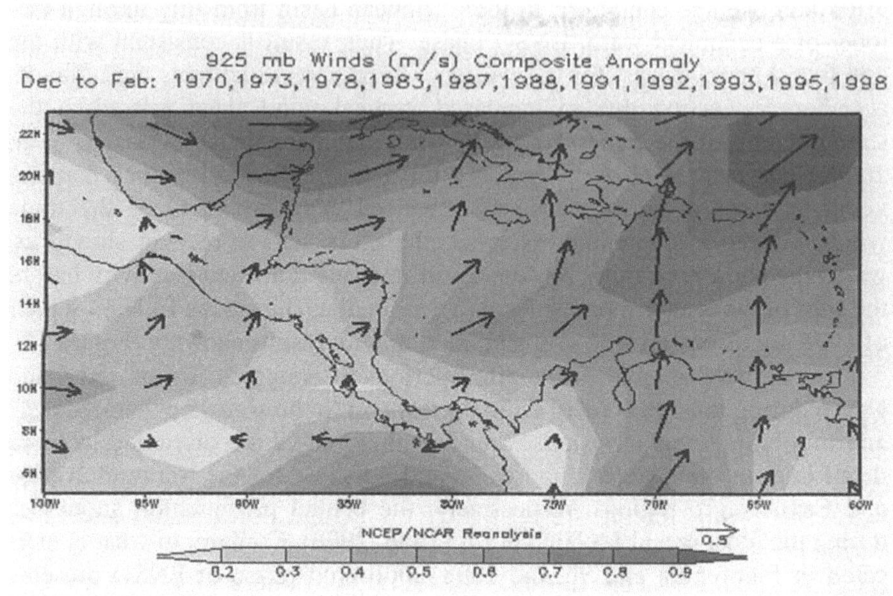


(b)

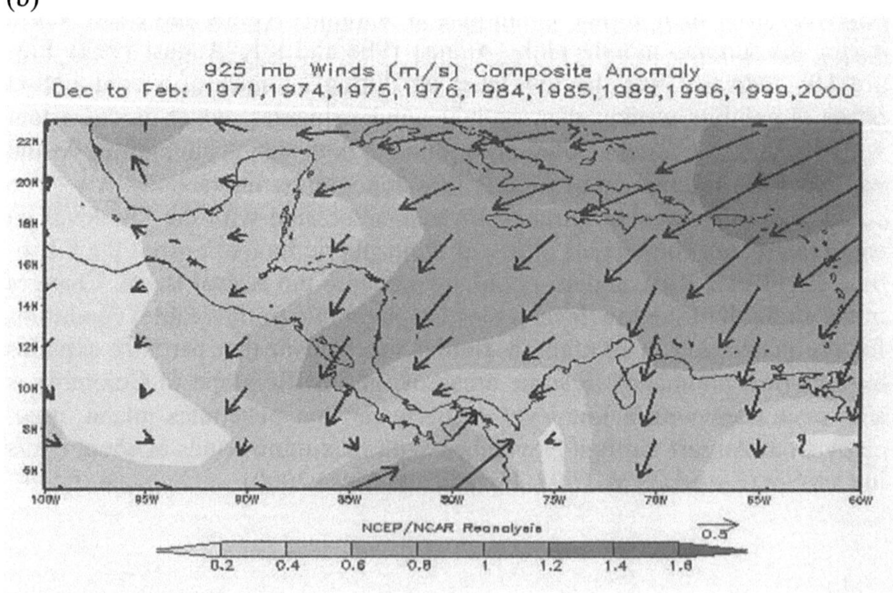


*Figure 11.* As in Figure 10 but for (a) El Niño winters (December through February), and (b) La Niña winters.

El Niño and La Niña composites for the 925 hPa vector wind anomalies for the Northern Hemisphere winter are shown in Figures 11a and 11b, respectively. The warm (cold) ENSO phase shows weaker (stronger) than normal wind velocities over the Caribbean region that are reflected in less (more) precipitation at both stations (Fig. 9a,b). The exception to the above statement is December 1998 through February 1999 for Caño Negro (Fig. 9d), which displayed near normal conditions. Noteworthy are the positive precipitation anomalies of the winter months of 1999–2000. A discussion of one of the extreme cases studied for the winter months of 1999–2000 follows.

Some case studies of short-term meteorological phenomena that hit the Arenal Basin are illustrated in Figure 12. The aim is to illustrate their relative importance to precipitation variability, and to identify these types of systems as fundamental components of the basin's hydrologic budget. Figure 12a shows the contribution to local precipitation of several extreme cases during the 1997–98 El Niño event, namely, the "temporales" of August 1–6, 1997, and of March 9–13, 1998.

The name "temporales" is the term used locally for a period of weak to moderate, nearly continuous rain affecting a relatively large region for several days. Hastenrath (1991) provides a definition of the "temporales" for the Pacific region of Central America close to the one used here; however, his concept includes a condition that winds are weak, while in some cases, as is shown below, winds could be intense and long lasting.

The frequency of these events shows a great deal of interannual and intraseasonal variability, and their relationship to ENSO or to other large-scale climatic signals is still unclear. As was discussed by Velásquez (2000), these perturbations not only originate in the Pacific as disturbances associated with the ITCZ (also discussed in Hastenrath 1991), but are also related with westward traveling low-level cloud systems over the Caribbean, which are not always associated with mid-latitude air intrusions. That is the case for the August 1997 episode and the beginning of the January 2000 event (the latter is illustrated in Fig. 12b). It can be noted in Figure 12a that Caño Negro experienced a dramatic increase in rainfall that surpassed 300 mm in about 5 days for the August 1997 temporal. Other stations (not shown) also measured important amounts of precipitation related to this temporal.

Note from Figure 12a that Naranjos Agrios, in the northwest lowlands of the basin, did not show an important response to this Caribbean temporal. Another case that confirms this difference in the response of the basin to Caribbean temporales is included in Figure 12a for March 1998. The January 2000 case was initially characterized by a rapid wind intensification over the Caribbean from January 5 to 9, 2000, as is illustrated in Figure 13b. Figure 13a demonstrates that this wind change was not initially related to cold air intrusions from mid-latitudes. This case corresponds to the period January 11–20, 2000, which affected almost the whole basin (Fig. 12b). Due to the anomalous rainfall of late 1999 in the southeast sector (see Fig. 9d), and to the temporal of January 2000, the reservoir level rose to an unprecedented level of nearly 548 m. A safe level of operation, according to ICE, is 546 m, and therefore emergency measures had to be taken regarding the water use and management for electricity generation. The photograph presented in Figure 14 illustrates the overall impact of the anomalous precipitation on the water level of the lake, with vegetation that is normally situated in the dry shoreline, submerged by the rapidly rising waters. Note that this temporal contributed more than 3 times the precipitation attributed to Hurricane Mitch at Caño Negro during October 26–30, 1998 (Fig. 13b), which occurred during a cold ENSO event. After this extreme event hit Costa Rica's northern region in January 2000, overall bean crop losses due to flooding and severe meteorological conditions were reported to be of the order of US$3 million.

The cases presented above provide evidence that the acute topographical features of the basin play a crucial role as a forcing mechanism for generating rainfall. Changes in low-level trade wind intensity are closely related to disruptions in precipitation by means of flow-topography interaction processes. The local dependence of precipitation amount upon elevation in areas of high relief in Costa Rica has been well established (Chacón and Fernández 1985; Fernández et al. 1996); however, the relationship of wind intensity to precipitation amount still requires further investigation.

## 5. CONCLUSION

The use of pentad precipitation data and the availability of a relatively dense station network to update aspects of the climatology of the Arenal River Basin, has helped to improve our understanding of the seasonal distribution of precipitation for water management purposes. It was also possible to identify the areas with strong spatial and temporal contrasts in precipitation patterns for water use management. A relative minimum in the annual cycle of precipitation, known as the midsummer drought, or "veranillo," weakly affects the westernmost portion of the Arenal reservoir during July–August. Intense trade winds, associated with the development of a low-level jet over the Caribbean during the summer months, have an effect opposite to that of the veranillo. Increases in the wind flow result in increased precipitation in the easternmost region of the reservoir, due mainly to the interaction of the wind flow with the basin's topography.

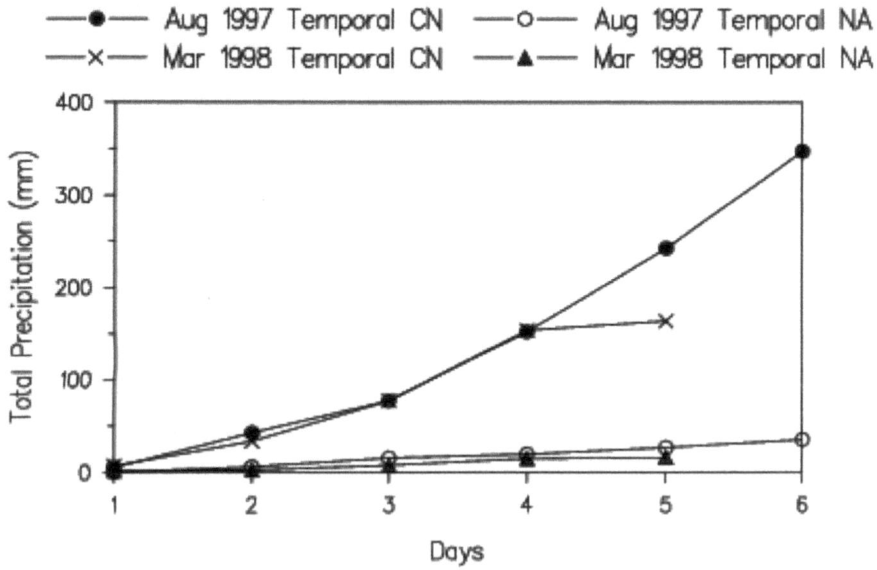


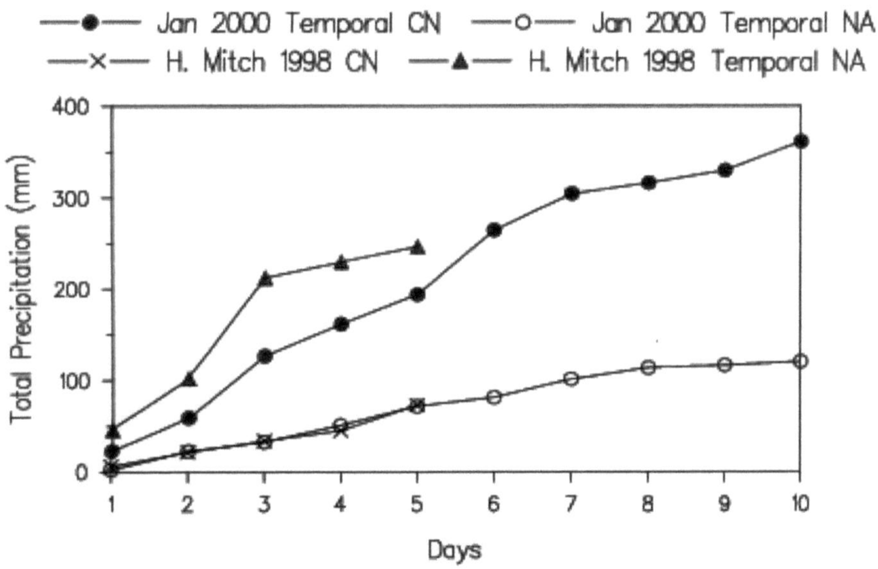


*Figure 12.* Total precipitation contribution (millimeters) of selected "temporales" and extreme meteorological events at Caño Negro (CN) and Naranjos Agrios (NA) during (a) El Niño 1997–98, and (b) La Niña 1998–2000.

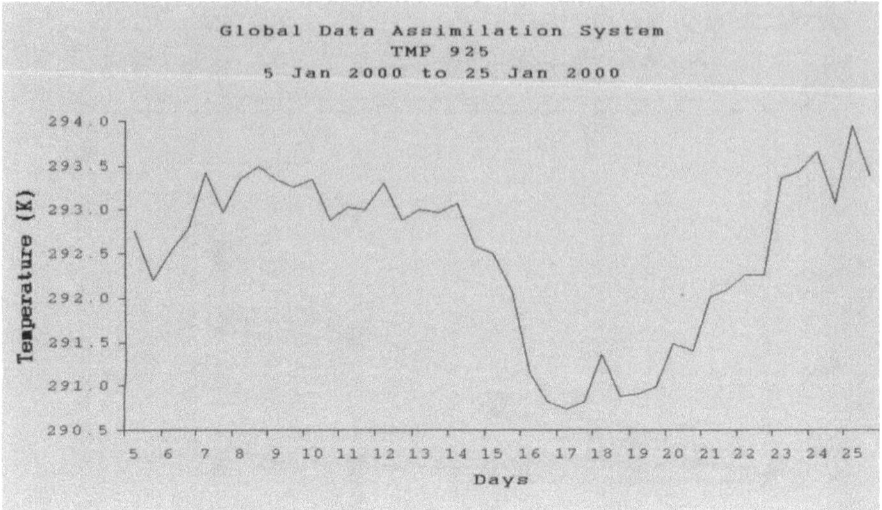


(a)

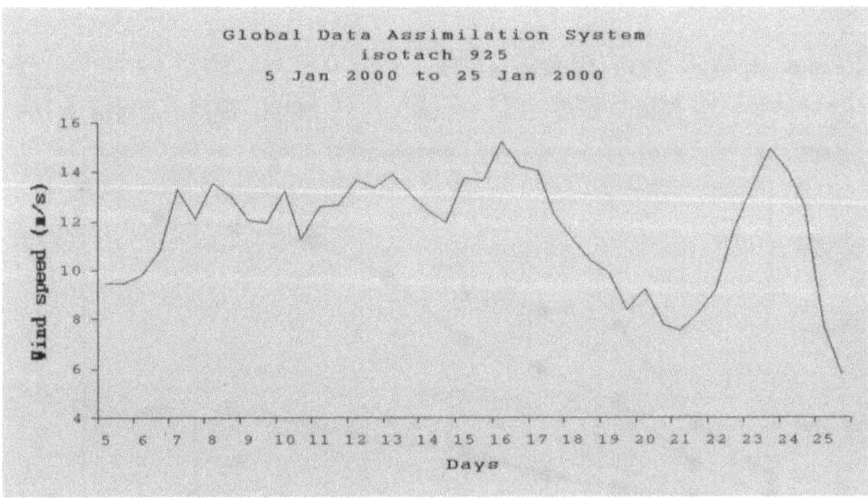


(b)

*Figure 13.* Average values at 925 hPa over the area 10°–15°N, 75°–80°W, from January 5 to 25, 2000, for (a) air temperature (°K), and (b) wind speed (meters per second).

*Figure 14.* Photograph of the Arenal Lake during the "temporal" of January 11–20, 2000. Note the location of an approximately 6 m high tree seen in the lake, which used to be along the lakeshore before the unprecedented water level rise due to the temporal and La Niña rainfall anomalies of late 1999 and early 2000. Photo taken by J.A. Amador on January 15, 2000.

The 1997–98 El Niño and 1998–2000 La Niña episodes have been closely studied in relation to their influence on precipitation distribution in the basin. Based on previous results by Amador et al. (2000a,b) and those discussed in this chapter, El Niño events are shown generally to be associated with a decrease in reservoir level, although some other factors also affect this parameter. La Niña events generally increase the precipitation in the basin, especially that in the northern and western parts, by means of disturbances originating in the Caribbean side, such as the so-called "temporales" and hurricanes.

The first-order climate disruption from the normal precipitation pattern in the Arenal Basin, as determined by principal component analysis of anomalies of monthly precipitation for two regions (northwest and southeast), appears to be weakly related to ENSO episodes and to a greater degree to low-level Caribbean wind anomalies, respectively. Although these low-level wind changes seem to be modulated partly by ENSO, results bring out the importance of other non-ENSO signals for understanding climate disruptions in the region.

Another factor that contributes to interannual precipitation variability in the basin is one related to the frequency of cyclone formation in the Caribbean Basin. It is generally accepted that during El Niño episodes, fewer tropical cyclones tend to form in the Caribbean region. Amador et al. (2000c) found, on the one hand, that the existence of anomalously warm SSTs in the eastern Pacific results in an enhanced low-level jet, and, therefore, in stronger than usual wind shear during summer. On the other hand, during La Niña years, the low-level jet weakens and the vertical wind shear decreases. Partly as a result of such changes in the wind shear environment, the number of

Caribbean hurricanes varies from one year to another. The fluctuations in the intensity of the low-level jet are also reflected in the SST anomaly field over the western Caribbean Sea, north of the Venezuelan coast. An intense (weak) jet results in negative (positive) SST anomalies over the western and central Caribbean, due to strong (weak) Ekman transport and upwelling. In this way, the low-level jet may also play a role relating the SST anomalies in the eastern Pacific during El Niño or La Niña events, to SST anomalies over some regions of the Caribbean.

The importance of identifying precipitation variability on interannual time scales, other than that related to ENSO, that is associated with regional climatic features such as the "veranillo" and the low-level jet, has led us to identify basic steps in the development of practical rainfall forecasting schemes for the Arenal Basin. The problem of climate variability in Costa Rica and other regions of Central America constitutes a challenge, since the region is surrounded by warm ocean pools in which convective activity is intense, and where the interaction of strong trade winds associated with regional climate systems in complex topography leads to large differences in precipitation over relatively short distances. Most forecasting schemes aimed at determining the spatial patterns of precipitation anomalies are based on statistical methods or on analogs, using the El Niño signal as primary input. As has been shown here, not all the anomalous climate patterns exhibit the same spatial characteristics, or are related to the ENSO signal directly. Even more, differences in large-scale climatic patterns associated with different ENSO events sometimes make the analogs or multiple regression models based only on tropical Pacific SST information of very limited use. Therefore, the use of more physically based prognostic tools, such as that provided by numerical models with a multiparameter output, should be incorporated as a more realistic alternative to traditional prediction schemes in the future.

## 6. ACKNOWLEDGMENTS

This work was partially funded by the University of Costa Rica (grants VI-805-94-204 / 98-506 / 112-99-305), and the Inter-American Institute for Global Change Research (IAI) grants ISP3-030 and CRN-073. Additional support from the National Oceanic and Atmospheric Administration, Office of Global Programs (NOAA-OGP) and Comité Regional de Recursos Hidráulicos through a project for regional modeling is also acknowledged. Thanks are due to Dr. Victor Magaña, National Autonomous University of Mexico, for many fruitful discussions on regional climate systems, and to Dr. Henry Diaz for his assistance and encouragement throughout this work. In addition, we would like to thank Dr. E. Alfaro and Dr. F.J. Soley, University of Costa Rica, for helpful talks over practical aspects of EOF. The authors are also indebted to E. Rivera, R. Velásquez, I. Mora, R. Madrigal, Z. Umaña, and other staff members of CIGEFI, for their assistance during the course of this work. Reviewers provided many valuable comments to improve the quality of the manuscript. ICE and Instituto Meteorológico Nacional kindly provided climate data for the Arenal River Basin and Costa Rica. Use of NCEP/NCAR public domain data facilitated part of this research.

## 7. LIST OF ACRONYMS

| | |
|---|---|
| AN | Arenal Nuevo Hydrometeorological Station |
| ARCOSA | Hydroelectric Complex Arenal-Corobici-Sandillal |
| CN | Caño Negro Hydrometeorological Station |
| EC | El Cairo Hydrometeorological Station |
| IAI | Inter-American Institute for Global Change Research |
| IGN | Instituto Geográfico Nacional (National Geographic Institute) |
| ICE | Instituto Costarricense de Electricidad (Costa Rica Institute of Electricity) |
| NA | Naranjos Agrios Hydrometeorological Station |
| NCAR | National Center for Atmospheric Research |
| NCEP | National Center for Environmental Prediction |
| NT | Nuevo Tronadora Hydrometeorological Station |
| PACS-SONET | Panamerican Climate Studies Sounding Network |
| PS | Presa Sangregado Hydrometeorological Station |

# 8. REFERENCES


Alfaro, E. 2000. Eventos Cálidos y Fríos en el Atlántico Tropical Norte. *Atmósfera* 13: 109–119.

Alfaro, E.J., and L. Cid. 1999. Ajuste de un modelo VARMA para los campos de anomalías de precipitación en Centroamérica y los índices de los océanos Pacífico y Atlántico Tropical. *Atmósfera* 12: 205–222.

Amador, J.A. 1998. A climatic feature of the tropical Americas: The trade wind easterly jet. *Tópicos Met. y Ocean.* 5(2): 91–102.

Amador, J.A., and V.O. Magaña. 1999. Dynamics of the low level jet over the Caribbean sea. Preprints 23$^{th}$ Conference on Tropical Meteorology, January 10–15, 1999, Dallas, Texas, American Meteorological Society, pp. 868–869.

Amador, J.A., S. Laporte, and R.E. Chacón. 2000a. Cuenca del Río Arenal: Análisis de los eventos El Niño de los años 1992–93, 1994–95 y 1997–98. *Tópicos Met. y Ocean.* 7(1): 1–21.

Amador, J.A., R.E. Chacón, and S. Laporte. 2000b. Cuenca del Río Arenal: Análisis de los eventos La Niña de los años 1988–89 y 1996. *Tópicos Met. y Ocean* 7(1): 22–42.

Amador, J.A., V.O. Magaña, and J.B. Pérez. 2000c. The low level jet and convective activity in the Caribbean. Preprints 24$^{th}$ Conference in Hurricanes and Tropical Meteorology, May 29–June 2, 2000, Fort Lauderdale, Florida, American Meteorological Society, pp. 114–115.

Chacón, R.E., and W. Fernández. 1985. Temporal and spatial rainfall variability in the mountainous region of the Reventazón River basin, Costa Rica. *Journal of Climatology* 5: 175–188.

Enfield, D., and E.J. Alfaro. 1999. The dependence of Caribbean rainfall on the interaction of the tropical Atlantic and Pacific Oceans. *Journal of Climate* 12: 2093–2103.

Fernández, W., and P. Ramírez. 1991. El Niño, la Oscilación del sur y sus efectos en Costa Rica : Una revisión. *Tecnol. Marcha* 11: 3–11.

Fernández, W., R.E. Chacón, and J.W. Melgarejo. 1986. Modifications of air flow due to the formation of a reservoir. *Journal of Climate and Applied Meteorology* 25: 982–988.

Fernández, W., R. E. Chacón, and J. W. Melgarejo. 1996. On the rainfall distribution with altitude over Costa Rica. *Revista Geofisica* 44: 57–72.

George, R., P. Waylen, and S. Laporte. 1998. Interannual variability of annual streamflow and the Southern Oscillation in Costa Rica. *Hydrological Science* Journal 43: 409–424.

Hastenrath, S. 1991. *Climate Dynamics of the Tropics*. Dordrecht: Kluwer Academic Publishers, 488 pp.

Hoerling, M.P., and A. Kumar. 2000. Understanding and predicting extratropical teleconnections related to ENSO. In: Diaz, H.F., and V. Markgraf (eds.), *El Niño and the Southern Oscillation: Multiscale Variability and Global and Regional Impacts*. Cambridge, UK: Cambridge University Press, pp. 57–88.

Kalnay, E., and Coauthors. 1996. The NCEP/NCAR Reanalysis 40-year Project. *Bulletin of the American Meteorological Society* 77: 437–471.

Kiladis, G.N., and H.F. Diaz. 1989. Global climatic anomalies associated with extremes in the Southern Oscillation. *Journal of Climate* 2: 1069–1090.

Magaña, V. 1999. *Los Impactos de El Niño en México*. Universidad Nacional Autónoma de México y Dirección General de Protección Civil, Secretaría de Gobernación, México, 229 pp.

Magaña, V.O., J.A. Amador, and S. Medina. 1999. The mid-summer drought over Mexico and Central America. *Journal of Climate* 12: 1577–1588.

Molinari, J., D. Knight, M. Dickinson, D. Vollaro, and S. Skubis. 1997. Potential vorticy, easterly waves and eastern Pacific tropical cyclogenesis. *Monthly Weather Review* 125: 2699–2708.

Portig, W.H. 1976. The climate of Central America. In, Schweidtfeger, W. (ed.), *World Survey of Climatology*, Vol. 12, *Climate of Central and South America*, Elsevier Publishing Company, pp. 405–478.

Poveda, G., and O.J. Mesa. 2000. On the existence of Lloró (the rainiest locality on Earth): Enhanced ocean-atmosphere-land interaction by a low-level jet. *Geophysical Research Letters* 27(11): 1675–1678.

Poveda, G., M. Gil, and N. Quiceno. 1998. El Ciclo Anual de la Hidrología de Colombia en Relación con el ENSO y la NAO. *Bull. Inst. Fr. d'Etudes Andines* 27(3): 721–731.

Schultz, D., W.E. Bracken, L. Bosart, G. Hakim, M. Bedrick, M. Dickinson, and K. Tyle. 1997. The 1993 Superstorm cold surge: Frontal structure, gap flow, and tropical impact. *Monthly Weather Review* 125: 5–39.

Schultz, D., W.E. Bracken, and L. Bosart. 1998. Planetary and synoptic-scale signatures associated with Central American cold surges. *Monthly Weather Review* 126: 5–27.

Srinivasan, J., and G. Smith. 1996. Meridional migration of Tropical Convergence Zones. *Journal of Climate* 9: 1189–1202.

Trenberth, K.E. 1997. The definition of El Niño. *Bulletin of the American Meteorological Society* 78: 2771–2777.

Velásquez, R.C. 2000. Mecanismos físicos de variabilidad climática y eventos extremos en Venezuela. Tesis de Licenciatura en Meteorología, Departamento de Física Atmosférica, Oceánica y Planetaria, Escuela de Física, Universidad de Costa Rica, 118 pp.

Wahler, W.A., and Associates. 1975. *Design Report for Arenal Dam*. Vol. 1, 2, *Project Summary*, California, U.S.A., 154 pp.

Waylen, P., C. Caviedes, and M. Quesada. 1996a. Interannual variability of monthly precipitation in Costa Rica. *Journal of Climate* 9: 2606–2613.

Waylen, P., M. Quesada, and C. Caviedes. 1996b. Temporal and spatial variability of annual precipitation in Costa Rica and the Southern Oscillation. *International Journal of Climatology* 16: 173–193.

15

# NONLINEAR FORECASTING OF RIVER FLOWS IN COLOMBIA BASED UPON ENSO AND ITS ASSOCIATED ECONOMIC VALUE FOR HYDROPOWER GENERATION

Germán Poveda,* Oscar J. Mesa,* and Peter R. Waylen**

**Programa de Posgrado en Approvechamiento de Recursos Hidráulicos,Escuela de Geociencias y Medio Ambiente, Universidad Nacional de Colombia, AA 1027 Medellín, Colombia*

***Department of Geography, University of Florida, Gainesville, Florida, 32611-7315 U.S.A.*

**Abstract** There is ample evidence of hydrologic variability at annual and interannual time scales over the northern regions of tropical South America. Hydroelectric power provides a cheap regional energy source, yielding over seventy percent of Colombia's national energy annually. The El Peñol scheme on the Nare River is the country's largest providing roughly 14% of national production. The region is particularly susceptible to droughts during warm phases of ENSO, however stream flow inputs are also subject to a variety of regional and local factors other than ENSO, which may makes forecasting difficult. The identification of a model which permits the reliable incorporation of readily available ocean-atmosphere variables, and variables derived from standard forecasts, to a potentially non-linear prediction of monthly stream flows is crucial to the optimal operation of the reservoir. The MARS (Multiple Adaptive Regression Splines) model is calibrated to provide forecasts of monthly stream flows over the period 1956–86. The applicability of the forecast technique is discussed by reference to comparisons between observed and forecasted flows in a separate model validation series (1987–92). Reliability of the stream flow forecasting methodology is investigated over "forecast horizons" ranging from 3 to 12 months, and the potential economic value of incorporating the methodology into the operation of the national power generating system is illustrated.



*Henry F. Diaz and Barbara J. Morehouse (eds.), Climate and Water*, 351-371.

# 1. INTRODUCTION

The last two decades have witnessed a marked increase in our knowledge of the causes of interannual hydroclimatic variability and our ability to delineate regions within which similar fluctuations are experienced. Such regions of coherent behavior frequently encompass the whole, or parts, of many nations. One of the most widely studied causes of such variability is the El Niño/Southern Oscillation (ENSO) phenomenon (Neelin et al. 1998) Many regions of Latin America, located as it is near the seat of the ENSO phenomenon, have been shown to experience variations in precipitation in concert with ENSO (Trenberth et al. 1998). These perturbations are transmitted through the land-based portion of the hydrologic cycle to the rivers, which are often an important source of regional and national energy. In regions where hydroelectric power contributes a large proportion of the national energy supply, this linkage may provide a direct connection between climatic fluctuations and national and regional economies. This research specifically investigates the possibility of improving the forecasts of streamflow into the largest reservoir used for hydropower generation in Colombia, and illustrates the potential economic savings of such forecasts.

Figure 1 displays the percentages of national energy derived from various sources throughout the countries of South and Central America. Fourteen of the twenty countries obtain more than half of their national power from hydroelectricity. It can be seen that Colombia's reliance on hydropower (approximately 75% of total energy production) is by no means atypical of the region. Despite potential environmental impacts, hydroelectricity provides a cheap, renewable form of energy. In most cases the only nationally viable alternative source of energy comes from thermal power plants fired by coal, gas, or oil. Currently there is little international trade of energy within the region; therefore, any shortfall in hydroelectric power generation (stemming from a drought) has to be compensated for by more expensive thermal sources. There are generally high costs associated with the start-up of these thermal plants, and expensive fuel has to be either imported or diverted from domestic supplies, which may result in the loss of valuable export revenues.

The level of uncertainty that this climatic linkage brings to the national supply of power is illustrated by annual energy generation in Colombia between 1977 and 1999 (Fig. 2). During this period, energy production increased threefold in a fairly steady fashion, with a marked break in 1992 corresponding to a warm phase ENSO event (El Niño, which brings drought to Colombia). The proportion of this energy derived from hydropower over the period varied from 66% in 1983 to a high of 87% in 1979. Low percentages correspond to the warm ENSO events of 1982–83, 1992–93, and 1997–98; however, the less pronounced warm event of 1986–87 is not so clearly marked. The cold phase of ENSO (La Niña) produces copious rains in Colombia (Poveda and Mesa 1997). Only one such event, that of 1988–89, occurred entirely during this period, although the beginnings of the 1999–2000 event may also be seen. Both correspond well with high percentages of national energy generated by hydropower. The national effects of such fluctuations are compounded when considered in conjunction with the simultaneous disruption of other sectors of the economy, such as agriculture, and government services, including health (Poveda and Rojas 1996; Poveda et al. 2001b) and transportation infrastructure.

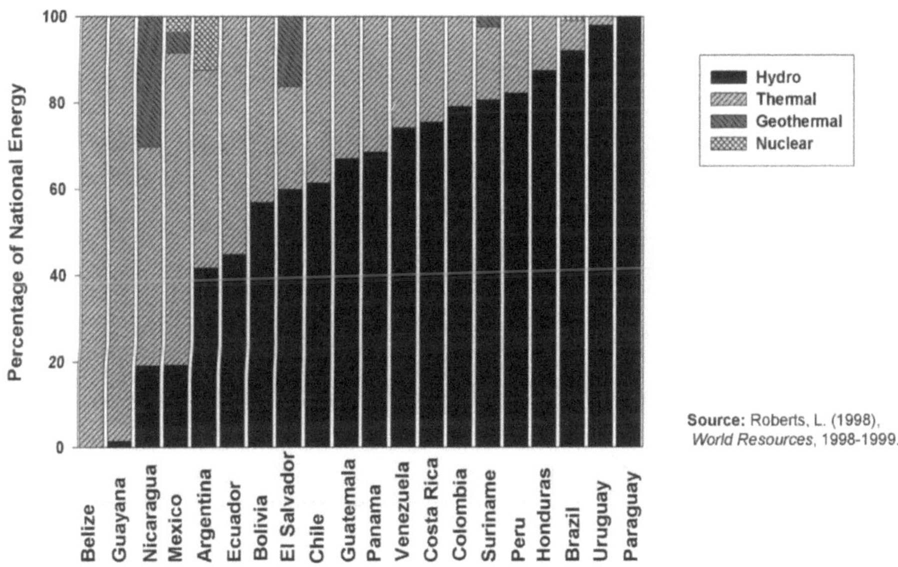


*Figure 1.* Percentages of national electricity generation from various sources in Latin American countries.

While nothing can be done to alter the natural variability of the climatic effects on the reservoirs, enhanced ability to forecast reductions in precipitation/streamflow would permit better operation of the reservoirs and the releases of water in anticipation of the low-flow season and expected tariff increases. For economic reasons the energy sector requires medium- to long-term planning. Hydrologically, the annual-to-interannual regulation capacity of the reservoirs— in combination with the necessity for slow,

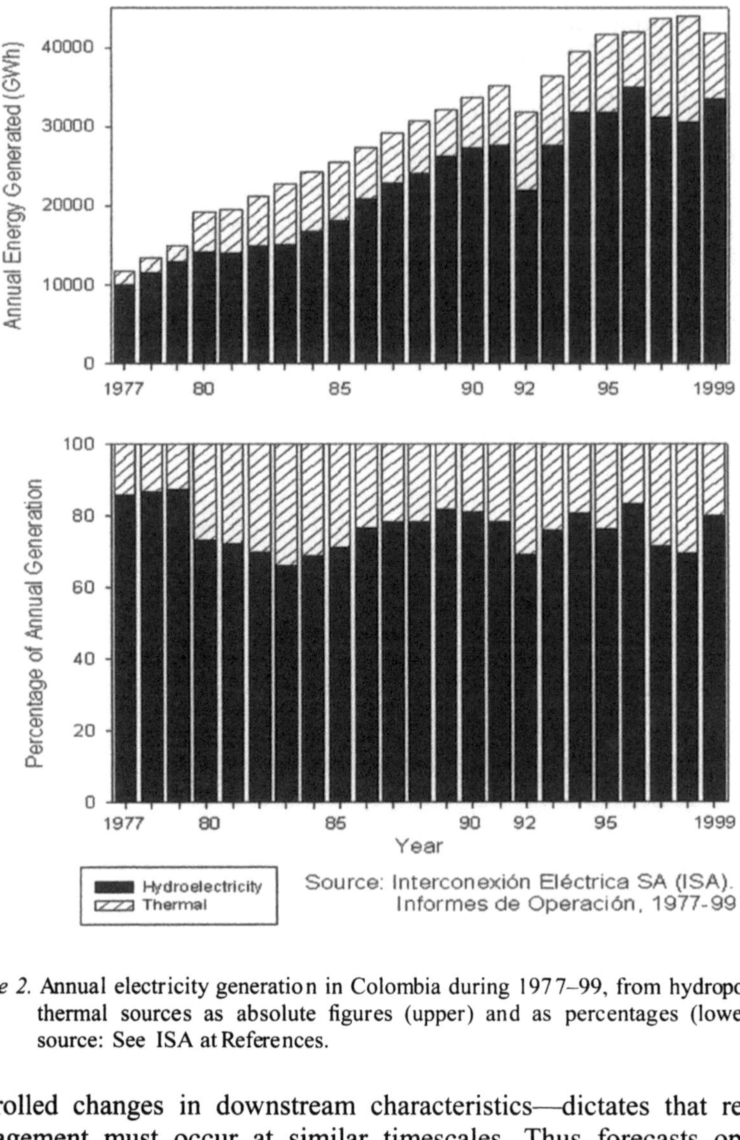


*Figure 2.* Annual electricity generation in Colombia during 1977–99, from hydropower and thermal sources as absolute figures (upper) and as percentages (lower). Data source: See ISA at References.

controlled changes in downstream characteristics—dictates that reservoir management must occur at similar timescales. Thus forecasts optimally must be made 6 to 12 months in advance. This study considers the use of Multivariate Adaptive Regression Splines (MARS) in combination with historic hydroclimatological records and readily available macroclimatic indices, to improve forecasts of monthly inflows to the reservoir of Colombia's most important power generating plant. Costs associated with the application of the model would be negligible, as it requires only existing streamflow and precipitation records, and macroclimatic indicators that are distributed through the Internet (http://www.cpc.noaa.gov/data/indices/index.html, http://www.atmos.washington.edu/~mantua/abst.PDO.html, http://tao.atmos.washington.edu/data_sets/tropicalraints/, and http://www.cdc.noaa.gov/~kew/MEI/). Yet the potential benefits are high.

## 2. THE STUDY SITE

The hydroelectric plant at Guatapé (6°20' N, 75°10' W) is part of the design on the Río Nare, which drains 1,250 $km^2$, and is supplied by the El Peñol reservoir. It is located at La Araña in the municipality of San Rafael, eastern Antioquia, and has an installed capacity of 560 megawatts (MW), contributing a fairly consistent average of 2,730 gigawatt-hours (Gwh), or about 14% of Colombia's annual energy production. This consistency is maintained by the high capacity of the impounded reservoir, which is the largest in the country, covering 6,240 hectares (ha) and possessing a storage capacity of $1{,}236 \times 10^6$ $m^3$, of which about 95% is usable. The historical time series of annual discharges on the Río Nare for 1956–98 is shown in Figure 3(upper). The project itself is located at 1,800 m above sea level; however, the contributing basin attains altitudes of over 3,000 m as it drains the Cordillera Central of the Colombian Andes. Mean basin precipitation is approximately 2,000 mm, varying considerably with elevation and orientation. Peaks in monthly precipitation (of almost equal size ~300 mm) occur in May and October (Fig. 4), separated by a slight minimum (180 mm) in July and an annual minimum of 80 mm in January. These peaks and troughs correspond to the migration of the Intertropical Convergence Zone (ITCZ) over northern South America; they are reflected in the monthly discharges, which average about 45 $m^3s^{-1}$, with the October peak being slightly larger (65 $m^3s^{-1}$) than the May peak (58 $m^3s^{-1}$). The periods of lower flow in July and February sustain 45 $m^3s^{-1}$ and 30 $m^3s^{-1}$, respectively.

There is growing strong evidence of an association between the phases of ENSO and the hydrology of the study area (Hastenrath 1990; Poveda and Mesa 1997; Trenberth et al. 1998; Poveda et al. 2001a; Gutiérrez and Dracup 2001), with droughts occurring in warm phase years (El Niño) and excessive rains during cold phases (La Niña). However, as was noted by Ropelewski and Halpert (1987), this signal is not always clear or consistent in this region, and other macroclimatic phenomena also affect the hydroclimatology there (Poveda and Mesa 1997).

Figure 3 illustrates some of the difficulties in establishing a relationship between streamflow and ocean atmosphere conditions in the Pacific.

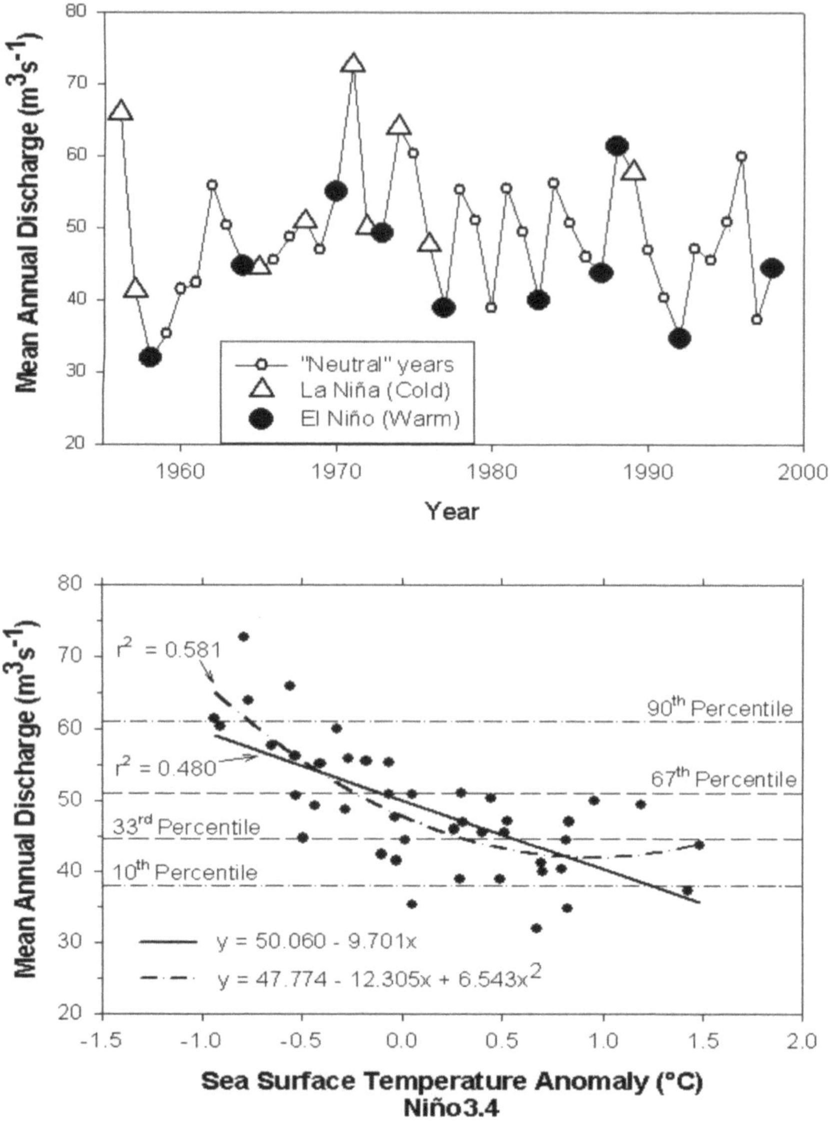


*Figure 3.* Time series of annual discharges on the Río Nare for 1956–98 (upper). El Niño and La Niña years are identified by using the classification scheme of the Center for Ocean-Atmospheric Prediction Studies, Florida State University. Relationship between annual sea surface temperature anomalies in the Niño3.4 region of the Pacific Ocean and annual discharges on the Río Nare (lower). Fitted functions are of the first and second order.

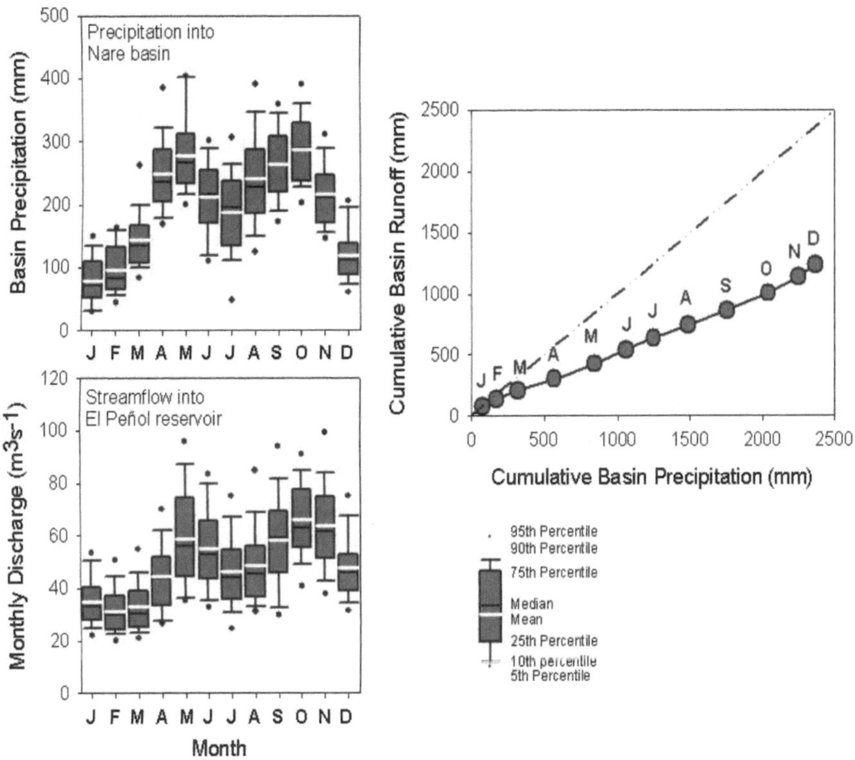


*Figure 4.* Summary of monthly characteristics of precipitation inputs to (upper left), and streamflow outputs from (lower left), the Río Nare basin and El Peñol reservoir, 1956–92. A comparison of the cumulative mean monthly inputs and outputs is shown in the form of a double mass curve (right), with a 1:1 line shown for reference.

The upper panel suggests that La Niña (cold phase of ENSO) is associated with higher discharges, and El Niño (warm phase) with lower flows. The relationship, however, which is almost completely effaced during the decade of the 1960s, seems to be nonstationary. The lower graph implies that the relationship is nonlinear, with discharges responding disproportionately to negative sea surface temperature (SST) anomalies, with a potential threshold at an anomaly of approximately 0°C.

The nature and strength of the association appears to vary over the period of historic records, and local topographic effects can also bring about marked differences in response. The ENSO phenomenon itself is cyclical, and the switch between warm and cold phases is currently believed to result from subsurface oceanic adjustments in the equatorial Pacific (Neelin et al. 1998). It also displays weak nonlinearity similar to the features of a complex nonlinear system (Vallis 1986; Hense 1987; Jin et al. 1994; Chang et al. 1994; Neelin et al. 1998). Neelin et al. (1998) suggest that this behavior may result from the interaction of the slow components of the atmosphere-ocean system with the annual cycle, or from the random influences of other weather-related phenomena, which may amplify or dampen the ENSO phenomenon. Poveda and Mesa (1997, 2000) illustrate the potential interactions and feedbacks between hydrologic and climatologic controls in the region. Further controls over regional hydroclimatology have been related to events and conditions in such diverse regions as the North Atlantic and Pacific Oceans (Enfield and Alfaro 1999; Giannini et al. 2000), North America (Schultz et al. 1998), and the Amazon (Poveda and Mesa 1997). The nature of the operation of the land-based portion of the hydrologic cycle is likely to produce further nonlinear responses, particularly through the effects of soil moisture storage and evapotranspiration (D'Odorico et al. 2000; Poveda et al. 2001b).

As the primary objective of this research is to provide forecasts of streamflow rather than a physical explanation of the phenomenon, a series of readily available macroscale atmosphere-ocean indicators are selected for use. These indicators are all correlated to ENSO and teleconnected to the hydrometeorology of the study area. They are also available in real time from the aforementioned Internet sources.

## 3. FORECAST MODEL

Due to the noted nonlinearities in the systems under study and the potential chaotic nature of the ENSO forcings, it is necessary to use a numerical forecasting procedure capable of accommodating such complexity. Friedman and co-workers (1989), and Friedman (1991) provided such a regression model in MARS, which is nonlinear in its explanatory variables. The technique is analogous to simple linear regression with the exception that all of the following may potentially vary: (1) the form of the model (additive/multiplicative), (2) the number of independent variables, and (3) the coefficients associated with those variables. In this regard, it may be viewed as similar to piecewise regression. Lall et al. (1996) have shown that the model performs particularly well in representing the changing relationships between variables derived from a low-dimensional chaotic system.

Traditional regression models may have limited utility when used to model time series that display quasi-periodic behavior (resulting from such phenomena as ENSO, nonlinear dependence, or a chaotic system). These parametric models try to adjust a parametric function, $\underline{g}(x)$, as an estimator of the real function $\underline{f}(x)$, usually through the use of least squares. However, it is possible that some predictor variables have greater importance only in certain ranges of predicted values, or that the underlying system is so complex that it cannot be represented by a simple function alone. This is particularly relevant for regional hydroclimatic variations (and their ENSO linkages), which often exhibit nonlinearity, nonstationarity, and thresholds in their relationships as exemplified by Figure 3. Nonparametric approximations do not try to adjust a single function throughout the entire global range of predictions. Instead, there exist two alternatives: (1) Adjust a global function, which may contain different model parameters (local parametric approximation), or (2) approximate $\underline{f}$ by various simple parametric functions (generally low-order polynomials), each defined within a subregion of the global domain of the predicted variable, and joined/smoothed by the use of splines. The MARS model has been recognized as an appropriate methodology for the prediction of hydroclimatological variables, which exhibit nonlinear dynamics and long-term oscillations (Abarbanel et al. 1996; Abarbanel and Lall 1996; Lall et al. 1996; Sangoyomi et al. 1996; Lewis and Ray 1997). In general it has the form:

$$f(x)=a_0+\sum_{K_m=1} f_i(x_i)+\sum_{K_m=2} f_{ij}(x_i,x_j)+\sum_{K_m=3} f_{ijk}(x_i,x_j,x_k)+\ldots..$$

Where $a_o$ is a constant, the second term represents a sum of all basic functions involving a simple variable, the third term is a sum of all functions involving the interaction of two variables, the fourth term involves interactions of three variables, and so forth.

Let

$$V(m)=\{v(k,m)\}_1^{K_m}$$

be the combination of variables associated with the $m^{th}$ basic function, $B_{m.}$ Each function in the first summation term of the equation above, can be expressed as

$$f_i(x_i)=\sum_{K_m=1} a_m B_m(x_i)$$

$$(i,j) \in V(m)$$

This is a summation of all the simple basic functions involving the variable $\underline{x}_i$, raised to the first power. Each bivariate function in the second term can be expressed as:

$$f_{i,j}(x_i x_j) = \sum_{K_m = 2} a_m B_m (x_i, x_j),$$

which is a summation of all the basic functions involving the interaction of two variables $\underline{x}_i$ and $\underline{x}_j$. This basic technique may be applied to successively larger combinations of variables. The MARS model identifies the nature of the basic functions, specific variables which are to be entered into the model, and whether the terms are simply additive or involve interactions with other variables. A more detailed discussion of model structure is provided by Lall et al. (1996).

The historic monthly streamflows are divided into two sections, one to be used to calibrate the model, and a second smaller section upon which the model may be validated. The MARS model "penalizes" the user for each additional predictor variable and the number of "segments" of regression used by reducing the degrees of freedom; therefore, it is desirable to have as long a calibration period as is possible. Similarly, because of the periods of the low-frequency fluctuations in some potential predictor variables, such as the decadal scale of North Atlantic sea surface temperatures (SST) (Latif 2001), it is preferable physically to have a calibration period that at least spans the similar periods of years. However, the larger the set of validation data, the greater the ability to test the applicability of the model, particularly under the variety of combinations of macroclimatic predictors that are likely to be encountered. To this end the larger portion (1956–87) of the available historic records are assigned to calibration and the shorter period (1988–92) to validation. Because of the noted importance of ENSO, it was ensured that both calibration and validation periods contained examples of warm and cold phase years.

Monthly values of the potential predictor variables are lagged up to 12 months and entered into the procedure. The final model is the one that minimizes the errors between observed and fitted streamflows during the calibration period. Various forms of the model, additive and multiplicative, are also employed with linear and quadratic terms.

Macroclimatic variables and the well-established persistence characteristics of rainfall and river discharges are used to implement the MARS methodology. The following hydroclimatic variables were used as predictors: Multivariate ENSO Index (MEI; Wolter 1987, see http://www.cdc.noaa.gov/~kew/MEI), the Southern Oscillation Index (SOI; Trenberth 1976), the North Atlantic Oscillation Index (NAO, Hurrell 1995; Jones et al. 1997), the Pacific Decadal Oscillation (PDO; Mantua et al. 1997), the Quasi-Biennial Oscillation (QBO; Marquardt and Naujokat 1997), and SST at the Niño3.4 region. Most of those variables can be obtained at http://tao.atmos.washington.edu/data_sets/.

## 4. RESULTS

In order to avoid spuriously high measures of the MARS model's ability to reproduce the observed patterns of interannual variability in monthly streamflows, all comparisons between the observed data and fitted model are carried out upon standardized monthly streamflows, $Q'$, which have previously been normalized by removing the long-term monthly mean and variance:

$$Q'_{i,j} = (Q_{i,j} - \mu_j) / \sigma_j ,$$

where
i = Year of record (in calibration period)
j = Month
Q = Observed monthly discharge
$\mu$ = Mean monthly discharge
$\sigma$ = Standard deviation of monthly discharge

Figure 5 shows scatter plots of the observed and fitted standardized discharges during the calibration period. A measure of the model performance for each month is also shown. Statistical significance of the following results is tested at the 0.05 level unless otherwise stated. The results indicate that the fitted MARS model explains 65% of the interananual variability of monthly flows on the Nare (statistically significant at less than the .0001 level). Ideally, the slope of the best-fit straight line should be 1 and the intercept zero. The intercept (0.001) is not significantly different from zero ($p = 0.9986$), while the slope (0.696) is significantly different from unity ($p < 0.0001$). This result implies that the fitted MARS model tends to underestimate interannual variability by approximately 30%. The monthly comparisons all yield statistically significant correlations, but the variance explained ranges from about 55% in June and July, to 80% in February through April. Performance seems to be best during the low-flow season and worst between the earlier and later high-flow seasons. For no month can the computed intercept be considered different from zero, but the slopes of the lines are all different from unity with the exception of that for March. All other months report slopes of less than one, ranging from 0.554 for July to 0.811 for February. Their seasonal pattern of variability naturally mirrors that of the correlation coefficients.

The model was validated against observed data over the period 1988–92 (Fig. 6), which includes the large cold phase ENSO event of 1988 and the warm phase of 1992. The fit for the first 3 years is very good. The model successfully reproduces the higher flows associated with the cold phase of 1988, as well as the elevated high-flow seasons of 1989 and the unusual dryness of the intervening months. For 1990 it successfully matches the decreased levels of flow in the early high-flow season and their higher levels in the latter high-flow season. In general, the protracted 1991–93 El Niño event is anticipated but the model frequently fails to match exactly the degree to which flows fall below the monthly mean. This is also reflected in the scatter plot, where there appears to be a greater variance of estimated values at below average flows, and a positive intercept of 0.242 (although not significantly different from zero). It appears that the overestimates of flows in the drought conditions of El Niño are forcing a lower overall slope (0.659, significantly different from unity) to the regression line. When monthly observations from 1991 and 1992 are omitted, the coefficient of determination and regression slope increase, while the intercept is almost halved, and the plot appears less heteroscedastic. The increase in slope is sufficiently great that it cannot be considered significantly different from unity; however, this observation should be tempered by consideration of the reduced sample sizes involved.

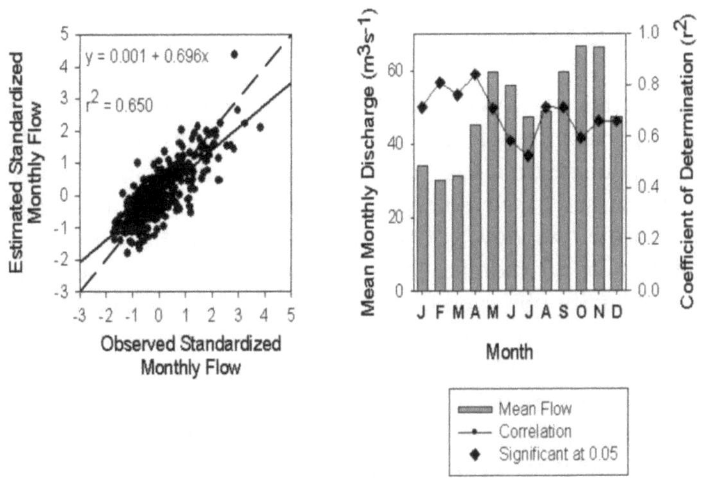


*Figure 5.* Graphical summary of the goodness of fit of the MARS model to observed monthly flows during the period of model calibration (1956–87). Observed and fitted monthly flows, standardized by the appropriate monthly means and variances, are shown to the left. The performance of the model relative to the mean monthly hydrograph is shown to the right.

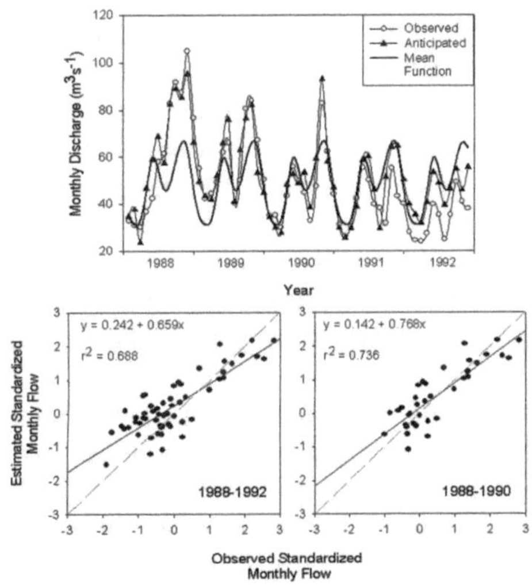


*Figure 6.* Summary of the MARS model's performance during the period of model validation (1988–92). Time series of observed and anticipated monthly flows are shown (upper) in comparison to the long-run mean monthly hydrograph. Comparisons of standardized monthly flows, observed and anticipated, during the entire period are shown in the lower left and those for the period excluding the 1991–92 warm phase ENSO event are shown in the lower right.

## 5. FORECASTS

Estimates of flows during the model validation period are made based upon perfect knowledge of the historic values of the predictor variables at the requisite time lags. Ultimately the model would be used to make real-time forecasts based solely upon current and former macroclimatic variables. As the "window" of the forecast extends further into the future, for example, 6 or 12 months, the likelihood of deviations of the forecasts from the eventually observed discharges might be expected to increase. On the other hand, the value of the forecasts to the operation of the reservoir increases as the forecast horizon expands. In order to replicate these circumstances, the MARS model was applied over the period 1985¯92 and forecasts were made at 6-month increments for the coming 6 months, based on the necessary variables available at the time of forecasting. Figure 7 provides a comparison of real-time forecasts of future monthly values of streamflow using this 6-month "forecast-horizon." The coefficient of determination drops to 0.412 (still significant), and again the model performs least well in forecasting the lower flows associated with the 1991–92 El Niño.

Model performance drops off dramatically ($r^2$ = 0.136) when the "forecast horizon" is extended to 12 months (Fig. 8). This explained variance is still statistically significant, however, and represents an almost 14% reduction in the interannual variability that had previously been unaccounted for. At this "horizon" it is possible to see the degree to which the forecasts can be in error as time since the forecast was made increases. For instance, at the end of 1987 the model "foresaw" the magnitude of the impact of the coming 1988 La Niña but was in error in forecasting the exact time of rise in streamflow, which took place halfway into the forecast window. Similarly, forecasts made at the end of 1990 appear to have totally missed the decline in flows at the end of that year resulting from the initiation of the 1991–93 El Niño event.

In order to provide some indication of the potential economic importance of these longer-term forecasts, the methodology, employing a 12-month forecast horizon, has been applied retrospectively to the major hydroelectric power generating schemes in Colombia over the period 1977–92 (Rendón, 1997). All reservoirs are operated in conjunction, under present operating constraints, in such a way as to maximize the production of national energy and to minimize the costs of operation and maintenance of the generating system, through adaptive dynamic programming. Figure 9 compares the observed operating costs of the national power generating system and the costs anticipated had the MARS streamflow forecasting procedure been used together with adaptive dynamic programming. Since both the forecast and historic data result from similar centralized control and optimization criteria, the primary difference between them and the modeled figures arises from the use of MARS to model streamflow as opposed to the extant Markov linear forecasting procedures. In the long run, savings in operating costs seem to be between 35% and 40%, although for some years, notably 1978, utilization of the combined methodologies leads to increased costs. These observations illustrate the potential economic benefits of incorporating ENSO and other macroclimatic variables associated with inter- and intra-annual variability, in hydrologic short-to-long term forecasting.

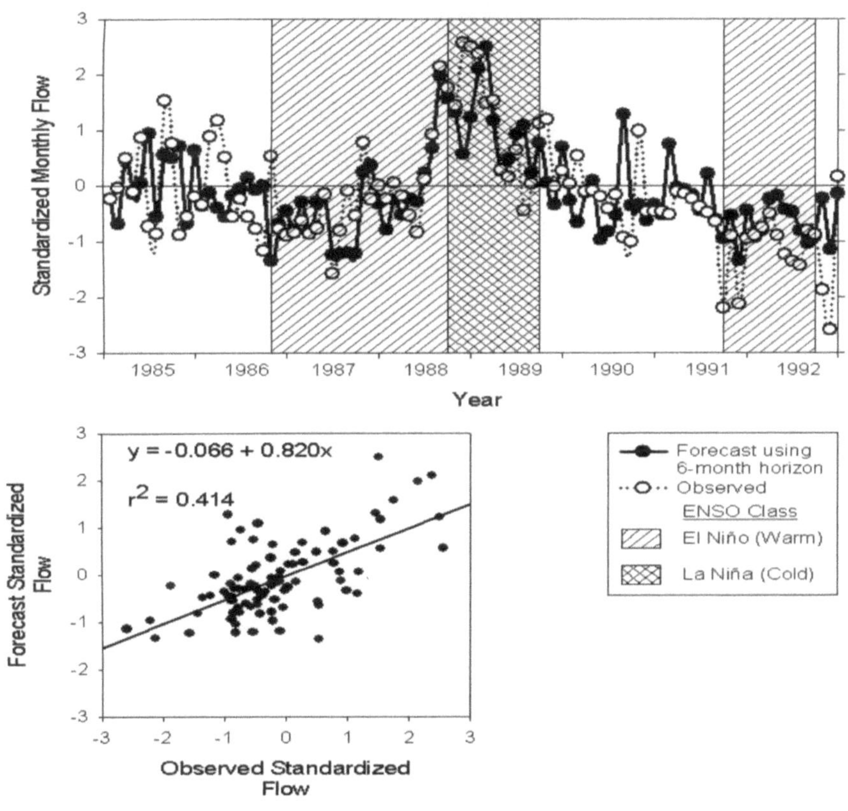


*Figure 7.* Time series of observed and forecasted standardized monthly flows during the period 1985–92 using a forecast window of 6 months (upper). Comparative plot of standardized observed and forecasted monthly flows using the same forecast window (lower). El Niño and La Niña years are identified by using the classification scheme of the Center for Ocean-Atmospheric Prediction Studies, Florida State University.

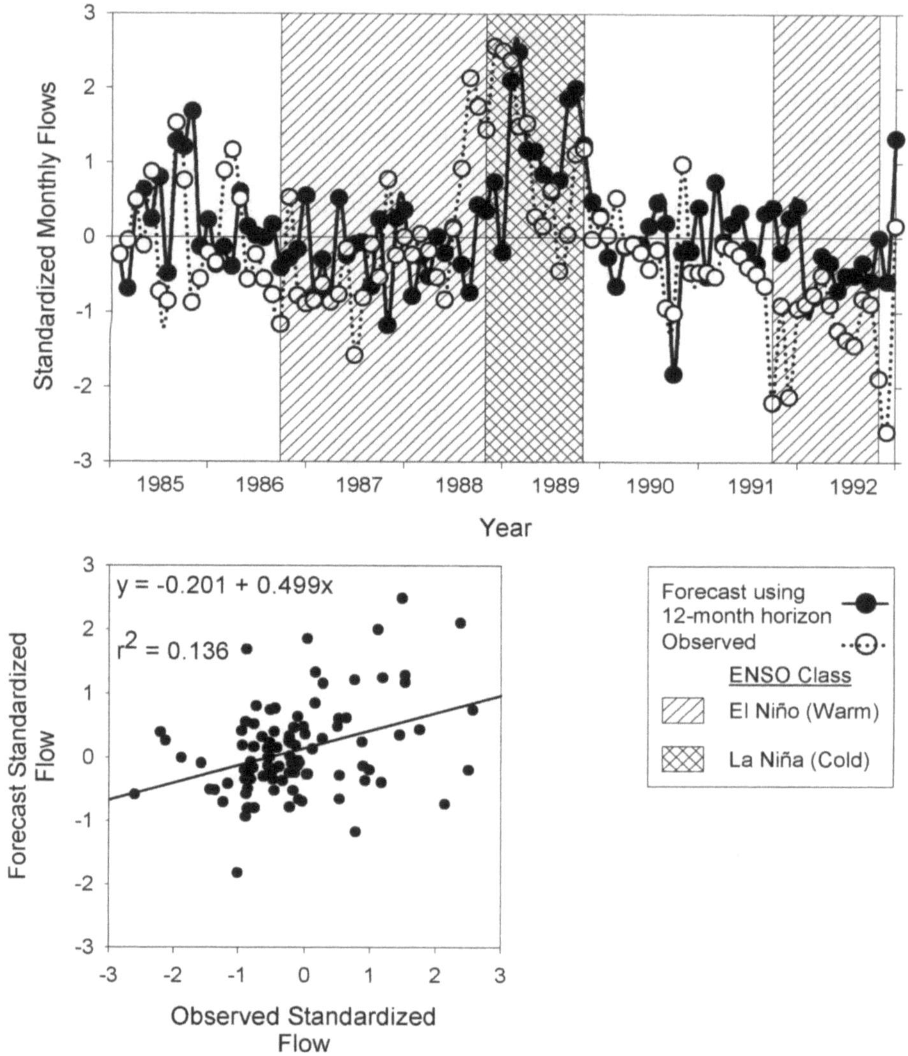


*Figure 8.* Same as Figure 7 except using a forecast window of 12 months. The solid symbols represent forecasts made with a 12-month lead time. The open circles pertain to the corresponding observations.

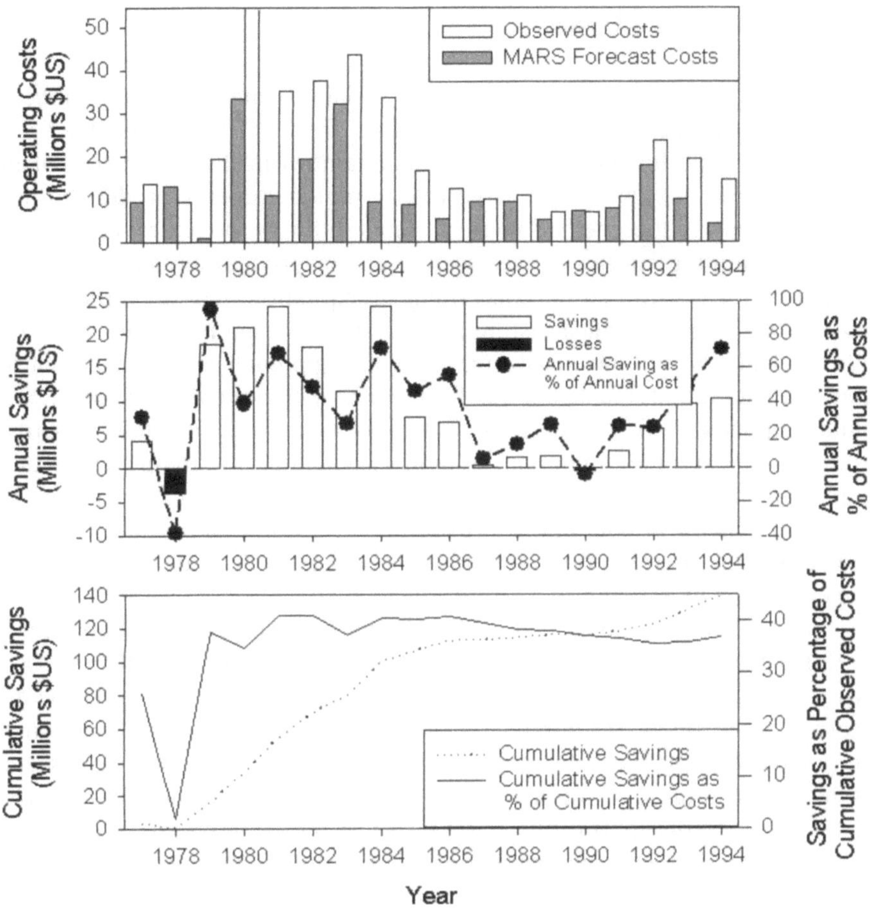


*Figure 9.* Summary of observed operating costs of the Colombian national electricity generating system, 1977–94, and those derived by using the modeling techniques discussed in the text (upper). Absolute annual savings and savings expressed as percentages of observed annual costs, which would have resulted from the use of the methodology (middle). Cumulative absolute savings and cumulative savings as percentages of cumulative absolute savings during the same period (lower).

# 6. CONCLUSIONS

Given the complexity of climatic variability and potential nonlinearity of the streamflow response in Colombia, the MARS models appears to offer some hope for the reduction of uncertainties in the forecasting of inflows to reservoirs that play a large role in regional energy production. Implementation of the methodology requires little or no extra investment. Power generating entities (governmental and/or private) already collect the necessary local information concerning precipitation and streamflow. The macroscale climatic indicators are readily available in "real time" from the World Wide Web. One disadvantage of the approach is that it has fairly sizeable data requirements and the form of the model would need to be constantly updated. This updating imposes only a small administrative burden and may be extremely important in light of potential low-frequency decadal signals in many climatic responses, as well as the potential for anthropogenic influences on both climate (particularly through global warming) and the responses of streamflow to land use changes such as deforestation and the expansion of agriculture.

It should be noted that in this particular case, the model appears to perform least well, although not intolerably so, during the high-flow season, when the influence of climatic inputs is at its greatest, as opposed to the dry season, when flows are dominated by the drawdown of hydrologic stores from the preceding wet season. Perhaps more significantly, the model seems to be better attuned to the excess streamflows associated with cold phase ENSO than the droughts of the warm phase. Although forecasts of excessive rainfalls may be useful in a number of other hydroclimatological fields, such as flood forecasting, there are few economic savings to be made in reservoir operation during years when water is abundant, as it was in 1988.

Improved long-term forecasts (up to 12 months in advance) of reservoir inputs may help create a more reliable supply of energy, an essential element of a healthy economy and regional stability. In the past, such as in 1991, lack of energy and blackouts have coincided with other socioeconomic consequences of ENSO in related fields of agriculture, livestock, human health, and forest fires, leading to considerable social disruption and political unrest. Although the modeling technique may well be applicable in many locations, the neighboring nations of Costa Rica, Panama, and Venezuela—all of which experience similar hydrologic responses to the ENSO signal and have a heavy reliance upon hydropower (Fig. 1)—would appear to have the most to gain from application of the methodology.

## 7. ACKNOWLEDGMENTS

We thank Carlos David Hoyos for his assistance in running the MARS model, and Luis Fernando Carvajal, Luz Adriana Cuartas, and Janeth Barco for their help with adaptive dynamic programming. The authors are grateful for the helpful comments and suggestions on the original manuscript by Shaleen Jain. The research was supported by Grant ISP3-22 from the Inter-American Institute (IAI).

## 8. REFERENCES

Abarbanel, H.D.I., and U. Lall. 1996. Nonlinear dynamics of the Great Salt Lake: System identification and prediction. *Climate Dynamics* 12: 287–297.

Abarbanel, H.D.I., U. Lall, M.E. Mann, Y. Moon, and T. Sangoyomi. 1996. Nonlinear dynamics and the Great Salt Lake: A predictable indicator of regional climate. *Energy* 21: 655–665.

Chang, P., B. Wang, T. Li, and L. Ji. 1994. Interactions between the seasonal cycle and ENSO-frequency entrainment and chaos in a coupled atmosphere-ocean model. *Geophysical Research Letters* 21: 2817–2820.

D'Odorico, P., L. Ridolfi, A. Porporato, and I Rodriguez-Iturbe. 2000. Preferential states of seasonal soil moisture: The impact of climate fluctuations. *Water Resources Research* 36: 2209–2219.

Enfield, D.B., and E.J. Alfaro. 1999. The dependence of Caribbean rainfall on the interaction of the tropical Atlantic and Pacific Oceans. *Journal of Climate* 12: 2093–2103.

Friedman, J.H. 1991. Multivariate adaptive regression splines. *The Annals of Statistics* 19: 1–141.

Friedman, J.H., and B.W. Silverman. 1989. Flexible parsimonious smoothing and additive modeling. *Technometrics* 31: 3–39.

Giannini, A., Y. Kushnir, and M.A. Cane. 2000. Interannual variability of Caribbean rainfall: ENSO and the Atlantic Ocean. *Journal of Climate* 13: 297–311.

Gutiérrez F., and J.A. Dracup. 2001. An analysis of the feasibility of long-range streamflow forecasting for Colombia using El Nino–Southern Oscillation indicators. *Journal of Hydrology* 246: 181–196 .

Hastenrath, S. 1990. Diagnostics and prediction of anomalous river discharge in northern South America. *Journal of Climate* 3: 202–215.

Hense, A. 1987. On the possible existence of a strange attractor for the Southern Oscillation. *Beitr. Phys. Atmosph.* 60: 34–47.

Hurrell, J.W. 1995. Decadal trends in the North Atlantic Oscillation: Regional temperatures and precipitation. *Science* 269: 676–679.

ISA, Informe de Operación. Annual Publication of Interconexión Eléctrica S. A., Medellín, Colombia. [Latest available on line at http://www.isa.com.co/publicaciones/InformeAnual/azul/index.htm].

Jin, F.-F., J.D. Neelin, and M. Ghil. 1994. El Niño on the Devil's Staircase: Annual subharmonic steps to chaos. *Science* 264: 70–72.

Jones, P.D., T. Jonsson, and D. Wheeler. 1997. Extension using early instrumental pressure observations from Gibraltar and SW Iceland to the North Atlantic Oscillation. *International Journal of Climatology* 17: 1433–1450.

Lall, U., T. Sangoyomi, and H.D.I. Abarbanel. 1996. Nonlinear dynamics of the Great Salt Lake: Nonparametric short-term forecasting. *Water Resources Research* 32: 975–985.

Latif, M. 2001. Tropical Pacific/Atlantic Ocean interactions at multi-decadal time scales. *Geophysical Research Letters* 28: 539–542.

Lewis, P.A.W., and B. Ray. 1997. Modeling nonlinearity, long-range dependence, and periodic phenomena in sea surface temperatures using TSMARS. *Journal of the American Statistical Association* 92: 881–893.

Mantua, N.J., S.R. Hare, Y. Zhang, J.M. Wallace, and R.C. Francis. 1997. A Pacific interdecadal climate oscillation with impacts on salmon production. *Bulletin of the American Meteorological Society* 78: 1069–1079.

Marquardt, C., and B. Naujokat. 1997. An Update of the Equatorial QBO and its Variability. 1st SPARC General Assembly, Melbourne, Australia, WMO/TD No. 814, Vol. 1, 87¯90.

Neelin, J.D., D.S. Battisti, A.C. Hirst, F-F. Jin, Y. Wakata, T. Yamagata, and S.E. Zebiak. 1998. ENSO theory. *Journal of Geophysical Research* 103(C7): 14261–14290.

Poveda, G., and O.J. Mesa. 1997. Feedbacks between hydrological processes in tropical South America and large scale oceanic-atmospheric phenomena. *Journal of Climate* 10: 2690–2702.

Poveda, G., and O.J. Mesa. 2000. On the existence of Lloró (the rainiest locality on Earth): Enhanced ocean-atmosphere-land interaction by a low-level jet. *Geophysical Research Letters* 27: 1675–1678.

Poveda, G., and W. Rojas. 1996. Impacto del fenómeno El Niño sobre la intensificación de la malaria en Colombia, Memorias XII Congreso Colombiano de Hidrología, Sociedad Colombiana de Ingenieros, Bogotá, 647 –654.

Poveda, G., A. Jaramillo, M.M. Gil, N. Quiceno, and R. Mantilla. 2001a. Seasonality in ENSO related precipitation, river discharges, soil moisture, and vegetation index (NDVI) in Colombia. *Water Resources Research* 37: 2169–2178.

Poveda, G., W. Rojas, I. D. Vélez, M. Quiñones, R. I. Mantilla, D. Ruiz, J. Zuluaga, and G. Rua. 2001b. Coupling between Annual and ENSO timescales in the Malaria-Climate association in Colombia. *Environmental Health Perspectives*, 109, 489–493.

Rendón, L.D. 1997. Beneficios de la Predicción Hidrológica en la Sector Eléctrico Colombiano, Considerando la Variabilidad Climática. Unpublished Masters Thesis, Programa de Posgrado en Approvechamiento de Recursos Hidráulicos, Facultad de Minas, Universidad Nacional de Colombia, Medellín, Colombia, 174 pp. (original in Spanish).

Roberts, L., 1998. World Resources 1998–99, Oxford University Press, Oxford, England, 369 pp.

Ropelewski, C.F., and M.S. Halpert. 1987. Global and regional precipitation patterns associated with El Niño/Southern Oscillation. *Monthly Weather Review* 115: 1606–1626.

Sangoyomi, T., U. Lall, and H.D.I. Abarbanel. 1996. Nonlinear dynamics of the Great Salt Lake: Dimension estimation. *Water Resources Research* 32: 149–160.

Schultz, D., W.E. Bracken, and L.F. Bosart. 1998. Planetary- and synoptic-scale signatures associated with Central American cold surges. *Monthly Weather Review* 126: 5–27.

Trenberth, K.E. 1976. Spatial and temporal variations of the Southern Oscillation. *Quarterly Journal of the Royal Meteorological Society* 102: 639–653.

Trenberth, K.E., G.W. Branstator, D. Karoly, A. Kumar, N.-C. Lau, and C. Ropelewski. 1998. Progress during TOGA in understanding and modeling global teleconnections associated with tropical sea surface temperatures. *Journal of Geophysical Research* 13(C7): 14291–14324.

Vallis, G.K. 1986. El Niño: Chaotic dynamical system? *Science* 232: 243–245.

Wolter, K. 1987. The Southern Oscillation in surface circulation and climate over the tropical Atlantic, eastern Pacific, and Indian Oceans as captured by cluster analysis. *Journal of Climate and Applied Meteorology* 26: 540–558.

16

# CLIMATE VARIABILITY AND CLIMATE CHANGE, AND THEIR IMPACTS ON THE FRESHWATER RESOURCES IN THE BORDER REGION

*A Case Study for Sonora, Mexico*

Víctor O. Magaña and Cecilia Conde

*Center for Atmospheric Sciences, National Autonomous University of Mexico (UNAM), Mexico City 04510, Mexico*

**Abstract** Climate change scenarios and recent studies on climate variability refer to important low-frequency variations of precipitation regimes for the northern regions of Mexico. Data on precipitation, streamflow, and water levels in storage reservoirs are used to analyze tendencies and changes in water availability during the last 30 years. El Niño/Southern Oscillation (ENSO) events become of particular interest since El Niño boreal winters generally result in more precipitation and increased water availability in northwestern Mexico. However, summer precipitation in the region does not show a clear El Niño signal. Furthermore, the interannual variability of precipitation associated with the Mexican monsoon does not appear to be modulated by sea surface temperature (SST) variations.

Historically, the economic development of the state of Sonora has been constrained by inequitable distribution of transboundary freshwater resources, a condition that may seriously worsen in the future. This issue will be particularly important in the design of strategies to adapt to or ameliorate the negative impacts of climate change. The predicted decline in water availability may exacerbate the trend toward increasing competition for water resources, namely, between productive activities, such as agriculture and industry (particularly *maquiladoras*), and domestic consumption by the increasing population in urban areas. These conditions make northern Mexico one of the most vulnerable regions in the country to climate changes.



*Henry F. Diaz and Barbara J. Morehouse (eds.), Climate and Water*, 373-391.

# 1. WATER RESOURCES AND WATER AVAILABILITY IN MEXICO

In Mexico, national and regional development, both rural and urban, is highly dependent on limited water resources. In contrast, political and/or economic regional plans or programs generally consider water as a constant, or as an unlimited resource that must satisfy an increasing population. However, current data contradict this view. A National Water Commission (Comisión Nacional del Agua, CNA, 1999) analysis of 96 aquifers showed that 14.1 x $10^5$ $m^3$ of water are extracted annually, while only 9.4 x $10^5$ $m^3$ are recharged. Moreover, scenarios of future climate and natural climate variability events indicate that rainfall could have important temporal (interannual) and spatial changes, even if the global annual precipitation increases. Water management, from a sustainable development perspective, should consider the mitigation of future negative climatic change impacts, in order to assure that development between regions or productive sectors is nondiscriminatory.

In 1999, the Minister of Environment and Natural Resources and Fisheries (Secretaría de Medio Ambiente, Recursos Naturales y Pesca, SEMARNAP, 1999) and the CNA, calculated the total water availability for the country at 463 $km^3$ per year. Nearly half of this total (43%) is now extracted from superficial or groundwater sources and used to satisfy the needs of the agriculture (60.5 $km^3$), public and industrial supply (13.5 and 4.1 $km^3$, respectively), fisheries (1.1 $km^3$), and thermoelectric (0.2 $km^3$) sectors. Approximately 48.9 $km^3$ of the total water resources available correspond to transboundary waters—regulated by international agreements from Guatemala (47 $km^3$) and the United States (1.9 $km^3$).

The geographic distribution of water availability per capita is not homogeneous. In the southern border region of the country, 28,400 $m^3$ per capita are available. In the central basins of the north and the Río Bravo region, the average has been calculated to be 2,200 $m^3$.

In 1998, more than 80 million people in the country had access to potable water, 80% of them living in urban areas. Increasing population (Fig. 1a), particularly in urban regions, has resulted in a greater demand for public services (Fig. 1b), and pressure to increase the water supply and to improve water quality. With the growth in the region's economy, the water demand has also increased for diverse economic sectors (Fig. 2).

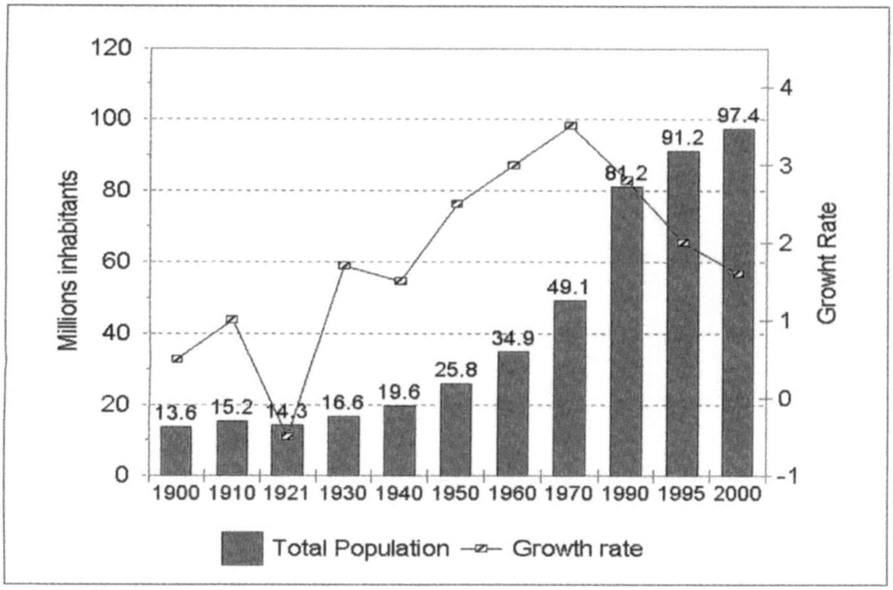


*Figure 1a.* Total population and population growth rate in Mexico, 1900–2000.

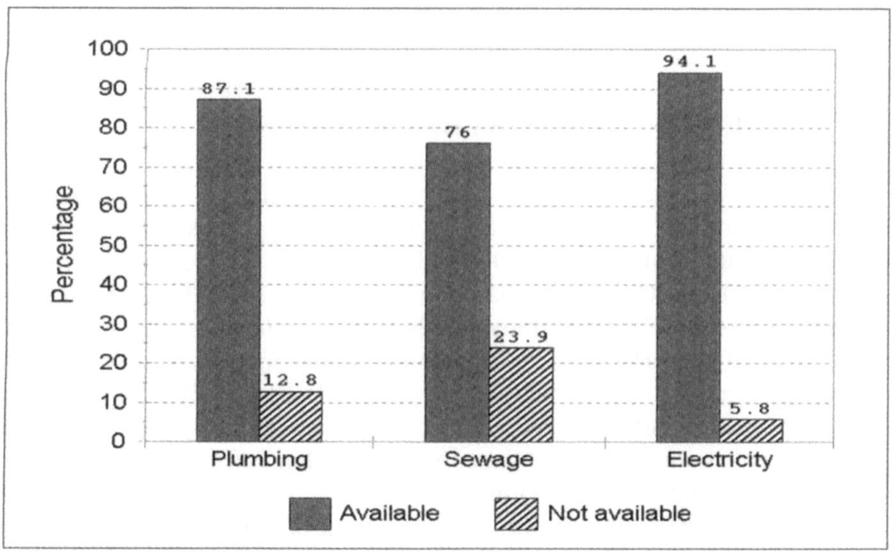


*Figure 1b.* Percentage of population in Mexico with access to potable water and public services, 1990–2000.

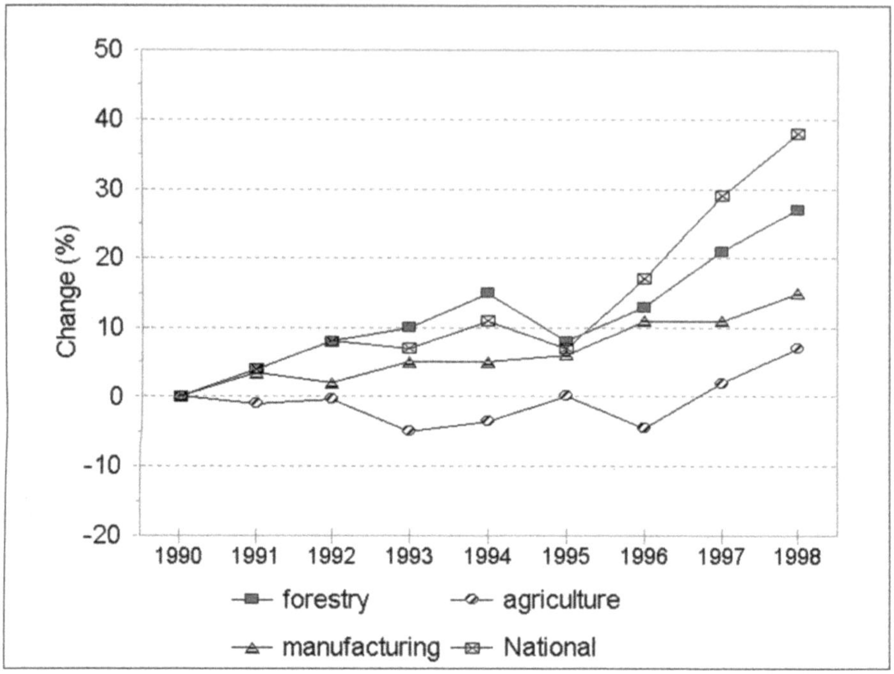


*Figure 2.* Percentage changes in the Mexican gross domestic product and for selected productive sectors: forestry, agriculture, and manufacturing, and national total.

## 2. THE MEXICAN NORTHERN BORDER

In 1983, Mexico and the United States signed the "La Paz Agreement," defining the "Border Region" as the area lying 100 km or 62.5 miles to the north and south of the 2,000 km long U.S.-Mexico border. Six Mexican states and four U.S. states share this border, its environment, and its resources. There is an intense transboundary movement of people and goods as a result of the rapid growth in population and industry (Table 1, Data from Demographic Data Viewer, 2000, and Border XXI Program web pages). Mutual benefits can be attained if greater political, economic, and scientific collaboration is established with consideration for the sustainable development of this critical region.

From 1950 to 1980, the border population almost doubled, and if the current rates of population growth continue, this population will double again by 2025 (Westerhoff 2000). In 1997, more than 10.5 million people lived along the 2,000 km border region (U.S. Environmental Protection Agency, USEPA, and SEMARNAP, 1997), 6.2 million in the United States and 4.3 million in Mexico. Almost 90% of them lived in urban areas (Table 1), in the so-called sister cities communities (Fig. 3), so named because of the proximity, and economic and social interdependence of the cities, on each side of the border. If the population trends continue, in the near future the competition for available water supplies will increase and water prices likely will also rise, forcing users and governments to apply stronger water conservation regulations. The Mexican minister of Environment, V. Lichtinger, has announced the gradual removal of subsidies for water consumption, especially for major agricultural and mining industries situated in the north of the country, "where there is no more water" (González 2001). This policy will severely increase production costs unless more efficient use and water distribution systems are developed.

*Table 1.* Population data for the Border States.

| Name | Total population, 2000* | Total population, 1997 | Urban (%), 1997 | Rural (%), 1997 | Population per square kilometer, 1997 |
|---|---|---|---|---|---|
| Baja California | 2,487,700 | 1,660,855 | 91.5 | 8.7 | 23.0 |
| Sonora | 2,213,370 | 1,823,606 | 81.5 | 18.6 | 10.1 |
| Chihuahua | 3,047,867 | 2,441,873 | 80.2 | 19.9 | 9.8 |
| Coahuila | 2,295,808 | 1,972,340 | 88.3 | 11.8 | 13.0 |
| Nuevo León | 3,826,240 | 3,098,736 | 93.0 | 7.1 | 47.5 |
| Tamaulipas | 2,747,114 | 2,249,581 | 83.3 | 16.8 | 29.0 |
| California | | 29,760,020 | 92.7 | 7.4 | 73.7 |
| Arizona | | 3,665,228 | 87.5 | 12.5 | 12.5 |
| New Mexico | | 1,515,069 | 72.9 | 27.1 | 4.8 |
| Texas | | 16,986,510 | 80.3 | 19.7 | 25.0 |

* Instituto Nacional de Estadística y Geografía e Informática (INEGI), 2001.

In the last decade, transboundary freshwater has become scarce, mainly because of the drought situation declared in the Mexican northern states, which allows them to have financial support from the national disaster prevention commission. In the last 4 years, Mexico has been unable to deliver its 430 million $m^3$/year obligation to the United States; the current Mexican government has agreed to deliver almost half of this debt before July 2001 (Restrepo 2001), a month when some northern regions typically experience the "canicula," or midsummer drought. This situation illustrates the necessity to develop binational agreements, which might prevent conflict between states or nations, and which could include climate variability and climate change conditions foreseen for the twenty-first century.

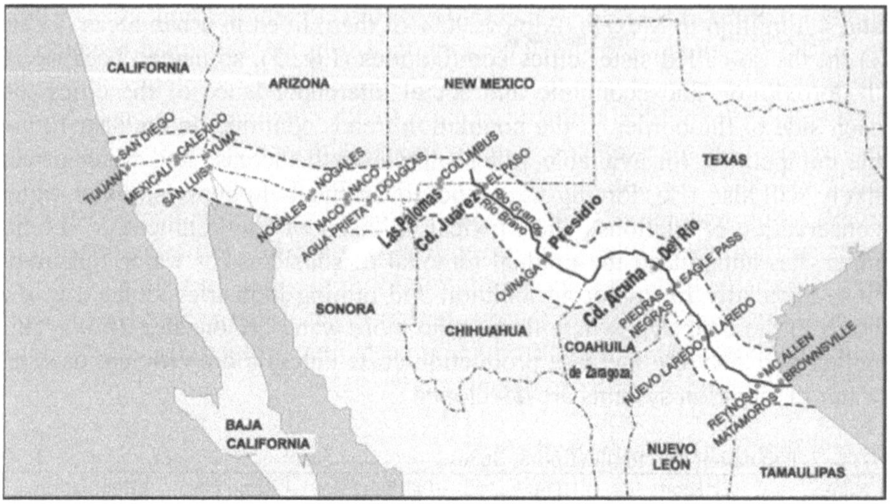


*Figure 3.* Sister cities along the Mexico-U.S. border. (Source: USEPA, SEMARNAP, 1997. The 1997 U.S.-Mexico Border Environmental Indicators report.)

The expected increase in population also indicates a possible shift in water use, now concentrated in agricultural irrigation, to domestic and industrial applications. Domestic uses may also imply an additional demand for higher water quality standards, which are higher for human consumption than for irrigation or industrial purposes.

Urban consumption and use of water is not homogeneous among municipios (municipalities) on the Mexican side of the border. Data on water use per capita in Mexicali city for 1995 shows 500 liters/day, while in Nogales, Sonora, the availability was only 183 liters/day, with only 64% of its urban population having access to potable water (Table 2).

*Table 2.* Water statistics for selected northern Mexico cities.

| City | Potable water available in 1996. In liters/day. | Population in 1996 with potable water. In percent. | Population in 1996 with sewage drainage. In percent. | Drinking water disinfected before delivery. In percent. |
|---|---|---|---|---|
| Mexicali* | 500 | 93 | 80 | 100 |
| Nogales, Son* | 183 | 64 | 81 | 100 |
| Ciudad Acuña | 372 | 89 | 39 | 100 |
| Piedras Negras | 419 | 95 | 80 | 100 |
| Matamoros | 262 | 72 | 47 | 100 |
| Reynosa | 294 | 92 | 57 | 100 |

*Sources: Comisiones Municipales de Agua Potable; Comisiones de Servicios Públicos 1995.

If we assume that standards of living will increase in the border region, then the Mexican municipios may tend to consume as much water as Mexicali in the future. Moreover, the average water consumption in the U.S. border counties in 1996 averaged 615 liters/capita/day (lpcd), 41% greater than in Mexico (Westerhoff 2000). This might indicate a future urban demand of at least twice the 1.3 x $10^6$ $m^3$ per year consumed in the U.S.-Mexico border region. All of these conditions in the region indicate that great efforts must be made to strengthen research and sustainable management of water resources. The Ministry of Environment (SEMARNAT 2001) has launched a "Crusade" for forest and water conservation in the country, as these issues are considered a matter of national security.

## 3. SONORA CASE

There are two main hydrologic water units in the state of Sonora (Celis 1992): the "sierra," which provides water for most of the state, and the costal valley, which delivers water to the southern region. In the state of Sonora the primary surface streamflows are mainly distributed from north-northwest to east-southeast. The Yaqui River drains 53% of the water toward the Gulf of California, and the Mayo River drains another 34% of the total. Runoff is stored behind 14 dams, with a total capacity of approximately $10^{10}$ $m^3$. Five of them have 98% of the total storage capacity (listed in Table 3, except Cuauhtémoc Dam). Plutarco E. Calles Dam is the largest, with a capacity of more than 3 x $10^9$ $m^3$, and is the only one in the list that is not near its limit of useful life, while the rest of them are more than 50 years old (Sánchez 2001). These dams have three main functions (Fig. 4): generation of energy; control of runoff; and water supply for livestock, agriculture, and urban purposes.

Sonora requires more than 7 x $10^{10}$ $m^3$ annually for agriculture (94% of the total demand of the state), urban uses, and mining and other industries. Storage can provide only 12% of this demand.

Dams like Angostura also are used in generating electricity. However, hydroelectric power generation represents less than a third of the total capacity installed in the country, which depends strongly on thermoelectric services. The Abelardo Rodriguez Dam has serious water quality problems. It receives solid waste from the urban areas, swine farms, and industries of Hermosillo.

The climate in the northwestern part of the state is dry, and therefore there are no permanent flows of surface water. Consequently, groundwater management becomes an issue of vital importance. Such management includes a strategy to reduce the rapid increase in the salinity of this resource.

Table 3. Principal dams in Sonora.

| Name of Dam (Name of nearest town) | Location | Latitude(°N) | Longitude (°W) | Altitude(m) |
|---|---|---|---|---|
| Presa Alvaro Obregón (Oviachic) | Yaqui River | 27.82 | 109.9 | 70 |
| Presa Lázaro Cárdenas (La Angostura) | Yaqui River | 30.45 | 109.38 | 965 |
| Presa Abelardo Rodríguez (Hermosillo) | Sonora River | 29.08 | 110.93 | 211 |
| Presa Adolfo Ruiz Cortines (Mocúzari) | Mayo River | 27.22 | 109.12 | 135 |
| Presa Cuauhtémoc (Sta. Teresa) | Concepción River-A. Cocóspera | 30.87 | 111.53 | 590 |
| Presa Plutarco E. Calles (El Novillo) | Yaqui River | 29.9 | 110.6 | 596 |

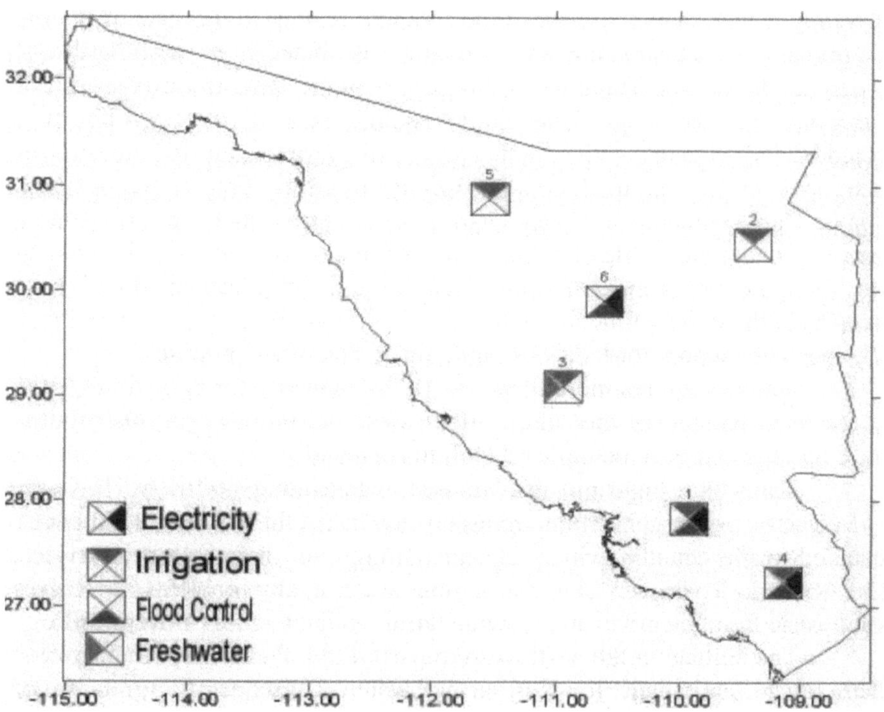


*Figure 4.* Principal dams in Sonora. Numbers match the first column in Table 3.

# 4. CLIMATE VARIABILITY, CLIMATE CHANGE, AND WATER RESOURCES IN SONORA

## 4.1. Mean Precipitation in Mexico

The U.S.-Mexico border is characterized by a semiarid climate, with higher precipitation towards both coasts (Magaña 2001). Evaporation exceeds precipitation almost all the year, causing soil moisture deficits.

Precipitation over the northern part of Mexico is relatively low compared to the average for the country, 772.7 mm (Fig. 5). The annual average precipitation amounts (1941–99) for the northern states were: Baja California, 448 mm; Sonora, 428 mm; Chihuahua, 423 mm; Coahuila de Zaragoza (or usually named just Coahuila), 314 mm; Nuevo León, 591 mm; and Tamaulipas, 764 mm. Low to very low precipitation rates exist in almost half of Mexico, especially in the northwest border regions of Sonora and Baja California, where annual mean precipitation ranges between 1 and 100 mm. The summer monsoon is one of the main sources of precipitation in northwestern Mexico, while winter storms originating in Alaska seem to play an important role in that season, particularly for the state of California (Diaz and Anderson 1995).

Temperatures ranging from 10° to 28°C can be found in the western end of the border, with the Arizona-Sonora border displaying a much more extreme and arid climate, with only 80 mm of precipitation and maximum temperatures reaching 45°C.

On a seasonal basis, 81% of the total precipitation in the country occurs from May through October, while the other 19% take place from November through April (Table 4). In the northwestern part of Mexico, summer rain is related to the North American monsoon (Douglas et al. 1993). The passage of mid-latitude systems, such as cold fronts, determines the winter rains.

*Table 4.* Monthly mean precipitation for Mexico.

| Month | Precipitation (mm) | Month | Precipitation (mm) |
|---|---|---|---|
| November | 32 | May | 40 |
| December | 30 | June | 103 |
| January | 27 | July | 138 |
| February | 18 | August | 137 |
| March | 15 | September | 141 |
| April | 19 | October | 73 |

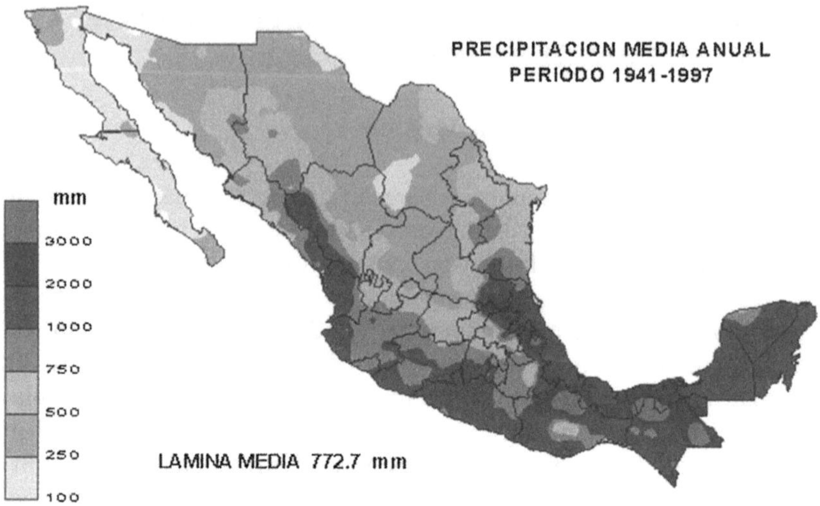


*Figure 5.* Annual mean precipitation for Mexico for the period 1941–97. The weighted average annual value for Mexico is approximately 773 mm. (Source: Comisión Nacional del Agua (CNA). http:/smn.cna.gob.mx/productos/map-lluv/GRAF-NAL.gif.).

In Sonora, the mean annual cycle precipitation exhibits a pronounced summer season maximum, with the highest amounts occurring during July and August (Fig. 6a). This annual cycle in water availability is reflected in the streamflow and in water levels at the Angostura Dam. Even though water availability is also determined by management policies, climate variability during the summer season can severely affect water consumption. Also, winter precipitation plays an important role in the recovery of the dam's water level, since processes like evaporation and irrigation reduce storage.

The precipitation regime described above can also be seen in San Luis Río Colorado (with county's meteorological station in Riíto, Sonora), and in Yuma, Arizona, two sister cities at the border. In Figure 6b, a 29-year monthly average of precipitation can be observed; it is clear that the rainfall in July and August plays an important role in determining seasonal totals. This behavior in the precipitation distribution can also be seen over a longer period of time (80 years) for Yuma. Nevertheless, the patterns for these two locations are not completely similar, particularly for the month of August, when Yuma seems to receive greater precipitation.

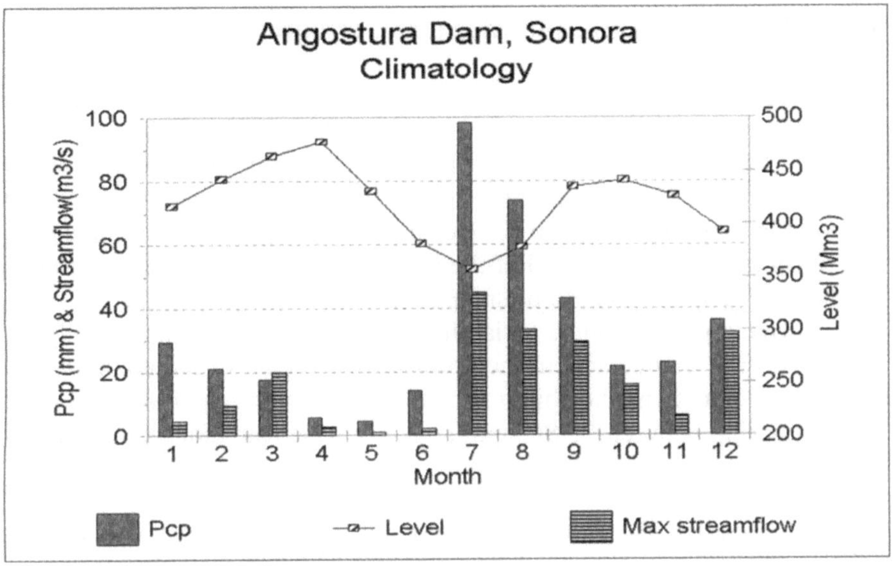


*Figure 6a.* Climatology of monthly precipitation (millimeters), streamflow (cubic meters per second), and dam level (million cubic meters) at La Angostura hydrological station.

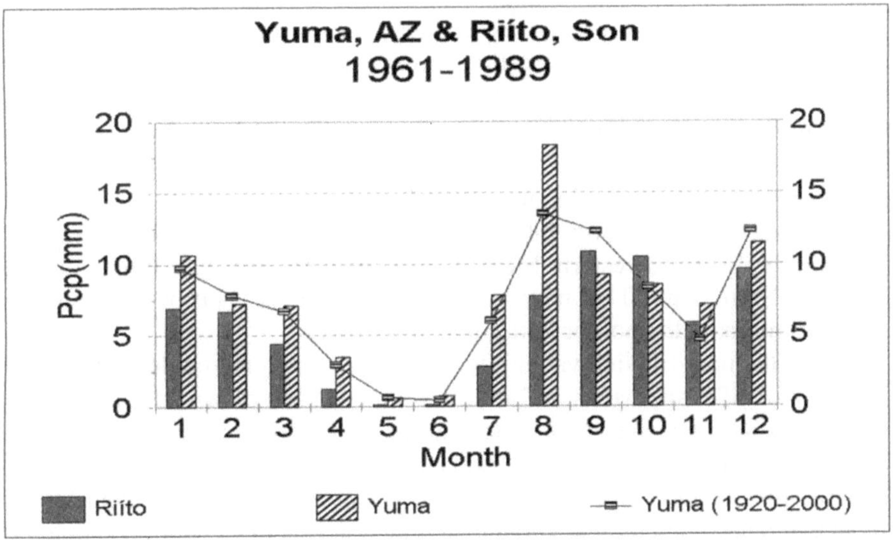


*Figure 6b.* Averaged monthly precipitation for Riíto, Sonora, and Yuma, Arizona, for 1961–89. The line corresponds to the averaged values for Yuma, for 1920–2000.

## 4.2. Interannual and Interdecadal Variability

One of the most challenging problems in water resources management is related to the interannual variability of precipitation. As was proposed by Magaña and Quintanar (1995), precipitation is modulated, on an interannual basis, by the occurrence of El Niño and La Niña events. These phenomena may explain up to 25% of the variability in monthly precipitation in some parts of Mexico. For the northern states, El Niño/Southern Oscillation (ENSO) impacts can be summarized as shown in Table 5.

The large interannual variability in winter precipitation is mainly related to ENSO, particularly in the northwestern region of the country. The problem of climate predictability for this region is an important one since the relationship between ENSO and the North American monsoon is still an open scientific question.

*Table 5.* El Niño and La Niña impacts for winter and summer along the northern Mexican states.

| ENSO event | Summer | Winter |
|---|---|---|
| El Niño (eastern Pacific sea surface temperature [SST] anomalies > 0°C) | Precipitation below normal | Precipitation above normal (in most cases) |
| La Niña (eastern Pacific SST anomalies < 0°C) | Precipitation above normal | Precipitation below normal |

In the northern part of the country, winter precipitation might even exceed the precipitation during the summer during strong El Niño events. This was the case during the 1982–83 El Niño, when the total amount of water in the six dams (listed in Table 3) was affected by a decrease in the total precipitation for the state in 1982 (Fig. 7) and was restored because of the heavy winter precipitation in 1982–83.

Although El Niño winters are associated in general with more water availability in northwestern Mexico, less precipitation has fallen during some El Niño years. This may be seen at the Huites Dam in the southern part of Sonora, at the border with Sinaloa (Fig. 8).

The inter–El Niño variability in Mexican precipitation depends on the intensity of SST anomalies and on the characteristics of winter atmospheric circulation. On the other hand, the impacts of La Niña events appear to be more consistent with less winter precipitation and normal or above normal summer precipitation and water availability. Current prediction schemes tend to take into account possible variations in the impacts of El Niño. It should be remembered that precipitation in northern Mexico also depends on hurricane activity, so regional ocean conditions should be taken into account in long-term prediction models.

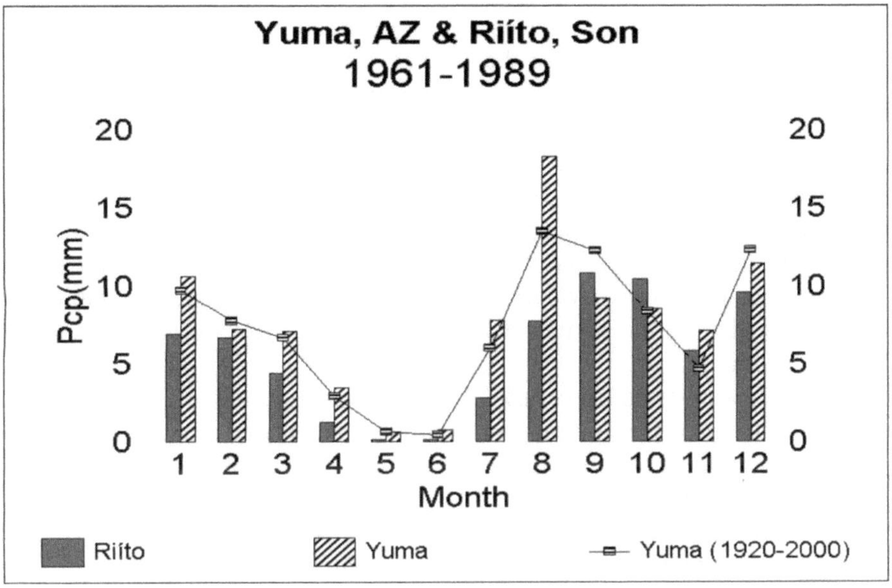


*Figure 7.* Standardized anomalies of the SST in the Niño3 region, the cumulative level of the six dams (Table 3), and the monthly precipitation for the state of Sonora.

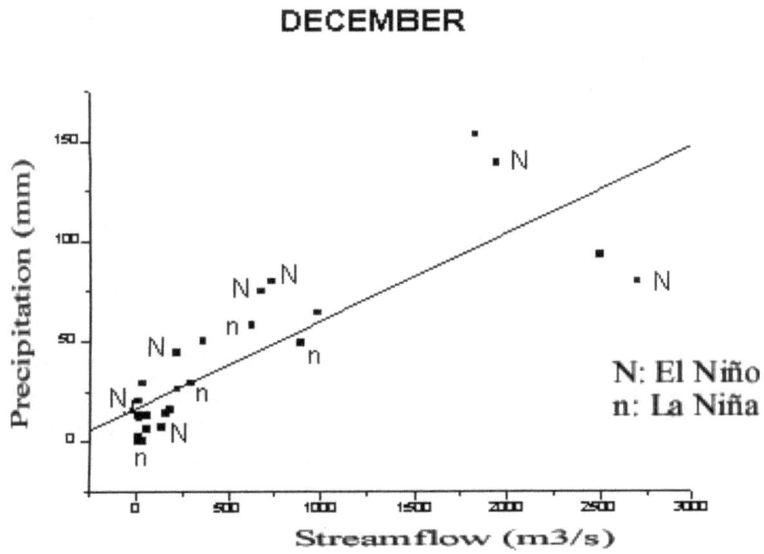


*Figure 8.* December precipitation (millimeters) vs. streamflow ($m^3/s$) at the Huites Dam (Magaña et al. 2000).

Water availability in northwestern Mexico exhibits large fluctuations on interdecadal time scales. As an example, this tendency can be observed in the precipitation for the state of Sonora since 1940 (Fig. 9a). Pronounced year-to-year fluctuations have occurred in recent decades, probably related to the more frequent occurrence of El Niño events (Fig. 9b). Even some dam level records for the same Sonoran region reflect part of this interdecadal variability (Fig. 10). This is the case for the Cuauhtemoc Dam, whose relative minima appear to be correlated to some extent to fluctuations in global surface temperature. In any event, there is a slight positive trend in the previous time series, which appears to suggest more water availability in recent decades, as the period of the 1950s to mid-1960s was exceptionally dry in the region.

As was stated previously, El Niño events during winter correspond generally to more precipitation in northwestern Mexico (Fig. 11a). The increases in streamflow at some locations, such as the Altar River (Fig. 11b), are mostly related to larger values of streamflow in winter, when the maximum values generally reflect the occurrence of extreme events, many of which are associated with the ENSO phenomenon.

More frequent and strong El Niño events have been experienced during the last two decades. Records indicate that there is a tendency for increased levels at some Mexican water reservoirs (Fig. 10). However, some very low water levels have been reached in recent years along Border States, associated with severe drought. Such large fluctuations in water levels, related to interannual variability in climate, must be an important element in water management planning.

## 4.3. Climate Change

Regional climate change scenarios for Mexico were generated by using statistical downscaling techniques applied to General Circulation Model- (GCM) outputs, for conditions of a possible duplication of carbon dioxide (2 x $CO_2$) in the future, associated to global climate change (Magaña et. al. 1997). The results indicate that Mexico will probably experience less or unchanged summer precipitation, but increased precipitation during winter. This conclusion is consistent with the recent climate record, which has exhibited more frequent strong El Niño events. Under these climate change scenarios, more water may be available along the border in winter.

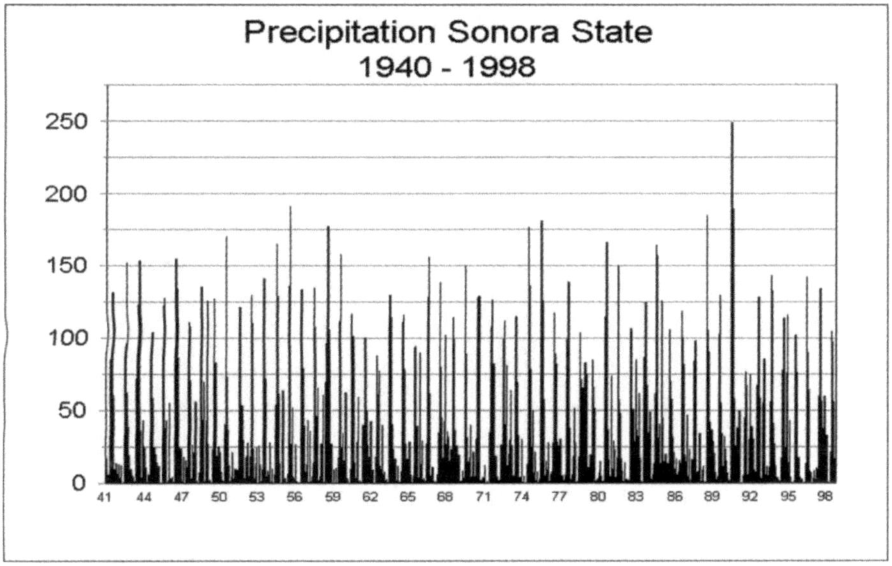


*Figure 9a.* Monthly precipitation (millimeters) for the period 1940–98 for the state of Sonora.

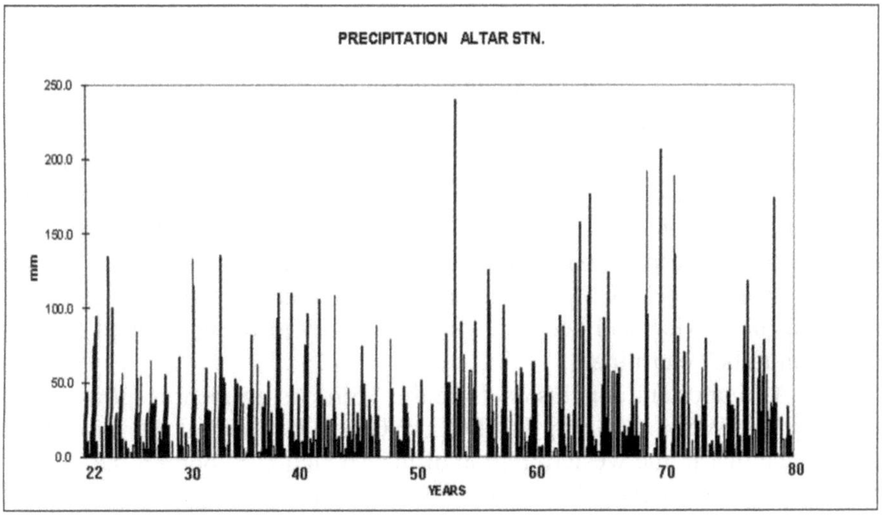


*Figure 9b.* Monthly precipitation (millimeters) for the period 1922–80 for Altar, Sonora.

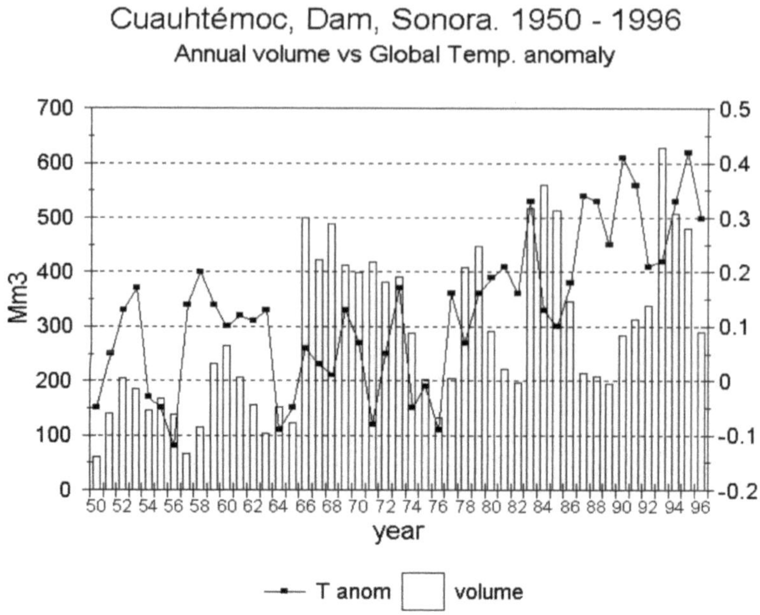


*Figure 10.* Annual volume (million cubic meters) for the period 1950–96 at Cuauhtémoc Dam, in northern Sonora, and global ocean and land surface temperature anomalies.
(http://www.ncdc.noaa.gov/ol/climate/research 1998/anomalies/annual_land.and.ocean.ts)

Climate change scenarios were also generated with a combination of a simple climate model, MAGICC, and a climate scenario database, SCENGEN (Hulme et al. 2000). The results indicate that a 10% to 20% increase in precipitation may occur after 2050 during winter in the northwestern regions of the country, under the emission scenario IS92a—defined by the Intergovernmental Panel on Climate Change (IPCC) in 1995—and applying simple interpolation techniques to the outputs of the Hadley Center Unified Model 2 (HadCM2). Whether this potential increase in freshwater resources in Mexico will keep up with the rapidly growing demand for water is still to be determined.

An important conclusion from the Third Assessment Report (TAR) of the IPCC, Working Group II (IPCC, WGII, 2001), is that there is a high confidence that in the western region of the Mexico-U.S. border, the "snowmelt-dominated watersheds... will experience earlier spring peak flows and reduction in summer flows...". Under these climatic change conditions, increased water storage during winter, for use during the summer low-flow season, would be a feasible adaptation measure, if management programs and infrastructure are planned accordingly.

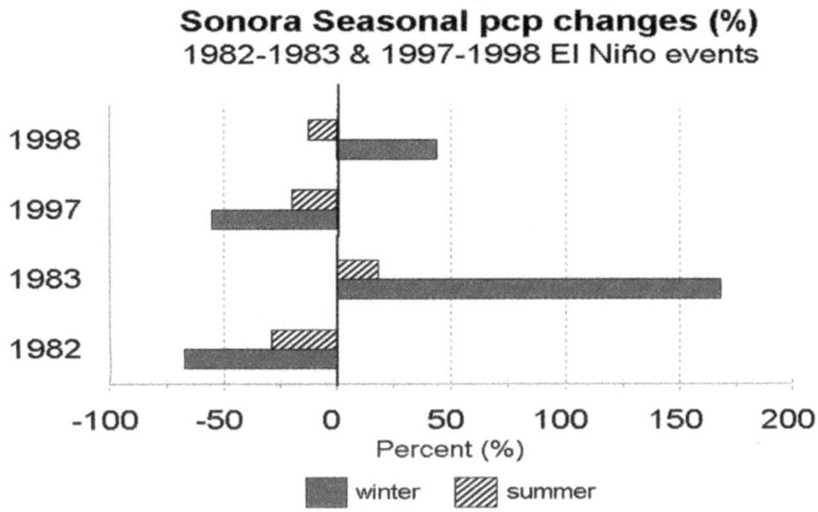


*Figure 11a.* Changes in precipitation during summer and winter in the state of Sonora for two very strong El Niño events. Winter changes are nearly double those in summer.

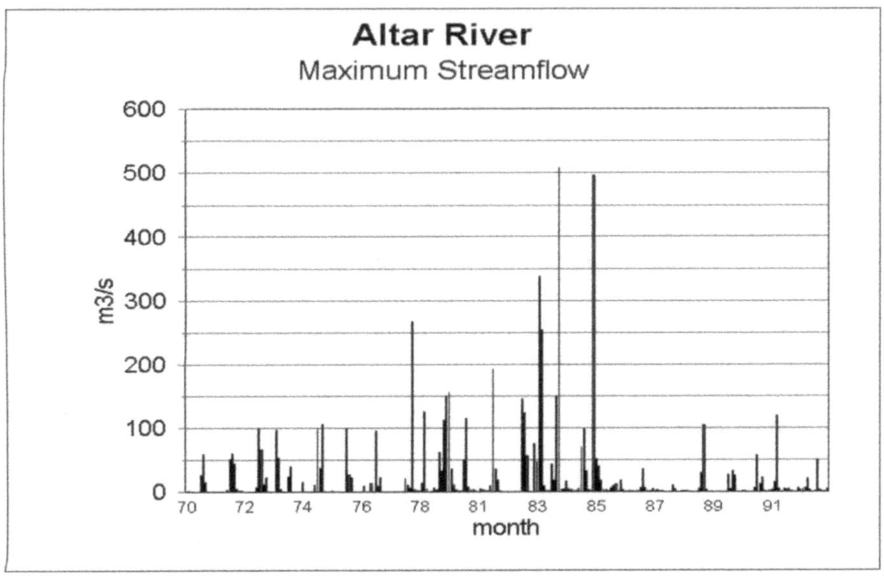


*Figure 11b.* Maximum monthly streamflow at Altar River, Sonora.

## 5. CONCLUSION

Given the nature of observed climatic variability and the range of possible future climatic scenarios, freshwater resources in the Mexico-U.S. border region are highly vulnerable. This vulnerability could be exacerbated because of increasing demand and competition, and declining water quality. Water management could be improved if social and governmental experiences during El Niño years are taken into account and integrated into binational strategies and future agreements. As part of these strategies, improved seasonal precipitation forecasts must be used to improve the decision-making processes for socioeconomic activities and plans. Also, new infrastructure, maintenance, and improvements of water treatment plants should be promoted, as well as more efficient technologies for increased water reuse.

Future research projects could extend the analyses of water availability in relation to the occurrence of ENSO events, and cooperative programs that monitor atmospheric and oceanic conditions in the region should be strengthened. More efficient use of water in the border regions should include new technologies but also more conscious and efficient use of water by the populations of both countries. Binational education activities related to this subject could be an important adaptation to possible climatic change in the future.

## 6. ACKNOWLEDGMENTS

This research was partially supported by the Interamerican Institute for Global Change Research (IAI) Collaborative Research Network 73 Grant to study climate variability in the Mexico, Central America, and Caribbean region. We also wish to acknowledge Henry Diaz and Carlos Gay for their comments on this chapter.

## 7. REFERENCES

Border XXI Program. http://www.epa.gov/usmexicoborder/ef.htm

Celis, P. 1992. Diagnóstico de la contaminación del agua en el estado de Sonora. In, *Ecología, Recursos Naturales y Medio Ambiente en Sonora.* José Luis Moreno (compiler). El Colegio de Sonora–Secretaría de Infraestructura Urbana y Ecología. Hermosillo, Sonora, pp. 155–187.

Comisión Nacional del Agua (CNA). 1999. *Compendio Básico del agua en México.* CNA, México. Secretaría de Medio Ambiente recursos Naturales y Pesca (SEMARNAP), México.

Demographic Data Viewer v3.1, U.S.-Mexico Version Copyright 2000 Center for International Earth Science Information Network (CIESIN) & The Trustees of Columbia University, New York. http://plue.sedac.ciesin.org /plue/ddviewer/ddv30-USMEX/

Diaz, H.F., and C.A. Anderson. 1995. Precipitation trends and water consumption related to population in the southwestern United States: A reassessment. *Water Resources Research* 31: 713–720.

Douglas, M.W., R. Maddox, K. Howard, and S. Reyes. 1993. The Mexican Monsoon. *Journal of Climate* 6: 1665–1667.

González, J. 18 April 2001. Busca Semarnat eliminar subsidios de agua para grandes industrias. *El Financiero*. p. 22.

Hulme, M., T.M.L. Wigley, E.M. Barrow, S.C.B. Raper, A. Centella, S. Smith, and A.C. Chiapanshi. 2000. *Using a Climate Scenario Generator for Vulnerability and Adaptation Assessments: MAGICC and SCENGEN*. Version 2.4 Workbook. Climatic Research Unit, Norwich, UK, 52 pp.

Instituto Nacional de Estadística y Geografía e Informática (INEGI). 2001. XII Censo General de Población y Vivienda, 2000. http://www.inegi.gob.mx/. Preliminary results.

Intergovernmental Panel on Climate Change, Working Group II, (IPCC, WGII). 2001. Summary for Policy Makers. Climate Change 2001: Impacts, Adaptation and Vulnerability. http://www.usgcrp.gov/ipcc/

Magaña, V. 2001. Contribution to Chapter 15: *North America*, of the IPCC Third Assessment Report, *Climate Change 2001, Impacts, Adaptations, and Vulnerability*. Cambridge, UK: Cambridge University Press, pp. 735–800.

Magaña, V., and C. Conde. 2000. Climate variability and freshwater resources in northern Mexico. Sonora: a case study. *Environmental Monitoring and Assessment,* 61: 167–185.

Magaña, V., and A. Quintanar. 1995. On the use of general circulation models to study regional climate. Universidad Nacional Autónoma de México (UNAM) CRAY Second Conference on Supercomputing, Mexico City, June 1995.

Magaña, V., C. Conde, O. Sánchez, and C. Gay. 1997. Assessment of current and future regional climate scenarios for Mexico. *Climate Research* 9(2): 107–114.

Restrepo, I. 23 Abril 2001. Sequía y mal uso del agua en la frontera. *La Jornada*. p. 20.

Sánchez, J.D. 2001. La Guerra que viene: por agua. *EPOCA* 517: 10–15.

Secretaría de Medio Ambiente, Recursos Naturales y Pesca (SEMARNAP). 1999. Estadísticas de Medio Ambiente. Información Estadística y Geográfica. Indicadores y Estadísticas Selectos.
http://beta.semarnap.gob.mx/ estadisticas_ambientales/estadisticas_am_98/agua/

SEMARNAT (Minister of Environment and Natural Resources). 2001. Crusade for Forests and Water. Available at http:// www.semarnat.gob.mx/bosque-agua/ vision.html. (Spanish).

U.S. Environmental Protection Agency (USEPA), SEMARNAP. 1997. The 1997 U.S.-Mexico Border Environmental Indicators report.
http://www.epa.gov/usmexicoborder/indica97/ic.htm

Westerhoff, P. 2000. Overview. In, Westerhoff, P. (ed.), *The U.S.-Mexican Border Environment: Water Issues Along the U.S.-Mexican Border*. Southwest Center for Environmental Research and Policy (SCERP) Monograph Series, No. 2. San Diego State University Press, pp. 1–8.

17

# LAND COVER CHANGES AND CLIMATE FLUCTUATIONS IN THE UPPER SAN PEDRO RIVER BASIN IN SONORA, MEXICO

Hector M. Arias

*Gabinete de Estudios Ambientales, A.C.,Blvd. Navarrete #125 Local 21, Colonia Valle Verde, Hermosillo, Sonora, Mexico 83240*

**Abstract** The San Pedro River is a transboundary river that originates in Sonora, Mexico, and flows north into Arizona in the United States. Previous studies using satellite imagery obtained for different years from 1973 through 1997 showed significant changes in land cover in the upper San Pedro region in Mexico. Here, we examine whether these land cover changes over the past few decades are related to climatic changes, to human impacts on the landscape, or both. In particular, this study examines rainfall fluctuations during the past 50 years for the San Pedro Basin, and contrasts those changes with land cover changes in the Upper San Pedro.

The results showed that natural vegetation types underwent significant reduction, especially during the period 1973–86. The most dramatic changes were the conversion of grasslands into mesquite and gallery forest into agricultural land. The precipitation record indicates the occurrence of a prolonged period of below average rainfall before the early 1980s, but then generally above average rainfall since then. It is possible that observed changes in land cover composition—mainly gallery forest, desert scrub, grassland, and mesquite—are at least partly related to a prolonged and severe dry period in the San Pedro River Basin, representing a delayed response to the multiyear episode of below normal precipitation. The higher rainfall in the past 20 years may have helped to eliminate negative climatic pressures on the landscape even though human impacts have continued.



*Henry F. Diaz and Barbara J. Morehouse (eds.), Climate and Water*, 393-402.

# 1. INTRODUCTION

The San Pedro River Basin (SPRB) is shared by the United States and Mexico. The river starts in the neighborhood of Cananea, Sonora, Mexico, and flows north, passing by Sierra Vista, Tombstone, Benson, and Winkelman, Arizona, where it joins the Gila River. Some characteristics and the location of the river are described in Arias (2000).

A study focused on the SPRB and dealing with land cover changes during 10-year periods, based on information available through the North American Land Cover (NALC) Program, showed significant changes in land cover from 1973 through 1997 (Kepner et al. 1998). On the Mexican side of the basin, grasslands have been displaced by mesquite, while in the United States, irrigation is displacing phreatophyte vegetation and urbanization is displacing shrubs.

Proposed factors responsible for those changes include rodents and jackrabbits, fire suppression and other human activities, climate, and a combination of all of them (Hastings and Turner 1980). There is enough evidence showing the effect of cattle in the spreading of some invasive plants like mesquite. Ignoring the impacts of rodents, and considering that cattle have been on the land for more than a century, and that fire suppression tends to be associated more with upland woodland areas, this study considers the possible impact of climate as a major factor in land cover changes.

Precipitation is a critical climate variable that determines natural vegetation types in different biogeographical regions.Therefore annual rainfall data were used as an indicator of the impacts of climate fluctuation on land cover changes. Six rain gage stations were selected for this study based on the length of the data record and its proximity to the study area. Annual rainfall changes were then compared with the estimated annual rate of change of land cover over the period of study, based on satellite images.

The conservation of natural resources, especially those that are rapidly disappearing in the San Pedro River Basin in the United States, such as grasslands, is a also a key consideration in the implementation of conservation methods for critical habitats in Mexico. The declaration of a reserve in the San Pedro River Basin (Yuncevich 1993) in the United States has significantly improved the riparian habitat, but species associated with grasslands appear to have a better opportunity for survival in the Mexican portion of the basin.

## 2. DATA AND METHODS

To analyze the influence of climate variables on vegetation, information from satellite images was used to estimate annual rates of change of land cover. Those data were compared with changes in annual rainfall at different sites in the area of study. Land cover changes were estimated from satellite images from 1973, 1986, 1992, and 1997. Land cover was separated into 14 classes (see Table 1): oak woodland, pine woodland, juniperus woodland, gallery forest, mesquite bosque, grassland, zacaton grass, desert scrub, barren land, urban land, irrigated cropland, bare cropland, ponds, and mines. Rainfall data were collected from stations that had at least 35 years of records and were analyzed as deviations from the mean annual rainfall. The changes in the land cover types deduced from satellite images obtained at different points in time were compared to changes in the observed rainfall data.

## 3. LAND COVER

Land cover changes were analyzed by Kepner et al. (1998) using satellite images from 1973, 1986, 1992, and 1997. The satellite images from 1973, 1986, and 1992 were from Landsat Multi-Spectral Scanner (MSS), while the 1997 image was from Landsat Thematic Mapper (TM). The images corresponded to Path 35 Row 38, and the dates were June 5, 1973; June 10, 1986; June 2, 1992; and June 8, 1997. Land cover was divided into 14 classes representing natural vegetation and human activities as described above. Natural vegetation consisted of oak woodland, pine woodland, juniper woodland, gallery forest, mesquite bosque, grasslands, zacaton grass, desert scrub, and barren land. Human activities were represented by urban land, irrigated agricultural land, barren cropland (abandoned irrigated agricultural fields, or fallow), ponds (stocking and irrigation ponds), and mines (including tailing ponds). Barren land comprised eroded land and calcareous soils, agricultural land is used for the production of forage and vegetables, and mines are almost entirely used for the extraction of copper.

## 4. CLIMATE VARIABLES

Rainfall records for this region are often widely spaced and show many data gaps in the record, particularly for Mexico. The most reliable rainfall measurement network was established in 1960 and started to be operational by 1965. Fortunately, the rain gage at Cananea was installed in 1918, and a private ranch, La Volanta, installed a rain gage in 1951. The Tombstone, Arizona, rain gage reports are available for years since 1931. Other stations in southern Sonora were installed in 1922 at Navojoa, and in 1927, at Quiriego, Sonora, both related to the development of irrigation projects. All other stations have records that started in the 1960s. Figure 1 shows the locations of some of the rain gages used for the study. As an independent check on the rainfall record in northern Sonora, additional station records were obtained from the Climate Diagnostics Center of NOAA in Boulder, Colorado, USA, and were used to compare the precipitation changes in the upper and lower San Pedro Basin.

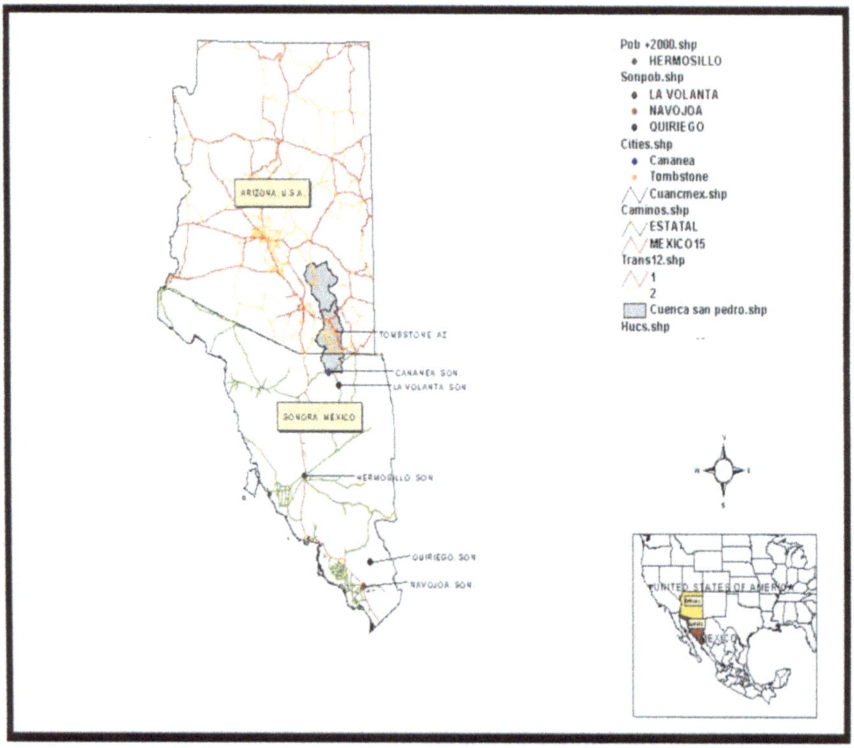


*Figure 1.* Locations of the study area in the Mexican State of Sonora and the U.S. State of Arizona, and the precipitation monitoring stations. Gray-filled area comprises the San Pedro River Basin.

*Table 1.* Land cover types used in the study.

| Land cover type | Characteristics |
|---|---|
| Oak woodland | Woodland conformed by *Quercus* sp. |
| Pine woodland | Vegetation conformed by different species of *Pinus* sp., widely distributed in the mountain ranges from 300 to 4,200 m |
| Juniperus woodland | Woodland conformed by *Juniperus* sp. |
| Gallery forest | Plant community associated with phreatophytic conditions where the dominant species are *Populus* sp., *Salix* sp., and *Alnus* sp. |
| Mesquite bosque | Plant community widely distributed, sometimes as secondary vegetation, mainly conformed by *Prosopis* sp. |
| Grassland | Plant community characterized by the dominance of different grasses. |
| Zacaton grass | Plant community associated with the floodplain of the river, mainly conformed by *Sporobolus wrightii* |
| Desert scrub | Vegetation associated with alluvial areas of deserts, mainly composed of *Prosopis* sp., *Larrea tridentata*, and *Atriplex* sp. |
| Barren | Barren areas or areas with calcareous soils that do not show any significant vegetation |
| Urban | Urban areas |
| Irrigated cropland | Cropland where moisture levels show recent irrigation activity |
| Bare cropland | Cropland that does not show recent activity |
| Ponds | Water storage construction for cattle watering and irrigation |
| Mines | Areas that show mining activity, including tailing ponds |

Cumulative deviations from the mean annual rainfall were used to highlight differences in precipitation on longer time scales. The records were analyzed for evidence of sustained dry periods and wet periods corresponding to time intervals with sustained annual rainfall below or above the mean. Inflection points on these curves identify years when the mean precipitation changes to a predominantly different regime.

## 5. RESULTS

### 5.1. Land Cover

Table 2 shows that in 1973 most of the basin was covered by grasslands (66.5%) and oak woodland (19.8%). Areas showing human activities, consisting of mines, cropland, urban areas, and ponds, occupied a little less than 1% of the area. By 1997, grasslands were significantly reduced (53.3%) and mesquite bosque was significantly increased (from 2.21% to 14.01%). Other significant changes during the 25-year period were the increase of all the human-related land uses from 0.94% to 3.95%, especially those activities related to agriculture, urbanization, and mining.

Table 3 shows the rate of change of each class for the periods 1973–86, 1986–92, and 1992–97, and for the full 24-year-period. The most dramatic change occurred from 1973 to 1986, especially the reduction of grasslands (1,679.5 ha/yr), desert scrub (144.7 ha/yr), and gallery forest (52.4 ha/yr). In contrast, mesquite bosque increased at almost the same rate that grassland decreased (1,726.6 ha/yr), and combined agriculture (bare and cropped) was 113.1 ha/yr.

From 1986 to 1992 the rate of change of land cover was considerably less than that for 1973–86; the rate of disappearance for grasslands was 10-fold reduced (197.6 ha/yr), the same, approximately, as desert scrub (21.4 ha/yr). Interestingly, the largest increase in land cover type came from abandoned cropland (157.9 ha/yr). The economic crisis of 1982–83 in Mexico influenced the reduction of agriculture in the area, since most subsidies to agriculture were reduced to almost zero.

*Table 2.* Land cover in the San Pedro River Basin in Sonora, Mexico (modified from Kepner et al. 1998).

| Cover type | 1973 | | 1986 | | 1992 | | 1997 | |
|---|---|---|---|---|---|---|---|---|
| | Area (ha) | % | Area (ha) | % | Area (ha) | % | Area (ha) | % |
| Oak woodland | 35586.4 | 19.76 | 35474.4 | 19.70 | 35474.4 | 19.70 | 35387.6 | 19.65 |
| Pine woodland | 1453.3 | 0.81 | 1444.7 | 0.80 | 1397.4 | 0.78 | 1350.3 | 0.75 |
| Juniperus woodland | 3037.7 | 1.69 | 2992.0 | 1.66 | 2983.0 | 1.66 | 2986.2 | 1.66 |
| Gallery forest | 1333.1 | 0.74 | 651.6 | 0.36 | 651.6 | 0.36 | 652.8 | 0.36 |
| Mesquite bosque | 3980.2 | 2.21 | 26425.4 | 14.68 | 25240.0 | 14.02 | 25230.3 | 14.01 |
| Grassland | 119737.4 | 66.50 | 97904.2 | 54.38 | 96221.4 | 53.44 | 95946.6 | 53.29 |
| Zacaton grass | 2952.0 | 1.64 | 2843.6 | 1.58 | 2767.0 | 1.54 | 2772.3 | 1.54 |
| Desert scrub | 7784.3 | 4.32 | 5902.6 | 3.28 | 5774.4 | 3.21 | 5737.8 | 3.19 |
| Barren soil | 2480.7 | 1.38 | 2820.6 | 1.57 | 2881.4 | 1.60 | 2881.5 | 1.60 |
| Urban | 106.2 | 0.06 | 414.0 | 0.23 | 546.8 | 0.30 | 911.1 | 0.51 |
| Irrigated cropland | 688.7 | 0.38 | 1101.2 | 0.61 | 1463.4 | 0.81 | 1456.4 | 0.81 |
| Bare cropland | 213.8 | 0.12 | 1271.9 | 0.71 | 3481.9 | 1.93 | 3488.9 | 1.94 |
| Ponds | 79.6 | 0.04 | 70.9 | 0.04 | 263.5 | 0.15 | 294.3 | 0.16 |
| Mines | 614.9 | 0.34 | 731.1 | 0.41 | 902.1 | 0.50 | 952.0 | 0.53 |
| Total | 180048.2 | 100.0 | 180048.2 | 100.0 | 180048.2 | 100.0 | 180048.2 | 100.0 |

During the period of 1992–97, grasslands continued decreasing in coverage but at a much slower pace (54.9 ha/yr). Oak woodland decreased, too, compared with other periods, as well as desert scrub, but this time urbanization increased (72.9 ha/yr), due to a boom in copper mining. Copper prices reached a peak in the middle of the period but then dropped by 1997, bringing more people into town. Also, the land occupied by mining increased due to modifications in the drainage system in 1984. An extraordinary storm produced spills from the tailing ponds into the San Pedro River in the United States and complaints were brought at an international level, which forced the mining company to change the direction of the tailing ponds into Rio Sonora instead of Rio San Pedro, increasing the tailing ponds' area in at least 115 ha.

Considering the 24-year period, the greatest changes were those associated with the disappearance of grasslands (991.3 ha/yr), desert scrub (85.3 ha/yr), and gallery forest (28.3 ha/yr). The land cover increased in mesquite (885.4 ha/yr), combined cropland (168.5 ha/yr), but with predominance in bare cropland due to abandonment (136.5 ha/yr), and urbanization (33.5 ha/yr).

*Table 3.* Rate of change of land cover in the San Pedro River Basin in Sonora, Mexico.

| Cover type | Rate of change of land cover (ha/yr) | | | |
|---|---|---|---|---|
| | 1972–86 | 1986–92 | 1992–97 | 1973–97 |
| Oak woodland | –8.6 | 0.0 | –17.4 | –8.3 |
| Pine woodland | –0.7 | –7.9 | –9.4 | –4.3 |
| Juniperus woodland | –3.5 | –1.5 | 0.7 | –2.1 |
| | | | | |
| Gallery forest | –52.4 | 0.0 | 0.2 | –28.3 |
| Mesquite bosque | 1726.6 | –197.6 | –1.9 | 885.4 |
| Grassland | –1679.5 | –280.5 | –54.9 | –991.3 |
| Zacaton grass | –8.3 | –12.8 | 1.1 | –7.5 |
| Desert scrub | –144.7 | –21.4 | –7.3 | –85.3 |
| Barren soil | 26.1 | 10.1 | 0.0 | 16.7 |
| Urban land | 23.7 | 22.1 | 72.9 | 33.5 |
| Irrigated cropland | 31.7 | 60.4 | –1.4 | 32.0 |
| Bare cropland | 81.4 | 368.3 | 1.4 | 136.5 |
| Ponds | –0.7 | 32.1 | 6.2 | 8.9 |
| Mines | 8.9 | 28.5 | 10.0 | 14.0 |

For all periods, the reductions of grasslands were equivalent to increases in mesquite. One may conclude then that grasslands were largely displaced by mesquite. Although most of the irrigated agricultural land had been displacing the gallery forest, that change was not as clear as for the grassland displacement, because some agricultural land was put into rest (fallow fields), which here is classified as bare cropland.

## 5.2. Climatic Changes

Changes in annual rainfall in the San Pedro River Basin (SPRB) were examined and compared to changes in land cover. Figure 2 illustrates the changes in annual precipitation over the SPRB, as well as for a somewhat larger geographic area. The data are presented in terms of cumulative annual deviation from the long-term mean. It can be seen from this figure that a prolonged dry period occurred in the SPRB from the late 1940s to the end of the 1970s. From the early 1980s, the climate of the region was wetter than normal, until about the last few years.

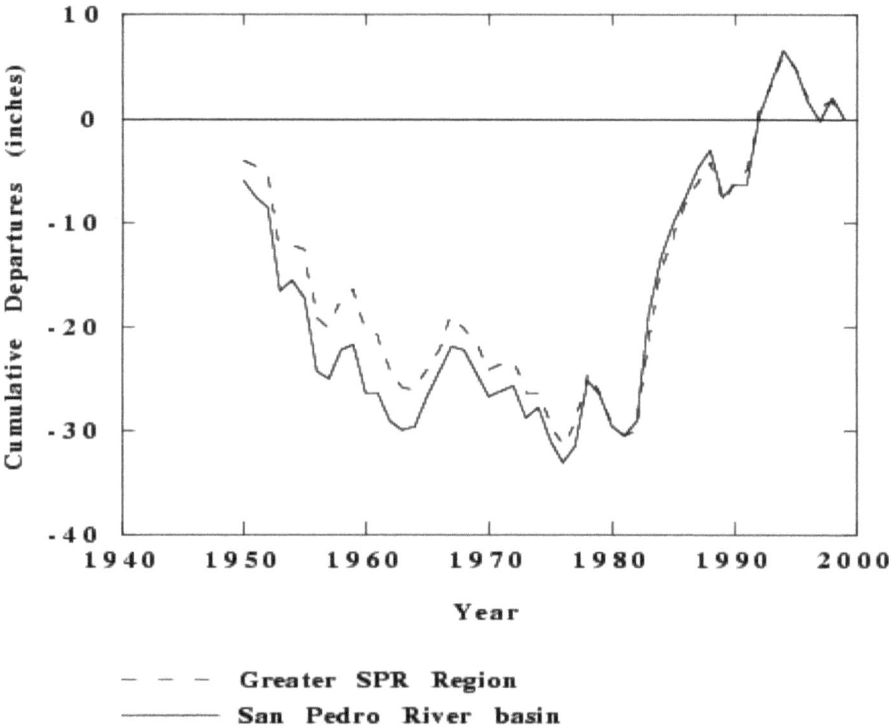


*Figure 2.* Cumulative departures of annual precipitation (in inches) in the San Pedro River Basin (solid line), and for a somewhat larger area representative of southeastern Arizona, denoted as the Greater San Pedro River region, shown by the dashed curve.

Long-term climate data for the Upper San Pedro River Basin in Mexico are available for Cananea and La Volanta. The annual rainfall for Cananea, which has the longest record, is given in Figure 3, by using the technique of cumulative departures as in Figure 2. The data for Cananea also show that from 1935 to 1984 the annual rainfall was lower than the mean annual rainfall. During the period 1953 to 1964 the dry period was worst, and dryness was also severe from 1973 to 1983. It can be seen also that starting around 1984, the annual rainfall has been above the average.

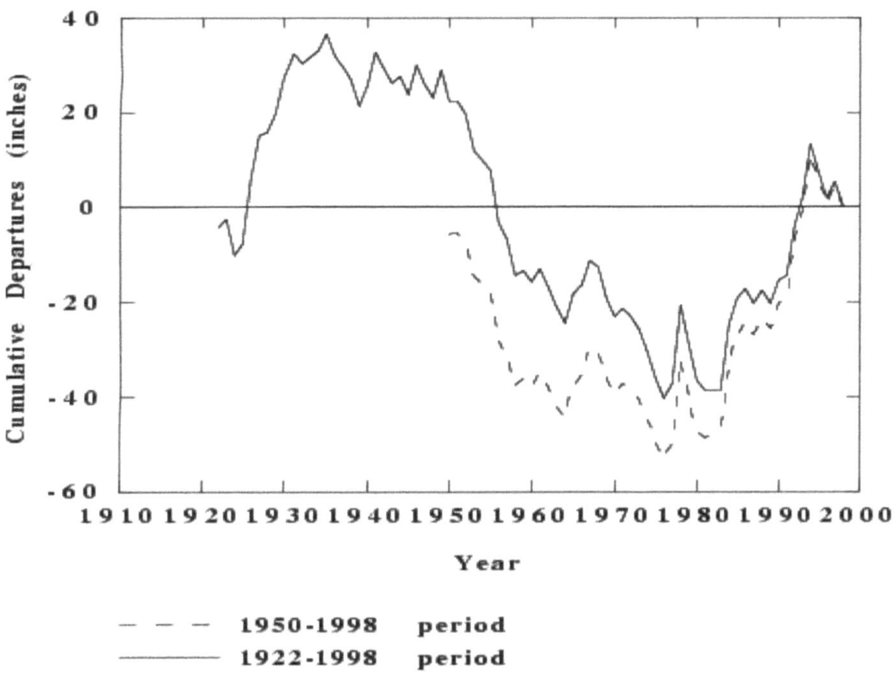


*Figure 3.* Same as Figure 2, but for Cananea in northern Sonora, Mexico, located in the Upper San Pedro River Basin. The changes from 1950–1998 are also shown for direct comparision with Fig. 2.

The satellite images for 1973 showed that the land was dry. The images for 1986 showed a mix of 10 dry years followed by 3 wet years; therefore, it could not be used as a dry period as could be the period 1973–86 versus 1986–92.

It is noteworthy to point out the replacement of grassland by mesquite. Although this study would benefit from an analysis of a 1983 Landsat MSS image, the rate of disappearance of grasslands during the period 1973–86 is taken here as evidence of the impact of a dry period on natural vegetation. Field reports of the impact of cattle in the spreading of mesquite seedlings (Bahre 1995) show the importance of mesquite control for the conservation of grasslands, a native ecosystem important not only to cattle growers, but to wildlife as well (Arias et al. 1999).

## 6. CONCLUSIONS

Land cover changes estimated from satellite images from 1973, 1986, 1992, and 1997 in the San Pedro River Basin in Sonora, Mexico, showed a dramatic reduction of natural vegetation, especially grasslands. The area covered by grasslands decreased more dramatically during the period 1972–83.

An analysis of the rainfall at stations within the basin, as well as south of the basin, showed that before 1983 the annual rainfall was well below average, and since then the annual rainfall has been above average. Substitution of mesquite for grassland could be accelerated in another dry spell, so measures to control mesquite should be taken to conserve grasslands, a major natural resource conservation objective of the San Pedro River Basin in Sonora.

## 7. REFERENCES

Arias, H.M. 2000. International groundwaters: The Upper San Pedro River Basin case. *Natural Resources Journal* 40(2): 199–221.

Arias, H.M., J. Bredehoft, R. Lacewell, J. Price, J.C. Stromberg, and G.A. Thomas. 1999. *Conservación y Enriquecimiento del Hábitat Ribereño de Aves Migratorias en los Altos del Río San Pedro.* Commisión para la Cooperación Ambiental, Montreal, Canada, 137 pp.

Bahre, C.J. 1995. Human impacts on the grasslands of southeastern Arizona. In McLaran, M.P., and T.R. Van Devender (eds.), *The Desert Grassland*, Tucson: University of Arizona Press, pp. 230–231.

Hastings, J.R., and R.M. Turner. 1980. *The Changing Mile: An Ecological Study of Vegetation Change with Time in the Lower Mile of an Arid and Semiarid Region.* Tucson: The University of Arizona Press, Fourth printing, 317 pp.

Kepner, W., C. Watts, C. Edmonds, and H. Arias. 1998. Landscape change in the Upper San Pedro watershed. Proceedings of the 11th Annual Symposium of the Arizona Hydrological Society. Tucson, Arizona, September 23–26, 1998. (http://www.epa.gov/crdlvweb/land-sci/fig3.htm, and http://www.epa.gov/crdlvweb/land-sci/san-pedro.htm)

Yuncevich, G.M. 1993. The San Pedro Riparian National Conservation Area. U.S. Dept. of Agriculture, Forest Service General Technical Report 226, pp. 369–372

MIX
Papier aus verantwortungsvollen Quellen
Paper from responsible sources
FSC® C105338

If you have any concerns about our products,
you can contact us on
**ProductSafety@springernature.com**

In case Publisher is established outside the EU,
the EU authorized representative is:
**Springer Nature Customer Service Center GmbH**
**Europaplatz 3, 69115 Heidelberg, Germany**

Printed by Libri Plureos GmbH
in Hamburg, Germany